Transport and Reactivity of Solutions in Confined Hydrosystems

NATO Science for Peace and Security Series

This Series presents the results of scientific meetings supported under the NATO Programme: Science for Peace and Security (SPS).

The NATO SPS Programme supports meetings in the following Key Priority areas: (1) Defence Against Terrorism; (2) Countering other Threats to Security and (3) NATO, Partner and Mediterranean Dialogue Country Priorities. The types of meeting supported are generally "Advanced Study Institutes" and "Advanced Research Workshops". The NATO SPS Series collects together the results of these meetings. The meetings are co-organized by scientists from NATO countries and scientists from NATO's "Partner" or "Mediterranean Dialogue" countries. The observations and recommendations made at the meetings, as well as the contents of the volumes in the Series, reflect those of participants and contributors only; they should not necessarily be regarded as reflecting NATO views or policy.

Advanced Study Institutes (ASI) are high-level tutorial courses to convey the latest developments in a subject to an advanced-level audience

Advanced Research Workshops (ARW) are expert meetings where an intense but informal exchange of views at the frontiers of a subject aims at identifying directions for future action

Following a transformation of the programme in 2006 the Series has been re-named and re-organised. Recent volumes on topics not related to security, which result from meetings supported under the programme earlier, may be found in the NATO Science Series.

The Series is published by IOS Press, Amsterdam, and Springer, Dordrecht, in conjunction with the NATO Emerging Security Challenges Division.

Sub-Series

A.	Chemistry and Biology	Springer
B.	Physics and Biophysics	Springer
C.	Environmental Security	Springer
D.	Information and Communication Security	IOS Press
E.	Human and Societal Dynamics	IOS Press

http://www.nato.int/science
http://www.springer.com
http://www.iospress.nl

Series C: Environmental Security

Transport and Reactivity of Solutions in Confined Hydrosystems

edited by

Lionel Mercury
Institute of Earth Sciences, University of Orléans, Orléans, France

Niels Tas
Department of Transducers Science and Technology, MESA+ Institute for Nanotechnology, University of Twente, Enschede, The Netherlands

and

Michael Zilberbrand
Israeli Water Authority, the Hydrological Service, Research Division, Jerusalem, Israel

Springer
Published in Cooperation with NATO Emerging Security Challenges Division

Proceedings of the NATO Advanced Research Workshop on
Alternative Water Resources in Arid Areas by Retrieving Water
from Secondary Sources
Ein Bokek, Israel
7–11 May 2012

ISBN 978-94-007-7543-5 (PB)
ISBN 978-94-007-7533-6 (HB)
ISBN 978-94-007-7534-3 (e-book)
DOI 10.1007/978-94-007-7534-3

Published by Springer,
P.O. Box 17, 3300 AA Dordrecht, The Netherlands.

www.springer.com

Printed on acid-free paper

Preface – Transport and Reactivity of Solutions in Confined Hydrosystems

The arid regions experience geopolitical and sociological problems which are related to the lack of water. Yet, it is well-known that some residual water remains inside the thinner part of the porous space in the soil, possibly being supported by a non-rainfall recharge linked to dew formation. This residual non-evaporating liquid in the unsaturated zone is occluded in thin pores and channels and can be stabilized in various ways: by capillary bridges concavely curved towards the atmosphere, by tiny filled porous spaces (solid-liquid-solid confinement) or by deposition on solid surfaces as adsorbed thin films (solid-liquid-air confinement). The properties of the liquid in each of these situations are of special interest for the management of the mobility and retention of this residual water. The pathway of recharge water downwards to aquifers through this unsaturated zone is very complex, particularly because of combining structural and textural heterogeneities with preferential flow coexisting with almost immobile water domains. Even deeper, in the saturated zone, it remains still a challenge to safely predict the reactive pathway of water-rock-pollutant interactions in natural settings, because of a similar complexity in the relationship between the structure of the hosting porous network, and the reactivity and migration properties of the guest fluids.

The scientific context is therefore wide enough, and the general aims of the workshop were:

- To define and discuss the different angles of view towards confinement and capillarity, with particular emphasis on the practical consequences in terms of mass balance and flux in real channels and pores, either of synthetic or of natural materials
- To bring together a number of leading scientists in the field of the theoretical description, experimental investigation, and/or phenomenological modeling and to have an inter-disciplinary exchange around the good, the bad, and the ugly in each conceptual frame

Altogether 30 participants from 11 countries attended the 5-day workshop. Some additional colleagues working at the Ben-Gurion University of the Negev, quite close to the workshop venue, attended some of the sessions. The main feature of this group was its large interdisciplinarity: the scientific fields encompassed nano-/micro-fluidics, fundamentals of surface forces, fluid physics, porous materials science, tree physiology, soil hydrodynamics and chemistry, and reactive transport in aquifers. This scientific scope allowed the group to tackle the "confinement-phases-network" triangle at different scales with very different approaches. The main goal of the present book of proceedings is to reflect the inter-disciplinary exchanges experienced by the participants: sharing common knowledge from such different points of view that, sometimes, it could be felt like different topics.

The book is organized thematically into four parts, preceded by a general introduction to the surface forces which are clearly the nutshell of the special physics arising around interfaces and inside restricted spaces. The first theme focuses on pore water flow at different scales, either to describe its special features in restricted spaces or to define the properties that cause the liquid to remain immobile ("Flow: from nano- to mega-scale"). The second theme deals with the corresponding topic of the solute movement and behavior under the same surface forces fields, introducing the strong effect of coulombic forces in the discussion ("Ions, hydration and transport"). The third theme starts the study of phase transitions with the liquid-vapor phase transition of superheated liquid. Superheating is a common event in natural thin channels and pores, especially under arid conditions, and the end of this metastable state is reached through an explosive boiling (known as cavitation), which is of fundamental and naturalistic value ("In-channels/pores cavitation"). The fourth and last part goes beyond to the solid-liquid phase transitions and their specific behavior when a growing crystal is confined in a thin channel or pore and brought in nanometric proximity to the host solid ("Crystallization under confinement"). Each part is briefly introduced by commenting the logics of the different contributions with respect to the workshop guidelines and the global questioning in the corresponding communities. In the rest of this foreword, we would like to describe how the workshop itself was organized and which lessons we learnt from this experience.

Scientific Program

The introductory lecture was delivered by Jacob Israelachvili, worldwide renowned for his in-depth researches on surfaces and interfacial forces, on both experimental and theoretical points of view. He presented the static (interaction) and dynamic (viscous) properties of water adjacent to single mineral surfaces (solid-liquid interface) as well as of water confined between two such surfaces. This topic allowed him to open the discussion on the main force fields able to modify the classic behavior of liquids, which are the direct or indirect causes of many phenomena, further elaborated during the workshop.

The seven other keynote lectures structured the program according to one thematic session per half-day or day. The Tuesday was devoted to the nano-scale (properties and mechanisms), in direct continuation of Israelachvili's lecture. The first half-day (keynote speaker: Sumita Pennathur) focused on nanofluidics and especially on the role of electrokinetics. In this and subsequent sessions the dynamic and static behavior of water confined in synthetic pores were discussed, and many direct consequences or associated phenomena were developed: dissolution in nanochannels, electrocavitation (induced by electro-osmotic flow), proton release and pH of the in-filling liquid, ion or molecule transport, elastocapillary pressure and the deformation of the channel host, seismoelectric effect. The second half-day (keynote speaker: O. Vinogradova) dealt with the surface forces, not as a general topic but related to diverse practical situations: slip on superhydrophobic surfaces, the stability of confined electrolyte solutions, and the computer simulations of solid/liquid interfaces with clay, cement, and polymer membranes.

The Wednesday (keynote speaker: Dani Or) focused on the pore scale processes, especially on the conceptual modeling of the gas-liquid (evaporation, water flow and retention, fingering) and solid-liquid equilibria (pressure solution) in natural pores. Key observation was that moving from the designed "pure" materials to the complexity and heterogeneity of natural samples requires to develop a simplified sketch of the studied volume, only capturing the relevant features to account for the evolution through time.

The Thursday was entitled "Pore/channel to network", progressively shifting to a global grasp of 3d materials and the role and properties of the in-filling fluids. The first half-day (keynote speaker: Michael Steiger) was essentially focused on salt crystallization in pores with the important issue of the location and the dynamics of the precipitation, but also the way salts can colonize in-pore spaces. The special behavior of hydrates in pores was studied, with the pore size effect but also the role of the metastable solids. The second half-day (keynote speaker: John Sperry) was related to the cavitation events in nature (trees xylem, soil channels, mud volcanoes). This topic was approached by measurements in situ and ex situ, by analogous experiments in labs and by thermodynamic calculations (equation of state).

The last half-day was devoted to the reactive transport issue, especially connected to the correct way to involve local heterogeneities (pore spaces, chemical composition, water state) in the modeling of extended natural media.

Discussions

The format of the workshop did deviate from conventional scientific congresses that mainly present validated results to its natural audience ("the community"). Here, the speakers tried to present a state-of-the-art of questions and methods in their area, while presenting their own approaches and originality. That resulted in the presentation of many ongoing investigations, making the debate less affirmative but more reflective for many participants, even those belonging to the speaker fields. The workshop also offered to many of us a chance to see phenomena closely related to our own, but from different angles of view, described in different vocabularies and studied by different methodologies.

The focus of the event was to encourage fruitful discussions hence all participants were introduced at the beginning of the symposium, with their expertise and their special interests. The lectures were held in an informal atmosphere making it easy to intervene during the talk with commentaries or questions. The timing was consequently difficult to follow, with lively exchanges of ideas and opinions amongst the participants, during the talks and the question times. It was especially noticeable that the general interest for each other did not fade through the whole period of the workshop despite the broad range of topics. The discussions in the technical sessions after the presentations did highlight some main topics, which will be summarized in the following.

Outcomes

During the entire symposium it became obvious that the different communities are speaking different "languages". The gap between the micro-/nanofluidics community and the natural pore sciences community for example was quite clear, despite the fact that they share work on channels filled with liquids. The real difference stems from their requisites and goals. The nanofluidics community works with well-designed systems and pure materials and targets a global understanding of these exact systems as well as the phenomena arising inside. The natural pores sciences try first to understand which features are essential (in which perspective?) to afterwards simplify the original (too complex) case studies in such a way that they can be quantitatively modeled. The game is then to test the initial assumptions by matching the results of simulations or modeling with practical observations or measurements. We could as well highlight the gap between the theoreticians and the naturalists, linked to the distinction between possible theoretical states (physical chemistry) and practical states (natural observations).

These two distinctions, and all the new concepts learnt when listening to colleagues from other scientific areas, made the workshop interesting for every participant, lively and interactive. Meanwhile, the multidisciplinarity caused a sort of limitation to define immediate practical outcome(s), though the cross-fertilization was felt very useful by all the participants. Upscaling the nano-processes and considering how they can act in natural settings at local (centimetric) and regional (megametric) scales were defined as one of the main challenges of further research. At the final round table, the participants expressed full satisfaction to have attended the workshop, predominantly because it broadened their scientific knowledge, and improved their networking. The informal setting, the opportunity to have discussions with colleagues from different, yet related, disciplines were often outlined in after-workshop messages, as well as the chance to gain new perspectives on research, especially by meeting potential collaborators. Despite this positive feeling, many participants were reluctant to build a proceedings book, because the workshop also highlighted the differences and the barriers from one community to the other.

Most of the participants agree that in the future the impact of confinement, superheating and capillarity on a better understanding of the natural phenomena should be developed. The studies that coupled different properties, like the cavitation processes (in nanochannels, membranes, glass capillaries, xylems, soil pores, etc.), the special nanoscale transport properties (related to the electrical double layer), the deformations due to the elastocapillary pressure (again, in different materials), the interplay between the chemical composition of the filling liquids and the surface forces, attracted much attention. Also critical is the use of well-defined nanochannels, not only to understand fundamental properties but also to serve as analogous materials to measure and quantify parameters specific to their scale. This knowledge could be embedded in renewed large-scale modelling. The nanochannels are known as key elements in "*Lab on a chip*" systems, but the workshop made clear to the participants that they could constitute "*Nature on a chip*" model systems as well.

The major outcome of the workshop was that the participants experienced the gap between fundamental science and research applied to natural sciences, and that the different communities talked to each other. By doing that, they started to learn each other's "languages" and were introduced to the problems and needs of the others. The event was an excellent start of a multi-disciplinary effort to develop cross-fertilization between physics, chemistry, biology, soil and earth sciences.

Conclusion

Eventually, we hope that the present proceedings book will reflect the valuable experience we lived through during this workshop, and will give a clear impression to readers of the richness gained by crossing disciplines to go into the *terra incognita* "where nobody feels comfortable", as one of the participants summed up.

This workshop was supported mainly by the NATO (grant ESP.MD.ARW 984162), but benefited as well of supports from the French ANR (Agence Nationale de la Recherche; grants CONGE BLAN-61001 & labex Voltaire ANR-10-LABX-100-01), the French embassy in Israel, the MESA+ Institute for Nanotechnology (The Netherlands), and the University of Orléans (France).

Contributors

Yuri Aristov Boreskov Institute of Catalysis, Novosibirsk, Russia

Shmuel Assouline Institute of Soil, Water and Environmental Sciences, Agricultural Research Organization (ARO), Bet Dagan, Israel

Mohamed Azaroual Institut des Sciences de la Terre d'Orléans, UMR 7327 Université d'Orléans - CNRS - BRGM, BRGM, Orléans Cedex 2, France

Isabelle Bergonzi Institute of Earth Sciences, University of Orléans, Orléans, France

Brian Berkowitz Department of Environmental Sciences and Energy Research, Weizmann Institute of Science, Rehovot, Israel

Nataliya Brunets Department of Transducers Science and Technology, MESA+ Institute for Nanotechnology, University of Twente, Enschede, The Netherlands

Jakob Buchheim Nanoscience for Energy Technology and Sustainability, Department of Mechanical and Process Engineering, Swiss Federal Institute of Technology Zurich (ETH Zurich), Zurich, Switzerland

Maria Dragila Department of Crop and Soil Science, Oregon State University, Corvallis, OR, USA

Chuanhua Duan Department of Mechanical Engineering, Boston University, Boston, MA, USA

Dag Kristian Dysthe Physics of Geological Processes, Department of Physics, University of Oslo, Oslo, Norway

Jan C.T. Eijkel BIOS/Lab on a Chip Group, MESA+ Institute for Nanotechnology, University of Twente, Enschede, The Netherlands

Rosa M. Espinosa-Marzal Environmental Engineering and Science, Department of Civil & Environmental Engineering, University of Illinois at Urbana-Champaign, Illinois, USA

Larissa Gordeeva Boreskov Institute of Catalysis, Novosibirsk, Russia

Avrami Grader INGRAIN, Houston, TX, USA

Jeroen Haneveld Department of Transducers Science and Technology, MESA+ Institute for Nanotechnology, University of Twente, Enschede, The Netherlands

Thomas Hankemeier Department of Analytical Biosciences, Leiden/Amsterdam Centre for Drug Research (LACDR), Leiden University, Leiden, The Netherlands

Teemu Hölttä Department of Forest Sciences, University of Helsinki, Helsinki, Finland

Attila R. Imre HAS Centre for Energy Research, Budapest, Hungary

Institute of Physical Chemistry, University of Cologne, Köln, Germany

Henri V. Jansen Department of Transducers Science and Technology, MESA+ Institute for Nanotechnology, University of Twente, Enschede, The Netherlands

Kjeld G.H. Janssen Department of Analytical Biosciences, Leiden/Amsterdam Centre for Drug Research (LACDR), Leiden University, Leiden, The Netherlands

Andrey G. Kalinichev Laboratoire Subatech, UMR 6457, Ecole des Mines de Nantes, Nantes, France

Department of Chemistry, Michigan State University, East Lansing, MI, USA

Rohit Karnik Department of Mechanical Engineering, Massachusetts Institute of Technology, Cambridge, MA, USA

Arnault Lassin Institut des Sciences de la Terre d'Orléans, UMR 7327 Université d'Orléans - CNRS - BRGM, BRGM, Orléans, Cedex 2, France

Manuel Marcoux Institut National Polytechnique de Toulouse (INPT), Institut de Mécanique des Fluides de Toulouse (IMFT), UMR 5502 CNRS-UPS, Toulouse, France

Lionel Mercury Institute of Earth Sciences, University of Orléans, Orléans, France

Uri Nachshon Department of Environmental Hydrology & Microbiology, Zuckerberg Institute for Water Research, Jacob Blaustein Institutes for Desert Research, Ben-Gurion University of the Negev, Beersheba, Israel

Now at: University of Saskatchewan, Saskatoon, Canada

Kfir Narkis Institute of Soil, Water and Environmental Sciences, Agricultural Research Organization (ARO), Bet Dagan, Israel

Hyung Gyu Park Nanoscience for Energy Technology and Sustainability, Department of Mechanical and Process Engineering, Swiss Federal Institute of Technology Zurich (ETH Zurich), Zurich, Switzerland

Marc Prat Institut National Polytechnique de Toulouse (INPT), Institut de Mécanique des Fluides de Toulouse (IMFT), UMR 5502 CNRS-UPS, Toulouse, France

Harvey Scher Department of Environmental Sciences and Energy Research, Weizmann Institute of Science, Rehovot, Israel

Vladimir N. Shilov Institute of Biocolloid Chemistry, Ukrainian Academy of Science, Kiev, Ukraine

Kirill I. Shmulovich Institute of Experimental Mineralogy, Russian Academy of Science, Chernogolovka, Russia

John Sperry Department of Biology, University of Utah, Salt Lake City, UT, USA

Michael Steiger Department of Chemistry, University of Hamburg, Hamburg, Germany

Niels R. Tas Department of Transducers Science and Technology, MESA+ Institute for Nanotechnology, University of Twente, Enschede, The Netherlands

Joost W. van Honschoten Department of Transducers Science and Technology, MESA+ Institute for Nanotechnology, University of Twente, Enschede, The Netherlands

Daniel S. van Schoot BIOS-Lab on a Chip Group, Mesa+ Institute for Nanotechnology, University of Twente, Enschede, The Netherlands

Stéphanie Veran-Tissoires Institut National Polytechnique de Toulouse (INPT), Institut de Mécanique des Fluides de Toulouse (IMFT), UMR 5502 CNRS-UPS, Toulouse, France

Noam Weisbrod Department of Environmental Hydrology & Microbiology, Zuckerberg Institute for Water Research, Jacob Blaustein Institutes for Desert Research, Ben-Gurion University of the Negev, Beersheba, Israel

Seul Ki Youn Nanoscience for Energy Technology and Sustainability, Department of Mechanical and Process Engineering, Swiss Federal Institute of Technology Zurich (ETH Zurich), Zurich, Switzerland

Michael Zilberbrand Israeli Water Authority, the Hydrological Service, Research Division, Jerusalem, Israel

Chapter 1
Interactions in Water Across Interfaces: From Nano to Macro-Scale Perspective

Rosa M. Espinosa-Marzal

Abstract In this work we first revisit the surface forces between two (model) mineral surfaces, mica, across an aqueous solution (KNO_3) over a broad range of concentrations. The significantly improved resolution available from the *extended* surface force apparatus (eSFA) allows the distinction of hydrated-ion structures. Above concentrations of 0.3 mM, hydrated-ion correlations give rise to multiple collective transitions (4 ± 1 Å) in the electrical double layers upon interpenetration. These features are interpreted as the result of hydrated-ion ordering (layering), and are responsible for hydration forces, in contrast to the traditional interpretation invoking water layering. At concentrations as low as 20 mM, attractive surface forces are measured in deviation to the DLVO theory. The estimated hydration number of the ions in the confined electrolyte is significantly below that of the bulk. A confined 1–3 nm thick ionic layer condensates at concentrations >100 mM, i.e. below bulk saturation. This study leads to new insights into crystal growth in nano-confinement that differs from the classical theory of crystallization. Finally, the impact of the properties of confined water or solution and in-pore crystallization on the macro-scale description of soil water distribution is discussed.

1.1 Introduction

Confined water, with dissolved organic and inorganic matter, is abundant in the narrow pores of soil, clay, rocks and biological systems. Surface interactions in narrow spaces across aqueous solutions are the driving force for many nano-scale phenomena such as ion exclusion desalination membranes, colloidal stabilization, or

---

R.M. Espinosa-Marzal (✉)
Environmental Engineering and Science, Department of Civil & Environmental Engineering, University of Illinois at Urbana-Champaign, Illinois, USA
e-mail: rosae@illinois.edu

L. Mercury et al. (eds.), *Transport and Reactivity of Solutions in Confined Hydrosystems*, NATO Science for Peace and Security Series C: Environmental Security,
DOI 10.1007/978-94-007-7534-3_1,

protein (un)folding and separation. Besides, a correct understanding of nano-scale interactions aids in elucidating large-scale phenomena; as in Seurat's pointillism the precise interaction of distinct single dots of pure color leads to the creating of a beautiful painting. Thus, the intrinsic properties of water either confined in narrow pores or as an adsorbed thin film (i.e. confined at the liquid/vapor interface), with dispersed clays and mineral grains, dissolved salts, organic matter, and living organisms play a significant role on the whole picture describing subsurface water distribution and flow, soil and rock weathering, and on keeping biological survival under arid conditions (e.g. water accumulations for plant uptake and microbiologic populations).

In fact, the mass balance of soil water at a macro-scale is influenced by the properties of confined water through both, transport and reactive term. Regarding the former, the effective soil permeability is pore-size dependent, which is described by the pore-network tortuosity. Although the permeability of nanopores is strongly reduced, there is still an exchange of water and solutes [1] that are confined in narrow spaces -e.g. in clay interlayers-, with those present in capillary and macro-pores, which affects the net water flow; viscosity of the confined solution, ion diffusivity in thin films, ion binding to the surface and surface potentials mainly determine this exchange. Concerning the reactive term, phase transitions involving water molecules and intermolecular interactions (e.g. capillary condensation, evaporation, salt hydration/dehydration, salt crystallization/dissolution) need to be included in the reactive term of the mass balance. Moreover, the properties of the confined thin film between a crystal growing in a pore and the pore wall determine the crystallization pressure and therefore they are directly involved in weathering and landscape shaping [14].

In this chapter I first revisit the surface forces acting between two atomically flat mineral (mica) surfaces in an electrolyte solution over a broad range of concentrations up to the saturation limit, and show new insights into the properties of the confined solution and into in-pore crystallization. The relevance of this fundamental study to the macro-scale description of soil water distribution and availability in arid areas is also discussed.

1.2 Theoretical Background

If natural minerals, such as calcite, clays, or quartz, are immersed in water or electrolyte solution they acquire a surface charge. An excess of counter ions exists in the proximity of the surface due to electrostatic interactions between dissolved ions and the charged surface. The Stern layer and the diffuse layer of counter ions form an electrical double layer (EDL) that significantly affects the way two such surfaces interact in solution. The DLVO theory [10, 24, 49] is commonly used as a benchmark to discuss the forces between surfaces that are slightly charged in low-concentration electrolyte solutions (e.g. in refs. [38, 39, 22]). It does not predict ion-specific effects [2], and, being a continuum theory, it must fail at the molecular scale [36]. Deviations from DLVO theory are reported in electrolyte solutions at concentrations as low as 0.3 mM [38]; oscillatory surface forces [39] with a periodicity of ~2.5 Å and an exponential repulsion below separation $D \sim 5$–10 nm [22, 44, 6, 25] are reported from direct force measurements with the conventional surface forces apparatus (SFA). These early measurements suggested that the oscillatory component of the so-called hydration force originates from a layered water structure in the vicinity of the surface ($D < 2$ nm), the layer thickness being equal to the diameter of a water molecule [26, 23]. Diverse mechanisms have been proposed to explain the short-range monotonic part of the hydration force, such as polarization of water molecules by surface dipoles, changes in the dielectric constant of the medium, breakage of hydrogen bonds at the liquid-solid interface, or the confinement of thermally mobile yet bound hydrated counter-ions on the surface and in the Stern Layer (see review in [50]). If the binding energy between hydrated counter-ions and the surface is higher than the energy released by replacing them with hydronium, the hydrated counter-ions reside in the Stern layer, their dehydration and steric effects leading to a repulsive surface force [38]. The experimental results by Pashley [38] show the ion specificity of these forces. Pashley's site-binding model was further improved by Miklavic and Ninham [35] and later by Shubin and Kekicheff [44], who included the steric restriction normal to the surface of the hydrated counter-ions in accordance with the triple-layer model [44, 6]. All models consider the hydration repulsion to result from diverse properties of water or (hydrated) ions within the Stern layer.

Moreover, it is accepted that high-frequency fluctuations in the polarization of ions give rise to dispersion forces between ions and between ions and a charged surface—so-called ion correlation interactions—that may lead to an extra attraction between surfaces. Various models have been proposed [28–31, 32] and they agree in as much as that this attractive force is relevant only at high concentration in divalent and trivalent electrolytes [37]. Further discrepancies between these models and experiments are expected, since water is considered as a dielectric continuum and the ion hydration structure is commonly neglected. In particular, excluded-volume effects of hydrated ions are ignored. Yet, Lee et al. [33] demonstrated by X-ray reflectivity the need to incorporate such ion hydration explicitly at the molecular scale to correctly describe the double layer. Molecular dynamics and Monte Carlo simulations of the confined electrolyte solution are the only modeling tools that can consider all these interactions (see [20] and the references mentioned there).

1.3 Materials and Experimental Method

Thin mica sheets were prepared by manual cleavage of optical-quality ruby mica blocks in a laminar-flow cabinet. Sets of thin mica pieces of uniform thickness (2–5.5 μm) were cut from the larger sheet into smaller (8×8 mm) rectangular pieces. The mica sheets were then coated with a silver film of 40 nm thickness by means of thermal evaporation in vacuum ($2 \cdot 10^{-6}$ mbar). The silvered mica sheets were glued, silver down, onto silica supports with cylindrical tops using epoxy resin glue (EPON 1004 F). After gluing, the samples were immediately (<5 min) inserted into the sealed measurement apparatus.

Potassium nitrate solution (KNO_3) has been used as electrolyte solution in this work. Solutions were prepared with ultrapure water (puriss. p.a., 0.2 mm membrane filtered; Fluka, Sigma-Aldrich) as received and potassium nitrate (Sigmaultra, >99.0 %; Sigma-Aldrich), at concentrations between 0.08 mM and 2.6 M, and filtered with a 0.2 μm-membrane. At a temperature of 22 °C the saturation concentration of KNO_3 is 3.1 M. The pH of the solutions was 5.6 ± 0.1 due to CO_2 dissolution.

The SFA [27, 48] is an instrument that directly measures forces occurring between two mica surfaces. A central element of this method is the optical measurement of the distance between the surfaces based on interferometry. The two surfaces are cylindrically curved ($R \approx 20$ mm) after being glued to a disk and facing each other such that the cylinder axes are crossed at an angle of 90°. One disk is supported by a spring and is approached towards the other using an approach actuator, of which the position P is known over time. If there is a surface force, this spring deflects by the amount D-P, with D the surface separation. Hence, the surface force F results from the distance measurement as $F = k(D\text{-}P)$, where k is the stiffness of the spring with known values between 800 and 2000 N/m in this work. The force, F, is normalized by the cylinder radius, R, to relate it to a thermodynamic meaningful change of energy (viz. Derjaguin approximation) [9].

The two transparent mica layers form an optical thin-film interferometer. The silver coating improves the reflectivity. White light is directed through the interferometer and the transmitted interference is analyzed in an imaging spectrograph.

A number of improvements in accuracy, resolution, speed, mechanical drift, thermal stability, automation and imaging were incorporated into the *extended* Surface Force Apparatus (*e*SFA) [18, 19]. Thus, the transmitted interference spectrum consisting of fringes of equal chromatic order are analyzed by a numeric fast-spectral correlation algorithm to determine both the distance D between the surfaces at the point of closest approach (PCA) with an accuracy better than 20 pm over a distance ranging from 0 to 10 μm, and the refractive index of the medium in *real-time*.

After placing the surfaces in the *e*SFA, the fluid cell was purged for 1 h with dry nitrogen and the mica thickness determined by means of thin-film interferometry in dry mica-mica contact. Then the surfaces were separated to about 10 μm and the *e*SFA cell filled with pure water (~80 ml) through an inlet that connects the SFA with a sealed container. The liquid was pre-conditioned in the container to 22 °C. Manual approach of the surfaces was then performed with a micrometer to a separation of ~1.5 μm. After thermal equilibration over about 3 h, approach and separation of the surfaces was performed in plain water at constant rate (0.5 nm/s). With a surface separation again of ~10 μm, water was drained from the *e*SFA cell, which was then filled with the salt solution that had been pre-conditioned in the closed and clean container. The direct force measurements were started after about 4 h of equilibration in solution and readjustment of the PCA. Two to four approach-separation cycles were measured with different solution concentrations at constant rates ranging from 5 nm/s to 0.5 Å/s. All experiments were performed at controlled temperature of 22.0 °C in the fluid cell. The total duration of an experiment series, in which an individual set of mica surfaces is submerged in different solutions, was limited to 100 h to preserve cleanliness.

1.4 Experimental Results and Discussion

The force curves measured across KNO_3 solution show a very clear qualitative evolution over the considered concentration range. A distinction between three phenomenological regimes: the "DLVO" regime at low concentrations, the "ordering" regime at medium concentrations and the "solidification regime" at high concentrations was introduced and described in detail in [11]. Here we only discuss some of the characteristic features.

1.4.1 Characteristic Features of the Surface Forces in the DLVO, Ordering and Solidification Regimes

The theory developed by Derjaguin and Landau [10], Verwey and Overbeek [49] (DLVO theory) describes the force between charged surfaces interacting through a fluid medium. It combines the effects of the van der Waals attraction and the electrostatic repulsion due to the so called double layer of counter ions assuming that the fluid medium is a continuum. The electrostatic part of the DLVO interaction is computed in the mean field approximation in the limit of low surface potentials.

The DLVO theory has been classically used to calculate surface forces between mica surfaces in electrolyte solution at concentrations below $\approx$ 100 mM using a Hamaker constant of $2.2 \cdot 10^{-20}$ J and the Poisson-Boltzmann equation assuming either a constant surface potential or a constant surface charge during approach [4]. A very good fit between DLVO theory and measurement can be obtained below 0.3 mM ("DLVO" regime, Fig. 1.1a), the agreement being better assuming a constant surface charge during approach. In the "DLVO" regime, the surfaces jump into contact due to the dominant effect of van der Waals attraction.

In a small concentration range around 0.3–1 mM we could observe two distinct transitions (Fig. 1.1a): the van der Waals instability (2 nm jump-in), followed by a single film-thickness transition of size $\Delta = 5.8 \pm 1$ Å that occurs at a moderate applied load of $F_t/R \sim 1.2$–12.9 mN/m. As in water, the mica surfaces also end up in close contact after the second transition. Further investigations show [11] that the potential associated with the hydrated ion layer at the interface can counterbalance the van der Waals instability, preventing from mica-mica contact until it is collectively expelled upon increasing the load. The occurrence of this second transition announces the crossover to the "ordering" regime.

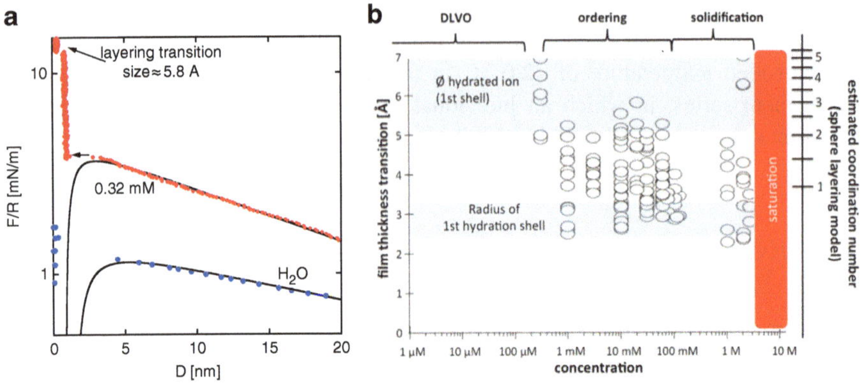

Fig. 1.1 (**a**) A jump into contact is measured in the DLVO regime, as shown for water ($\Delta \approx 4.6$ nm). The DLVO theory predicts well the surface interactions in this regime. In 0.32 mM KNO_3 solution, a jump is measured at $D = 3.1$ nm ($\Delta \approx 2.7$ nm) but intimate mica-mica contact is prevented. The *arrow* points at a film thickness transition $\Delta = 5.8 \pm 0.3$ Å during further confinement at $F/R = 12.8$ mN/m. (**b**) Film-thickness transitions measured at different salt concentrations. No transitions are observed in the DLVO regime where the van der Waals instability masks detection of any weaker structural correlation forces. The multiple layering in the ordering regime has a characteristic size of $\Delta = 4 \pm 1$ Å; only the final transition is smaller (2.9 ± 0.3 Å) (Figures from [11])

At intermediate concentrations, in the "ordering" regime, the picture becomes more complex, although the DLVO theory can still be used to fit the experimental results at surface separations larger than ~5–10 nm (Fig. 1.2a). At shorter distances the repulsive hydration force avoids the jump into mica-mica contact. The abovementioned $\Delta = 5.8 \pm 1$ Å transition is replaced by multiple film-thickness transitions of sizes 4.1 ± 1 Å at surface separations $D \leq 3$ nm, followed by one characteristic final transition of smaller size $\Delta = 2.9 \pm 0.3$ Å, occurring at high external load. The multiple film-thickness transitions are reminiscent of *layering*.

For each cycle, the external load was always increased up to 600 mN/m while the last measured transition was generally observed below 200 mN/m. Clearly, we cannot rule out that further structural transitions would occur at even higher loads. As shown in [11] the reduced adhesion measured in the ordering regime suggests that the mica-mica contact is no longer an intimate one and that ions must still be present at the interface. Thus, this last transition is interpreted as forced surface-association of hydrated ions.

The crossover to the "solidification" regime is characterized by an extra weakening of the double-layer repulsion at small separations (Fig. 1.2a, b). The reduction of repulsion is already observed at 20 mM and becomes very clear in 30 mM KNO_3 solution. This reduction is not predicted from a simple DLVO model, assuming invariant Hamaker constant. A long-range structuring of the solution is also detected, typically at $D < 8$ nm, but these film-thickness transitions are rather irregular and therefore different from those observed in the "ordering" regime.

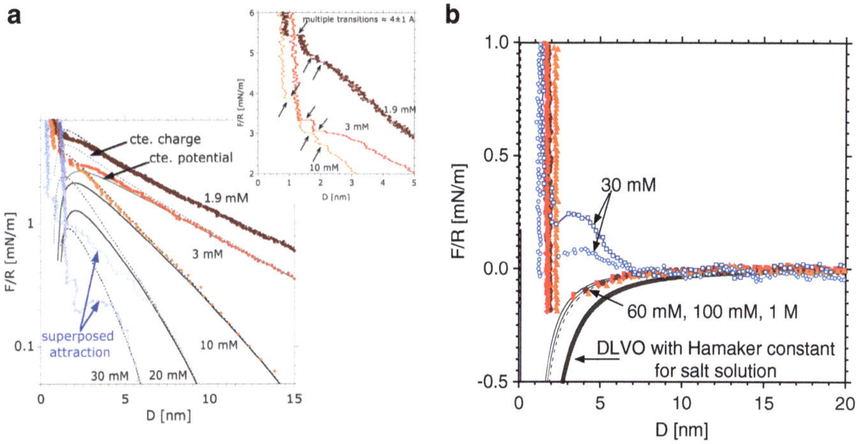

Fig. 1.2 (**a**) Measured forces during approach of mica surfaces in the ordering and crossover to solidification regime. The *dotted lines* give the calculated forces according to the DLVO theory assuming a constant surface charge and the full *lines* assuming a constant surface potential (ψ_s:−88 mV for 1.9 mM KNO_3, − 75 mV for 3 mM, −57 mV for 10 mM and − 40 mV for 20 mM). The *arrows* in the inset show the resolved transitions in the measured surface forces. (**b**) Measured force-distance curves in 30 mM (empty *squares* and *circles*), 60 mM, 100 mM and 1 M (full symbols) KNO_3 solution during approach at 3 Å/s. The attractive force that was identified in the transition to the solidification regime becomes dominant. An attractive force at surface separations smaller than 5 nm leads to an energetic minimum characteristic of the solidification regime. Upon further approach, a strong hard-wall repulsion is measured although no mica-mica contact has been achieved. The *thin lines* give van der Waal forces at the three selected concentrations calculated with the Liftshitz theory and neglecting retardation effects, fitted to the experimental results with a Hamaker constant equal to $0.78 \pm 0.13 \cdot 10^{-20}$, while the thick line was calculated with a Hamaker constant equal to $2.2 \cdot 10^{-20}$ J characteristic of mica-solution-mica system at low ionic strength (Figures from [11])

Above ∼100 mM, the surface force becomes attractive. This attractive surface force was already reported by Israelachvili and Adams [25] and attributed to a secondary minimum caused by van der Waals forces and recently by Perkin et al. [40] but not further elucidated nor discussed. Figure 1.2a shows that it cannot be described by the DLVO theory with an invariable Hamaker constant (thick line in Fig. 1.2b).

The surface force exhibits a minimum at ∼2 nm in the attractive regime, which is reminiscent of the minimum reported by Kjellander et al. [32] as predicted from ion-dispersion forces and measured for divalent counter ions. Attractive interactions due to ion-ion correlations are predicted by Kjellander and Marcelja [31] for divalent cations and at high concentrations and surface charges. However, an ordering of hydrated ions is neither considered nor predicted in the Kjellander model. An important simplification of the Kjellander model is to consider water as a continuum and to neglect ion hydration [33] and associated excluded volume. In contrast to his predictions, we find that attractive forces are significant already at moderate concentrations, in monovalent salts at concentrations as low as 20 mM, and, with relatively low surface charge of mica (max. 1 per 48 $Å^2$).

We finally press against a hard solid with a thickness of 1–3 nm. This interfacial solidification is characteristic for the concentration regime above 100 mM. Only one single final structural transition is measured; it is variable in nature, i.e. the transition size Δ ranges from 2.5 to 6.5 Å at applied loads from 10 to ~100 mN/m (Fig. 1.1b). It should be noted that a relatively low ionic strength, which can be expected for water in both saturated and unsaturated zone, is required for ordering and solidication regimes.

1.4.2 Correlations Between Hydrated Ions in Overlapping Electrical Double Layers

Molecular layering is a known hard-sphere correlation phenomenon, which can be discussed in terms of an oscillatory surface potential of periodicity close to that of the size of the underlying molecules [21]. The observation of the involved structural force for spherical fluids necessitates atomically smooth confining surfaces. It can be observed by SFA as a series of mechanical instabilities or film-thickness transitions, as found in this study.

Our high-resolution data, as well as new simulation results, allow a reinterpretation of the accepted theory for hydration forces. The size of the observed transitions (4 ± 1 Å) is significantly larger than a single water layer and therefore in contradiction to [26]. We thus conclude that the hydration (oscillatory exponential) force is due to correlations and excluded volume effects that lead to ordering or layering of hydrated ions in the overlapping double layers.

The range of structural forces increases with salt concentration and is detected up to a separation of at least 3 nm. At such separations and concentrations, the sum of charge of the counter-ions (here, cations) must be balanced by the presence of anions and we speculate that these layers therefore consist of both hydrated anions and hydrated cations alike. The physics of ordering is quite different in the case of hydrated ions than it is for layering of hard-sphere molecular liquids, due to hydration effects, and involved ion-ion, ion-water and ion-surface correlations, and can only be studied by molecular dynamic simulations.

1.4.3 Concentration of the Nano-Confined Solution

The first hydration shell of potassium ion is a sphere with a diameter of 5.66–7.4 Å with 5 water molecules [7]. The size of many film thickness transitions observed in the ordering regime falls between the known sizes of the radius of a hydrated potassium (or of a water layer) and the expected diameter of the first-shell hydrated ion (Fig.1.1b). The irregular size of the transitions is attributed to the soft-sphere behavior of the hydrated ions, i.e. to the lack of a sharp energetic distinction between hydration water and free water.

Since the size of a layering transition is an estimate for the size of the layering objects, at least in the case of hard spheres, we can roughly estimate the effective coordination numbers from the transition sizes. Both, coordination number and size, decrease with increasing concentration in bulk solution. Coordination numbers that are 5–8× smaller than the coordination number in bulk saturated solution (~8.25) (see right Y-axis in Fig. 1.1b) are obtained. This confinement-induced dehydration may be overestimated (different layering physics), but it is nevertheless in qualitative agreement with the MD simulations of Malani et al. [34] for nano-confined NaCl solution. At such low coordination numbers, a bulk solution would be supersaturated, and consequently crystallization should occur immediately.

However, the saturation concentration of nano-confined solution is significantly higher than that of the bulk solution due to the influence of the solid-liquid interfacial energy of small crystals (i.e. large surface to volume ratio) on their chemical potential [3]. Therefore, in solutions confined in a gap smaller than 10 nm, the influence of the interfacial energy, γ, on the saturation concentration must be considered. This leads to an increase of the saturation concentration $c_{sat,D}$ with decreasing gap distance, according to $c_{sat,D} = c_{sat,\infty} \exp\left(\frac{2\gamma V_m}{RTD/2}\right)$, where γ is the solid-liquid interfacial energy (36.4 mN/m for KNO_3) and V_m the molar volume of the crystal (47.9 cm^3/mol for KNO_3), R the gas constant, and T the temperature. Accordingly a relevant decrease of the coordination number of the ions for solution saturation in nano-confinement is obtained. For example, in a gap of 5 nm, the coordination number for saturation is expected to decrease from 8.25 to 5.6 and in a gap of 2 nm to 2.4. Hence, the observed transition sizes indicate that the solution might be near saturation already in the "ordering" regime. Interestingly, this confinement-induced dehydration could explain the nucleation of calcium carbonate between mica surfaces that Alcantar et al. [1] observed with the SFA at concentrations below bulk saturation.

1.4.4 Nucleation in Nano-Confinement

It is reasonable to postulate that ions lose their individual mobility in the solidification regime (i.e. condensation or clustering). Our estimated coordination numbers are less than unity, i.e. several ions share a water molecule. They are closely packed and largely dehydrated. Figure 1.2b shows a potential minimum for the mica-mica surface force in 60 mM, 100 mM and 1 M KNO_3 solutions at a surface separation of ~1.78–2.1 nm. If the attractive force is interpreted as a van der Waals interaction between the surfaces, we obtain an effective Hamaker constant of $0.78 \pm 0.13 \cdot 10^{-20}$. Such a small Hamaker constant could be explained with a dielectric constant of $\varepsilon \sim 10$–15 and a refractive index $n \sim 1.42$–1.48 for the medium between the mica surfaces, which is comparable, for example, with those of several imidazolium-based ionic liquids with diverse anions [45]. Thus, the confined solution in this regime has structural and dielectric characteristics of a condensed ionic liquid with strong attractive ion interactions. As obtained from the measured solid-liquid interfacial energy (adhesion energy), it does have a smaller cohesive energy than that of potassium nitrate crystal; we can depict an ionic condensate of anions and cations held together by Coulombic, Van der Waals and a decreasing amount of hydration interactions (and water content).

The formation of pre-nucleation clusters in unsaturated solutions has been reported, e.g. for calcium carbonate, and other biominerals [17] under non-confined conditions. This *non-classical* nucleation or crystallization theory considers that ion clusters form in unsaturated solution as a result of entropic effects, which are not considered in the classical theory and already have some structural features. It is speculated that the release of the hydration shells of the ions to form clusters of ionic polymers increases the degrees of freedom of the system, while in classical nucleation only enthalpic terms are considered. The size of the clusters is of the same order of magnitude as the solid layer observed in our experiments ($\approx$2 nm), but there is still no evidence that relates the solid layer with the observed ionic clusters in unconfined solution.

Since the ionic condensate exhibits elastic hard-wall properties, we cannot further elucidate its internal structure with the eSFA. Recently we have measured the electron-density distribution of diverse confined-solutions in the "DLVO" and "solidification" regimes by synchrotron X-ray reflectivity to confirm the multi-layering of hydrated ions [5, 41–43] and the ordered structure of these condensates. This work is still on progress.

1.5 From Nanoscale to Macroscale Phenomena

Water retained in the pores of the Earth's crust, in both saturated and unsaturated zones, can be considered as an alternative source of water e.g. for plant growth and microbiological populations, and therefore, it may contribute to the biological survival of arid areas. To correctly assess the potential soil water capacity and availability in arid areas, it is necessary to correctly establish a mass balance at a large-scale, including the responses to changes in climate and landscape. The system is indeed complex. The host, the soil, is subjected to weathering. In pore water, salts, minerals, organic matter and living organisms are present. Molecular interactions and resulting surface forces between the different species and at the interfaces affect transport and reactive properties of water, which needs to be considered in the net mass balance. This is discussed by means of selected examples in this section.

One goal of the previous section was to emphasize on the advances on understanding nano-scale properties of confined salt solutions in a pore. This can be applied to the properties of interlayer water in clays and to clay-salt interactions. Nano-scale properties rely on the force field across the thin film and lead to structural features of the solution and a confinement-induced dehydration of ions, among others. These characteristics strongly influence the retained water content in clays, and with it, that of flowing (or free) water. Moreover, retention of interlayer water is relative: a narrow path (of the range of a few water molecules) does not prevent the diffusion of water and ions into/out of the film even of hydrated ions larger than the thickness of the film [1], although the transport is retarded compared to that in bulk water. This leads to a continuous exchange of confined water (and ions) with capillary and free water as a function of solute concentration and temperature that effectively affect the transport term.

While water is necessary for microbiologic population, living organisms and organic matter also affect substantially water properties (see [12] and references mentioned there). We have recently reported the measured interactions between selected biomolecules (glycoproteins) adsorbed onto mica surfaces across aqueous salt solutions. The presence of these amphiphilic biomolecules influences notably the structural surfaces forces and shifts the long-range repulsion onset from a few to several tens of nanometers. These organic molecules need to be highly hydrated and accordingly it is to expect that determined biofilms strongly affect water distribution in the soil pore network. Besides, the presence of organic matter in clays might lead to larger interlayer spaces and higher water contents.

Long drying periods lead to strong evaporation and higher concentration of solutes in soil water in the unsaturated zone, and possibly in the saturated zone, too. When reaching a critical super saturation, salts precipitate in pores. As discussed in the previous section, a concentration gradient between thin film and free water can be expected due to structural forces, and it needs to be known to predict crystallization. On one hand, precipitated crystals reduce soil permeability by pore-clogging [15] influencing the transport term in the net mass balance of soil water. If the phase transition involves water (e.g. as part of a hydrated salt crystal), it induces a sink/source in the mass balance. Besides, crystallization might also lead to a retained solution thin film between the crystals and the pore wall even under strong drying conditions [13].

In this chapter it was shown that crystallization in a thin film does not follow the classical theory of nucleation/crystallization, due to acting structural forces. This is a fundamental result towards the elucidation of clay-salt interactions. But also for the crystal growth in large pore spaces, these results are of relevance: if a repulsive force acts between the growing crystal and the pore wall, a thin layer of aqueous remains between the crystal and pore wall, which permits diffusion of the ions to the growing crystal surface [8, 16]. Due to the high ionic strength of the solution, EDL forces are screened, and the repulsion results from structural forces in the electrolyte solution, while the van der Waals force is attractive at salt-mineral interfaces. The pore wall hinders the growth of the crystal in the supersaturated solution, which leads to the action of the crystallization pressure or force of crystallization, dependent on the solution super saturation [46, 47]. Crystallization pressure leads to stresses and eventually to material failure. Hence, the described crystal-pore wall interaction is involved in weathering of soil and rocks, and crack formation, which might open new paths and pores for water accumulation and distribution. The concentration in this thin film (and super saturation) can be (temporary) higher than that of the free water, which might explain "unexpected" salt weathering in some laboratory experiments and in the field.

In contrast, if this thin layer of solution did not exist due to an attraction between the crystal and the pore wall, the crystal would come into contact with the pore wall, the growth would stop, and no crystallization pressure would be exerted [14]. Thus, mineral and crystal surface properties (such as potential, and dielectric properties) and resulting surface forces are key for the understanding of in-pore crystallization and therefore, for the description of soil weathering and landscape shaping induced by salt crystallization. Moreover, soil weathering directly affects soil water distribution and flow.

Thus, surface interactions are strong at nano-scale and are the driving force for many nano-scale phenomena and responsible for the properties of the confined solution. This was investigated by measuring surface forces across a thin film of salt solution in a narrow gap at different concentrations. By means of several examples it was shown that a correct description of the reactive transport behavior of soil water in arid areas and of its distribution, both in the non-saturated and saturated zones, requires understanding the properties of water and solutions in confinement.

Acknowledgements I would like to acknowledge M. Heuberger, N.D. Spencer, J.F. van der Veen, and S. Chodankar for scientific discussions. Technical support for the eSFA was provided by J. Vanicek, M. Elsener and G. Cossu. This work was supported by the Swiss National Science Foundation. Selected figures from [11] reproduced by permission of the PCCP Owner Societies.

References

1. Alcantar N, Israelachvili J, Boles J (2003) Forces and ionic transport between mica surfaces: implications for pressure solution. Geochimica et Cosmochimica Acta 67(7):1289–1304
2. Boström M, Ninham BW (2004) Dispersion self-free energies and interaction free energies of finite-sized ions in salt solutions. Langmuir 20(18):7569–7574. doi:10.1021/la049446+
3. Cahn JW (1980) Surface stress and the chemical equilibrium of small crystals–I. The case of the isotropic surface. Acta Metall 28(10):1333–1338. doi:10.1016/0001-6160(80)90002-4
4. Chan DYC (2002) A simple algorithm for calculating electrical double layer interactions in asymmetric electrolytes–Poisson-Boltzmann theory. J Colloid Interface Sci 245(2):307–310. doi:10.1006/jcis.2001.7942
5. Chodankar S, Perret E, Nygard K, Bunk O, Satapathy DK, Marzal RME, Balmer TE, Heuberger M, van der Veen JF (2012) Density profile of water in nanoslit. EPL (Europhys Lett) 99(2):26001
6. Claesson P, Horn RG, Pashley RM (1984) Measurement of surface forces between mica sheets immersed in aqueous quaternary ammonium ion solutions. J Colloid Interface Sci 100(1):250–263. doi:10.1016/0021-9797(84)90433-8
7. Clementi E, Barsotti R (1978) Study of the structure of molecular complexes. Coordination numbers for Li+, Na+, K+, F- and Cl- in water. Chem Phys Lett 59(1):21–25. doi:10.1016/0009-2614(78)85605-x
8. Correns CW (1949) Growth and dissolution of crystals under linear pressure. Discuss Faraday Soc 5:267–271
9. Derjaguin BV (1934) Untersuchungen über die Reibung und Adhäsion, IV. Kolloid Zeitschrift 69(2):155–164. doi:10.1007/BF01433225

10. Derjaguin BV, Landau L (1941) Theory of the stability of strongly charged lyophobic sols and of the adhesion of strongly charged particles in solution of electrolytes. Acta Physicochim URSS 14:633–662
11. Espinosa-Marzal RM, Drobek T, Balmer T, Heuberger MP (2012) Hydrated-ion ordering in electrical double layers. Phys Chem Chem Phys 14(17):6085–6093. doi:10.1039/C2cp40255f
12. Espinosa-Marzal RM, Fontani G, Reusch F, Spencer ND, Crockett R (2013) Sugars communicate through water: oriented glycans induce long-range water structuring. Biophys J 104(12):2686–2694
13. Espinosa-Marzal RM, Scherer GW (2009) Crystallization pressure exerted by in-pore confined crystals. In: Poro-mechanics IV: proceedings of the fourth biot conference on promechanics, New York, pp 1013–1018
14. Espinosa-Marzal RM, Scherer GW (2010) Advances in understanding damage by salt crystallization. Acc Chem Res 43(6):897–905. doi:10.1021/Ar9002224
15. Espinosa-Marzal RM, Scherer GW (2012) Impact of in-pore salt crystallization on transport properties. Environ Earth Sci. doi:10.1007/s12665-012-2087-z
16. Flatt R, Steiger M, Scherer G (2007) A commented translation of the paper by C.W. Correns and W. Steinborn on crystallization pressure. Environ Geol 52(2):187–203. doi:10.1007/s00254-006-0509-5
17. Gebauer D, Völkel A, Cölfen H (2008) Stable prenucleation calcium carbonate clusters. Science 322(5909):1819–1822. doi:10.1126/science.1164271
18. Heuberger M (2001) The extended surface forces apparatus. Part I. Fast spectral correlation interferometry. Rev Sci Instrum 72(3):1700–1707. doi:10.1063/1.1347978
19. Heuberger M, Vanicek J, Zach M (2001) The extended surface forces apparatus. II. Precision temperature control. Rev Sci Instrum 72(9):3556–3560
20. Ho TA, Argyris D, Cole DR, Striolo A (2011) Aqueous NaCl and CsCl solutions confined in crystalline slit-shaped silica nanopores of varying degree of protonation. Langmuir 28(2):1256–1266. doi:10.1021/la2036086
21. Horn RG, Israelachvili JN (1980) Direct measurement of forces due to solvent structure. Chem Phys Lett 71(2):192–194. doi:10.1016/0009-2614(80)80144-8
22. Horn RG, Smith DT, Haller W (1989) Surface forces and viscosity of water measured between silica sheets. Chem Phys Lett 162(4–5):404–408. doi:10.1016/0009-2614(89)87066-6
23. Israelachvili J, Wennerstrom H (1996) Role of hydration and water structure in biological and colloidal interactions. Nature 379(6562):219–225
24. Israelachvili JN (1991) Intermolecular and surface forces, 2nd edn. Elsevier, New York
25. Israelachvili JN, Adams GE (1978) Measurement of forces between 2 mica surfaces in aqueous-electrolyte solutions in range 0–100 Nm. J Chem Soc Faraday Trans 1 74:975–1001
26. Israelachvili JN, Pashley RM (1983) Molecular layering of water at surfaces and origin of repulsive hydration forces. Nature 306(5940):249–250
27. Israelachvili JN, Tabor D (1972) The measurement of van der Waals dispersion forces in the range 1.5 to 130 nm. Proc R Soc Lond A Math Phys Sci 331(1584):19–38. doi:10.1098/rspa.1972.0162
28. Kjellander R, Marcelja S (1984) Correlation and image charge effects in electric double layers. Chem Phys Lett 112(1):49–53. doi:10.1016/0009-2614(84)87039-6
29. Kjellander R, Marcelja S (1985) Inhomogeneous Coulomb fluids with image interactions between planar surfaces. I. J Chem Phys 82(4):2122–2135. doi:10.1063/1.448350
30. Kjellander R, Marcelja S (1986) Double-layer interaction in the primitive model and the corresponding Poisson-Boltzmann description. J Phys Chem 90(7):1230–1232. doi:10.1021/j100398a006
31. Kjellander R, Marcelja S (1986) Interaction of charged surfaces in electrolyte solutions. Chem Phys Lett 127(4):402–407. doi:10.1016/0009-2614(86)80304-9
32. Kjellander R, Marcelja S, Quirk JP (1988) Attractive double-layer interactions between calcium clay particles. J Colloid Interface Sci 126(1):194–211. doi:10.1016/0021-9797(88)90113-0

33. Lee SS, Fenter P, Park C, Sturchio NC, Nagy KL (2010) Hydrated cation speciation at the muscovite (001) water interface. Langmuir 26(22):16647–16651. doi:10.1021/la1032866
34. Malani A, Ayappa KG, Murad S (2006) Effect of confinement on the hydration and solubility of NaCl in water. Chem Phys Lett 431(1–3):88–93. doi:10.1016/j.cplett.2006.09.071
35. Miklavic SJ, Ninham BW (1990) Competition for adsorption sites by hydrated ions. J Colloid Interface Sci 134(2):305–311. doi:10.1016/0021-9797(90)90140-j
36. Ninham BW (2002) Physical chemistry: the loss of certainty. In: Nylander T, Lindman B (eds) Lipid and polymer-lipid systems, vol 120, Progress in colloid and polymer science. Springer, Berlin/Heidelberg, pp 1–12. doi:10.1007/s02882002
37. Ninham BW, Yaminsky V (1997) Ion binding and Ion specificity: the hofmeister effect and onsager and lifshitz theories. Langmuir 13(7):2097–2108. doi:10.1021/la960974y
38. Pashley RM (1981) DLVO and hydration forces between mica surfaces in Li+, Na+, K+, and Cs + electrolyte solutions: a correlation of double-layer and hydration forces with surface cation exchange properties. J Colloid Interface Sci 83(2):531–546. doi:10.1016/0021-9797(81)90348-9
39. Pashley RM, Israelachvili JN (1984) Molecular layering of water in thin films between mica surfaces and its relation to hydration forces. J Colloid Interface Sci 101(2):511–523. doi:10.1016/0021-9797(84)90063-8
40. Perkin S, Goldberg R, Chai L, Kampf N, Klein J (2009) Dynamic properties of confined hydration layers. Faraday Discuss 141:399–413
41. Perret E, Nygard K, Satapathy DK, Balmer TE, Bunk O, Heuberger M, van der Veen JF (2009) X-ray reflectivity reveals equilibrium density profile of molecular liquid under nanometre confinement. EPL Europhys Lett 88(3):36004–36010. doi:Artn 36004, doi:10.1209/0295-5075/88/36004
42. Perret E, Nygard K, Satapathy DK, Balmer TE, Bunk O, Heuberger M, van der Veen JF (2010) X-ray reflectivity theory for determining the density profile of a liquid under nanometre confinement. J Synchrotron Radiat 17(4):465–472. doi:10.1107/S0909049510014858
43. Perret E, Nygard K, Satapathy DK, Balmer TE, Bunk O, Heuberger M, van der Veen JF (2010) Molecular liquid under nanometre confinement: density profiles underlying oscillatory forces. J Phys Condens Matter 22(23):235102
44. Shubin VE, Kekicheff P (1993) Electrical double layer structure revisited via a surface force apparatus: mica interfaces in lithium nitrate solutions. J Colloid Interface Sci 155(1):108–123. doi:10.1006/jcis.1993.1016
45. Singh T, Kumar A (2008) Static dielectric constant of room temperature ionic liquids: internal pressure and cohesive energy density approach. J Phys Chem B 112(41):12968–12972. doi:10.1021/jp8059618
46. Steiger M (2005) Crystal growth in porous materials – I: the crystallization pressure of large crystals. J Cryst Growth 282(3–4):455–469. doi:10.1016/J.Jcrysgro.2005.05.007
47. Steiger M (2005) Crystal growth in porous materials – II: influence of crystal size on the crystallization pressure. J Cryst Growth 282(3–4):470–481. doi:10.1016/J.Jcrysgro.2005.05.008
48. Tabor D, Winterton RHS (1969) The direct measurement of normal and retarded van der Waals forces. Proc R Soc Lond Ser A Math Phys Sci 312(1511):435–450
49. Verwey EJ, Overbeek JTG (1948) Theory of the stability of lyophobic colloids. Dover Publications, Amsterdam
50. Xu L, Salmeron M (1998) Effects of surface ions on the friction and adhesion properties of mica. Langmuir 14(8):2187–2190. doi:10.1021/la9713659

Part I
Flow, from Nano- to Mega-Scale

Transport processes in soil occur on a dazzling range of scales: the processes that determine molecular water absorption at the molecular level occur on a length scale that is some 12 orders of magnitude smaller than the depth of some aquifers. The processes at these vastly different scales can be studied in separation but will finally have to be studied in their interrelationship. The contributions in this part deal with analyses at various length scales somewhere in this tremendous range: they reach from the scale of molecular flow via nanoscale and microscale flow processes to meter-scale systems. We will see that the processes at the molecular scale can still be studied in isolation in single channels and pores, but that the processes at larger scales typically are many-channel problems that add the problem of connectivity. In soils there even is an added complexity: here water transport is coupled to air transport and dried-out regions can occur that isolate patches of water-filled soil. Even at the very large scales of soils however, the molecular-scale forces will still be present and play a highly important role.

Different length scales will bring about different types of flow. At the largest length scales unbounded water will generally flow in a turbulent manner. In rivers or lakes we find this type of flow, as well as in underground streams. When we go down to the scale of the channels in between grains of sand, the wall is never far away and the increasing influence of the friction with the wall will strongly determine the path that the flow chooses, giving rise to so-called viscous or laminar flow. When the length scale is further decreased, the influence of the wall will become increasingly larger. At typical length scales of 100 nm, the surface charge of chemical origin that is present on almost all materials, makes itself felt. This surface charge attracts ions of the opposite sign forming the so-called electrical double layer. When the ions in this electrical double layer are transported by solution flow, an electrical field is generated which pushes back the ions against the flow direction. A seemingly increased viscosity is then the result, and for this reason the phenomenon is called the electroviscous effect. At even smaller scales finally, the viscosity between hydrophilic surfaces really increases because the molecularly thin layer of water directly adjacent to the wall actually does have a higher viscosity. Between walls that have little affinity for water however, such as the walls of carbon nanotubes, water can almost flow frictionless. This stark contrast precisely demonstrates the point that the surface instead of the bulk determines flow behaviour at these smallest of scales. At the smallest length scales furthermore any description of water as a continuum with bulk properties breaks down and must give way to a molecular description, vastly increasing theoretical complexity.

In the above paragraph we considered single phase flow of aqueous solutions through conduits or pores. When soil dries out, water transport occurs simultaneously with air transport. Water transport then often occurs only in thin adsorbed films and along corners, where the capillary forces determine the shape of the adsorbed volumes of water. In dried-out or half dried-out regions vapour diffusion furthermore will carry the water across dried-out regions. The variable connectivity between different regions of the soil will now vastly increase the complexity of the water transport above the complexity already observed in channel networks entirely filled with water.

Under part I we will see examples of all these length scales mentioned above, and of both single-phase and two-phase flow in both artificial channels and real soil samples. Coming to a description of the various contributions in this part, the contribution of Youn et al. focuses on water transport in carbon nanotube membranes, which represents the smallest length scale of single water molecules. At the smallest scales surface properties dominate the bulk properties, and in this case the low affinity of the carbon nanotubes for water leads to the surprising and highly important phenomenon that the hydrodynamic flow resistance of carbon nanotubes practically collapses. The authors show that this fact, together with the exclusion of ions from thc tubcs by stratcgically placcd ionic 'guard molcculcs' at the pore entrances, can lead to very promising low-energy cost desalination membranes.

One scale up we find the contributions of Tas et al. and Bergonzi et al. Both research teams use silica glass chips with artificial enclosed one-dimensional nanochannels of dimensions from five up to a few hundreds of nanometers. The contribution of Tas et al. describes two-phase flow in such channels. At the small dimensions used the interfacial forces are huge, again due to the enormous surface to volume ratio. The authors show that these enormous interfacial forces cause mechanical deformation of the environment, an effect that will also be relevant in soils. Another surface-related effect is the influence of surface charge leading to an added resistance to solution flow because the generated charge transport generates backward transport. Their contribution analyses this so-called electroviscous effect and tries to separate it from the effect of the molecularly thin layer with decreased viscosity directly adjacent to the capillary wall.

Bergonzi and Mercury use the same silica glass nanochannels and perform FTIR measurements on the water confined therein. They analyse these measurements to acquire information on the thermodynamic properties of the confined water over a wide temperature range. Ultimately their analysis leads to knowledge on the forces that keep the water attached to surfaces in soils and as such is highly relevant for the description of soils.

The next contribution comes from Berkowitz. We now scale up to the microscale level and with this growing scale the added complexity of connectivity arises. Berkowitz' contribution thus addresses problems that are related to the transport of both water and air in interconnected networks. By careful experiments using a glass-etched porous 2D micropore model he demonstrates that porous media don't dry in a simple isotropic fashion. Instead the connectivity together with the capillary forces often causes water to stay behind in isolated clusters. At subsequent refilling of the system pockets of 'old water' can still remain, even after the 'new water' ran in. This study is of considerable importance to understand the fate of solutes (pollutants) in soil.

Processes at the largest scale finally are addressed by Assouline et al. By careful experiments and theoretical analysis on artificial soils systems they show that heterogeneities in the soil cause permeability gradients, that will strongly influence evaporation from the soil. They point at the importance of a continuous hydraulic connection between water at different locations in the soil. When the permeability is for example such that water in deep layers for a long time remains connected by a water column to the water at the soil surface, very rapid drying can occur. It should be pointed out that even though the processes observed here typically occur at the meter-scale, one of the important forces in the system has its origins at the pore-scale, namely the capillary force.

Summarizing this part we see a number of studies at different scales that illustrate the different physics at play at different length scales. The physics at the small length scales however does not become irrelevant at the longer scales, but continues to be present and exert its influence. This causes the extreme richness of this field and forms its largest challenge.

Jan C.T. Eijkel

Chapter 2
Confined Water in Carbon Nanotubes and Its Applications

Seul Ki Youn, Jakob Buchheim, and Hyung Gyu Park

Abstract Unique nanoscale transport properties of carbon nanotubes (CNTs) have inspired researchers for over a decade, initially with their analogies to various biological pores and later with the potential impact on water purification. Water can permeate through a nanometer-wide pipe of the CNT interior at rates far exceeding those predicted by Hagen-Poiseuille formulation and measured in nano conduits of different material, attributed to nano confinement, hydrophobicity, and smooth potential energy landscape. Also, chemical addition to the nanotube ends was found effective in electrostatic exclusion of ions without much loss of water permeability, suggesting the emergence of CNT membranes for desalination and purification of water resources. This article introduces *Carbon Nanotube Nanofluidics* by capturing important findings and progresses made in the early developments of the area.

2.1 Introduction

Limited access to clean water defines one of the global sustainability issues facing humanity today. Growing concerns of freshwater feeding to a significant portion of the world population are calling for immediate attention from science and technology [1, 2], pertaining to efficient seawater desalination and wastewater purification. Membrane-based water purification is a popular, promising and widely spread technological option today. Nanofiltration, reverse osmosis, and forward osmosis using polymer membranes have been considered actively as a means for clean water supply. Recent advances in nanotechnology and nanomaterial synthesis

S.K. Youn • J. Buchheim • H.G. Park (✉)
Nanoscience for Energy Technology and Sustainability, Department of Mechanical and Process Engineering, Swiss Federal Institute of Technology Zurich (ETH Zurich), Sonneggstrasse 3, Zurich CH-8092, Switzerland
e-mail: parkh@ethz.ch

L. Mercury et al. (eds.), *Transport and Reactivity of Solutions in Confined Hydrosystems*, NATO Science for Peace and Security Series C: Environmental Security,
DOI 10.1007/978-94-007-7534-3_2,

have allowed us to approach to scales comparable with individual membrane pore, thereby enabling rational design and manufacturing of membranes and opening up new avenues to processing water energy- and cost-efficiently.

Carbon nanotubes provide unique structures in studying nanoscale mass transport and flow phenomena that can be employed in a variety of important applications. Having nanometer-scale diameters of innermost walls whose graphitic surface is atomically smooth, they give rise to newly discovered phenomena of ultra-efficient transport of water through these ultra-narrow molecular pipes [3]. These findings lead to a vision of permeable membranes incorporating carbon nanotube pores for a filtration that could possibly address the global water stress issue. Important findings and progresses in the emerging field of so called "*Carbon Nanotube Nanofluidics*" are presented in this article.

2.2 Predictions of Water Transport in Carbon Nanotube

The task of observing and understanding water flows in CNT pores raises a set of unique fundamental questions. First, it is surprising that polar liquids such as water can fill narrow hydrophobic CNTs. Two arguments against water infilling of the CNT interior are based on considerations of energetics and capillarity. In order for a water molecule to move from bulk water to such a narrow CNT permitting only a single-file water configuration, it should be given average energy of approximately two hydrogen bonds. This unfavorable energetics was expected to prohibit the entrance of water in narrow CNTs. The capillarity argument is based on the Young-Laplace equation [4], which predicts that a very large pressure force would be needed to fill the narrow CNT pore with water. However, counterarguments can also exist. At a given temperature, water molecules share hydrogen bonding energy as a statistical distribution. It is unclear whether this energy distribution will take the same shape in bulk water and under extreme confinement, so called nanoconfinement. In addition, the capillarity argument is based on a continuum assumption in a way that a single contact angle value can be defined between water and CNT mouth. Nevertheless, it is possible that thermal fluctuation of a confined system would increase the variation in the contact angle values statistically, complicating the applicability of the continuum-based prediction. The first answer to this debate was presented by a molecular dynamics (MD) simulation. Hummer and colleagues [5] used MD simulations to observe the infilling of an extremely narrow CNT (0.81 nm in diameter and 1.34 nm in length) with water molecules. According to their simulation result it is surprising that water molecules filled the empty cavity in a CNT within a few tens of picoseconds and resided there for the entire simulation time of 66 ns. More importantly, the water molecules under such a high degree of nanoconfinement formed a single-file configuration, unseen in bulk water but observed in some biological channels [5]. The simulation allowed for comparison between binding energy distributions of water at bulk and under nanoconfinement of CNT (*i.e.*, nanotube water). Average binding energy is favorable for bulk water, but it is the weakly bound states that are reportedly less populated in nanotube water owing to the narrowed distribution of water molecules in the binding energy space under the nanoconfinement. As the result chemical potential of the nanotube water is lowered, and the probability of finding water molecules in the open CNT interior increases. This counterintuitive result could yet be a possible one partly because CNT is hydrophobic but not superhydrophobic [6]. For example, a slight (25 %) increase in hydrophobicity in the carbon-water interaction potential could turn the stable water infiltration to bi-stability between water-filled and empty states, while 40 % more hydrophobic interaction completely emptied the CNT.

The second fundamental question is regarding the dynamics of water under the CNT nanoconfinement. If water molecules can reside preferably in the CNT interior, how would they diffuse along the nanotube? This question about dynamics could be addressed by considering the shearing interaction between water molecules and graphitic channels. Driven in an axial direction, each water molecule atop the CNT wall would experience an energy landscape set by the overall potential interaction between water molecules and carbon graphitic crystals. It has been shown that the energy landscape can be smoothened if the graphitic surface takes a positive curvature, as known as an egg carton analogy, like in the case of water confined in a CNT [7]. Each water molecule right near the CNT wall could then experience less energy consumption to overcome the energy hills and valleys, hinting that atomic-scale smoothness of the wall may help water reduce the friction. Each rolling event of a water molecule over a pair of the potential hill and valley can then be modeled as a chemical reaction process. Babu and Sathian [8] has modeled this process by the transition state theory to suggest that wall shear stress, or dynamic friction, of water that shears the CNT channel can be expressed as a hyper sine function of the axial transport velocity, which suggests that the water transport rate will increase in proportion to the applied pressure drop, linearly for moderate flow rates and nonlinearly for high speed flows.

The single-file water wire is unique in that they move collectively. When mechanically biased by applied pressure drop in an MD domain the water wire showed a fast conduction through the nanotube, a speed comparable to the transport rate in aquaporins (4–6 H_2O/ns) [9]. In a hydrophilic channel as narrow as these CNTs, in contrast, water molecules cannot diffuse very fast because of a strongly attractive interaction with the constituent atoms of the channel wall, causing a strong friction mechanism. As the CNT dimension enlarges and the confinement is relaxed it becomes possible that the nanotube interior accommodates more "water wires," and at large diameters, the confinement-induced hydrogen bond ordering disappears. As the result, the water hydrogen bond network takes a bulk configuration at the core part of the nanotube while remaining somewhat ordered in the very vicinity of the wall [6].

Obviously, dynamics of the water transport in CNTs will hinge primarily on the hydrogen bond configurations at various graphitic nanoconfinements. MD simulations have attempted to provide an insight of the enhancement of water transport with reference to continuum flow formalism, or Hagen-Poiseuille flow [7, 10]. These models agree qualitatively to each other in that the enhancement factor,

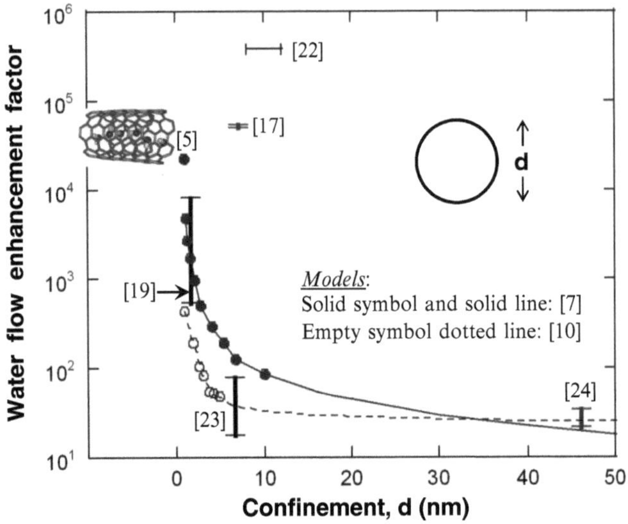

Fig. 2.1 Water flow enhancement over the Hagen-Poiseuille flow with respect to the CNT diameter. Comparison among different models and experiments

defined as the ratio of a water flow rate in a CNT to that of Hagen-Poiseuille formula at the identical tube diameter, varies from quite high values of 10^2–10^3 orders of magnitude at sub- to few nanometers of the CNT diameter to nearly unity at large diameters. For the detailed relationship between the enhancement and the nanotube diameter, nevertheless, they do not agree by exhibiting quantitatively distinct functions (Fig. 2.1). Such disparity between models hints that, although much progress has been made, the clear understanding of the water transport in CNTs is not yet achieved completely. For example, detailed friction mechanism demands careful investigations from experimental and theoretical ends. In the current understanding the water transport enhancement in CNTs is attributed to the nanotube properties including atomic level smoothness and hydrophobicity [11], as well as smoothening of the potential energy landscape related to nanotube curvature [7]. According to this frame of understanding, water transport in a strong nanoconfinement of the CNT is described as parallel trains of several hydrogen-bonded molecular wires traveling quite concertedly aloof from the wall (Fig. 2.2). This picture may allow drawing a plug-like velocity profile for water flows in a CNT [12].

2.3 Experimental Verifications

Despite a number of efforts to verify the aforementioned predictions with some breakthrough achievements, the field of *Carbon Nanotube Nanofluidics* today is still featured by imbalance between many simulations and rare experiments. The absence

Fig. 2.2 Trains of hydrogen-bonded water wires transporting in CNTs

of sufficient experimental data has hampered our detailed and comprehensive understanding of the water transport under graphitic nanoconfinement. Nonetheless, there exist several outstanding achievements worthy of notice.

Most of the early experimental investigations of CNT fluid structure and transport have been inspired by Hummer and colleagues' MD modeling studies. The first experimental evidence of water infiltration of the CNT interior was obtained by Kolesnikov et al. [13] and Gogotsi et al. [14, 15] Using neutron scattering methods, Kolesnikov et al. have probed the structure of water in 1.4-nm-diameter, single-walled CNTs to confirm the residence of water molecules inside the CNTs. At cryogenic temperatures they could observe a single-file water wire surrounded by a water nanotube of square lattices. Later, Gogotsi and colleagues observed liquid molecules trapped in CNTs by transmission electron microscopy, which provides direct evidence that liquid water can favorably reside inside CNTs at room temperature.

The next experimental task was to measure directly the water transport dynamics in CNTs. The challenge of the direct counting of water molecules is evident. Currently available direct flow measurement techniques may find limited applicability to molecular transport through a single CNT, illustrated by the range of flow rates predicted by MD simulations. For instance, Kalra et al.'s prediction [9] of *ca.* 6 H_2O/ns per 0.8-nm-diameter CNT at around 430 atm converts to *ca.* 12 H_2O/ns-nm^2. Assuming the bulk density of liquid water, this water flux is converted to a volume flow rate lower than one atoliter per second, extremely small for the present liquid flow measurement techniques.

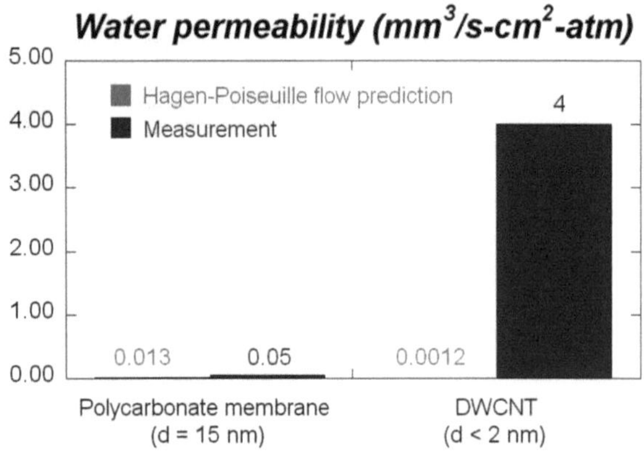

Fig. 2.3 Water permeation through sub-2-nm double-walled CNTs (*DWCNT*) and a 15-nm-diameter polycarbonate membrane relative to Hagen-Poiseuille flow prediction

One strategy of overcoming this challenge is to manufacture a manifold device out of a large number of aligned CNTs, called a CNT membrane. The manufacturing strategy is rather simple. First, CNTs are grown in a vertically aligned way with the end tip of each being capped. Gap space among these aligned CNTs is filled with a matrix material such as polymer or ceramic. Finally, the nanocomposite film embedded with aligned CNTs spanning across the film is etched to cut open the hollow interior of CNTs. Thus obtained structure is a thin film porous membrane where the hollow interior of CNTs serves as graphitic pores. Two groups have pioneered this membrane manufacturing avenue. Hinds and colleagues [16, 17] have fabricated a CNT membrane out of vertically-aligned multiwalled CNTs of 6–7 nm inner diameters spanning a polystyrene thin film. They have reported orders of magnitude rapid water transport relative to classical hydrodynamics predictions. This membrane was later used to take a biomimetic approach of "gate keeping" applied to macromolecules separation *via* hindered diffusion [18]. Water transport under a stronger graphitic nanoconfinement was measured by another type of CNT membrane employing narrow (1.6-nm-diameter) CNTs. Bakajin and colleagues adopted a MEMS-compatible microfabrication process to produce this CNT membrane [19]. This membrane allowed a closer comparison to the MD simulation results. Indeed they obtained two-to-three orders of magnitude enhanced water permeation rates relative to the continuum hydrodynamics prediction (Fig. 2.3), in great agreement with MD predictions [9]. Later, Fornasiero et al. [20] attached electrostatic charge to the membrane pore by tethering carboxyl groups at the CNT pore entrance. At pH values of the membrane feed side higher than the carboxylic pKa value (*ca.* 4.8), the deprotonated carboxyl groups generated a negatively charged electric double layer that could screen the CNT pore from anions entrance at feed concentrations around those of brackish water (a few tens of millimolar) [21]. It is then the feed-side electroneutrality requirement that could also impede the permeation of cations, thereby completing the charge exclusion mechanism. It was turned out that difference in ionic strength and mobility between anion and cation determines the ion exclusion or permeation by the CNT

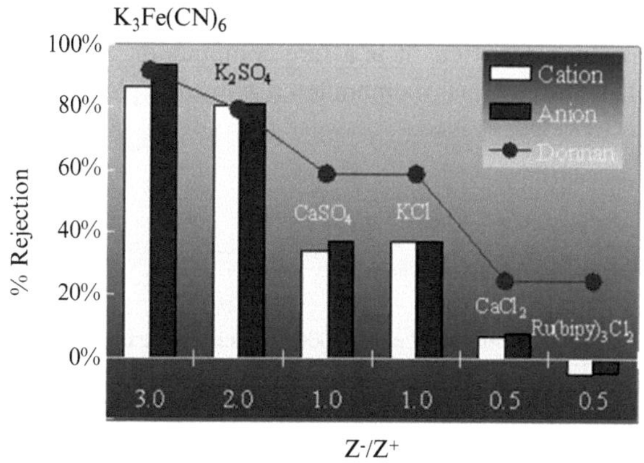

Fig. 2.4 Rejection coefficients measured for two CNT membranes and six salt solutions with the same equivalent concentration but different ion valence (Z)

membrane: Donnan equilibrium mechanism (Fig. 2.4). For example, KCl rejection by laboratory manufactured CNT membranes is around 40 % [20, 21], rather small but posing huge room for improvement by further studies.

2.4 Perspective

The technological impact of such rapid water flows established above involves a concept of an ultrapermeable membrane. This is a membrane that can be operational at far smaller pumping power than a few hundred bars in usual membrane applications including seawater desalination of a reverse osmosis type. For instance, water permeation through sub-2-nm CNT membranes has excelled the performance of commercial membranes [19]. Additionally established, Coulombic exclusion of ions in aqueous solutions suggests the use of the CNT membrane in water desalination, purification, waste water treatments, and extra-purification for special purposes.

Findings from MD simulations suggest that CNTs can be used in applications in which single-file water conduction is useful. For example, aquaporins almost exclusively permeate water molecules (and no ions) at speeds not low at all. Hence, one could envision seawater desalination and wastewater purification by membranes to which highly narrow CNTs serve as pores.

2.5 Conclusion

Nanoscale confinement by molecularly smooth hydrophobic graphitic walls gives rise to the unique phenomenon of ultra-efficient water transport, by some orders of magnitude faster than through other pores of comparable size. MD simulations and experimental measurements of water transport and ion exclusion are underscoring the great promise of CNT membranes in water desalination, water purification, and nanofiltration. It is a rare combination of transport efficiency and selectivity that makes CNT membranes a powerful platform for the next generation water technology.

References

1. Schwarzenbach RP et al (2006) The challenge of micropollutants in aquatic systems. Science 313:1072–1077
2. Shannon MA et al (2008) Science and technology for water purification in the coming decades. Nature 452:301–310
3. Noy A, Park HG, Fornasiero F, Holt JK, Grigoropoulos CP, Bakajin O (2007) Slippery nanopipes: nanofluidics in nanotubes. Nano Today 2:22–29
4. Whitby M, Quirke N (2007) Fluid flow in carbon nanotubes and nanopipes. Nat Nanotechnol 2:87–94
5. Hummer G, Rasaiah JC, Noworyta JP (2001) Water conduction through the hydrophobic channel of a carbon nanotube. Nature 414:188–190
6. Werder T, Walther JH, Jaffe RL, Halicioglu T, Koumoutsakos P (2003) On the water-carbon interaction for use in molecular dynamics simulations of graphite and carbon nanotubes. J Phys Chem B 107:1345–1352
7. Falk K, Sedlmeier F, Joyl L, Bocquet L (2010) Molecular origin of fast water transport in carbon nanotube membranes: superlubricity versus curvature dependent friction. Nano Lett 10:4067–4073
8. Babu JS, Sathian SP (2012) Combining molecular dynamics simulation and transition state theory to evaluate solid-liquid interfacial friction in carbon nanotube membranes. Phys Rev E 85:051205
9. Kalra A, Garde S, Hummer G (2003) Osmotic water transport through carbon nanotube membranes. Proc Natl Acad Sci U S A 100:10175–10180
10. Thomas JA, McGaughey AJH (2008) Reassessing fast water transport through carbon nanotubes. Nano Lett 8:2788–2793
11. Joseph S, Aluru NR (2008) Why are carbon nanotubes fast transporters of water? Nano Lett 8:452–458
12. Hanasaki I, Nakatami A (2006) Structure and stability of water chain in a carbon nanotube. J Chem Phys 124:144708
13. Kolesnikov AI et al (2004) Anomalously soft dynamics of water in a nanotube: a revelation of nanoscale confinement. Phys Rev Lett 93:035503
14. Naguib N et al (2004) Observation of water confined in nanometer channels of closed carbon nanotubes. Nano Lett 4:2237–2243
15. Maniwa Y et al (2007) Water-filled single-wall carbon nanotubes as molecular nanovalves. Nat Mater 6:135–141
16. Hinds BJ, Chopra N, Rantell T, Andrews R, Gavalas V, Bachas LG (2004) Aligned multiwalled carbon nanotube membranes. Science 303:62–65
17. Majumder M, Chopra N, Andrews R, Hinds BJ (2005) Nanoscale hydrodynamics: enhanced flow in carbon nanotubes. Nature 438:44
18. Majumder M, Chopra N, Hinds BJ (2005) Effect of tip functionalization on transport through vertically oriented carbon nanotube membranes. J Am Chem Soc 127:9062–9070
19. Holt JK, Park HG, Wang YM, Stadermann M, Artyukhin AB, Grigoropoulos CP, Noy A, Bakajin O (2006) Fast mass transport through sub-2-nanometer carbon nanotubes. Science 312:1034–1037

20. Fornasiero F, Park HG, Holt JK, Stadermann M, Grigoropoulos CP, Noy A, Bakajin O (2008) Ion exclusion by sub-2-nm carbon nanotube pores. Proc Natl Acad Sci U S A 105:17250–17255
21. Fornasiero F et al (2010) pH-tunable ion selectivity in carbon nanotube pores. Langmuir 26:14848–14853
22. Du F, Qu L, Xia Z, Feng L, Dai L (2011) Membranes of vertically aligned superlong carbon nanotubes. Langmuir 27:8437–8443
23. Park HG (2007) Mass transport through carbon nanotubes. PhD dissertation, University of California Berkeley
24. Whitby M, Cagnon L, Thanou M, Quirke N (2008) Enhanced fluid flow through nanoscale carbon pipes. Nano Lett 8:2632–2637

Chapter 3
Static and Dynamic Capillarity in Silicon Based Nanochannels

Niels Tas, Nataliya Brunets, Joost W. van Honschoten†, Jeroen Haneveld, and Henri V. Jansen

Abstract In this chapter we review the fabrication of silicon based nanochannels and their use in capillarity studies. Static capillarity measurements of the pressure in isolated liquid plugs confined in hydrophilic nanochannels, confirm the existence of capillary negative pressure, quantitatively in accordance with the Young-Laplace equation. The negative pressure can be quantified through measurement of the elasto-capillary deformation of the channel capping due to the pressure difference with the atmospheric pressure. By measuring the capillary filling dynamics in nanochannels of uniform and accurately defined height, different (apparent) viscosity effects in confinement have been revealed. One effect (visible in insulating sub-100-nm channels) is likely to be related to the influence of the electrical double layer (an electroviscous effect), while the other effect (visible in conductive sub-50 nm channels) seems to be related to a decrease in the effective channel due to a thin immobile layer close to the polar or charged channel wall.

3.1 Introduction

We define capillarity as a change in the chemical potential of the liquid in a two phase system, due to the curvature of the meniscus in confinement. Capillary action plays an essential role in the retention of water in the unsaturated zone (the capillary groundwater). Related phenomena, as for example cavitation and bubble formation by enclosure of gas will have a large influence on transport in the unsaturated

---

†Deceased January 12th 2011

N. Tas (✉) • N. Brunets • J.W. van Honschoten† • J. Haneveld • H.V. Jansen
Department of Transducers Science and Technology, MESA+ Institute for Nanotechnology, University of Twente, P.O. Box 217, 7500 AE Enschede, The Netherlands
e-mail: n.r.tas@utwente.nl

L. Mercury et al. (eds.), *Transport and Reactivity of Solutions in Confined Hydrosystems*, NATO Science for Peace and Security Series C: Environmental Security,
DOI 10.1007/978-94-007-7534-3_3,

zone. Silicon-based nanotechnology offers techniques for precise fabrication of artificial nano-pores. Often these channels are planar (in the plane of the substrate wafer) facilitating microscope observation of liquid inside of these channels. The combination of well-defined dimensions and observability makes silicon-based nanochannels an ideal model system for fundamental studies in capillarity. In Sect. 3.2 we will briefly review the main procedures to fabricate these nanochannels. We will then move on to discuss measurements of static capillarity in nanochannels and introduce the concept of capillary negative pressure (Sect. 3.3). Section 3.4 will be devoted to discuss experiments of dynamic capillarity, i.e. the capillary filling of nanochannels. These experiments give information on the (apparent) viscosity of water and aqueous salt solutions in confinement, a topic which is relevant to water transport in both the saturated and unsaturated zone.

3.2 Fabrication of Silicon Based Nanochannels[1]

Figure 3.1 illustrates the three main techniques for the creation of (1D-confined) nanochannels: Sacrificial layer etching (left), bulk etching (middle), and spacer layer technique (right). The sacrificial layer technique to produce nanochannels is based on the thin film encapsulation of a strip of sacrificial material and the subsequent selective removal of the sacrificial layer through access holes [1, 2] Typical combinations of materials are silicon oxide as the sacrificial layer and polysilicon as the structural material, or polysilicon as the sacrificial layer and silicon oxide or silicon nitride as the structural material. The latter combination provides optically transparent nanochannels. A practical issue in this fabrication scheme is the finite selectivity of the etchant during removal of the sacrificial layer,

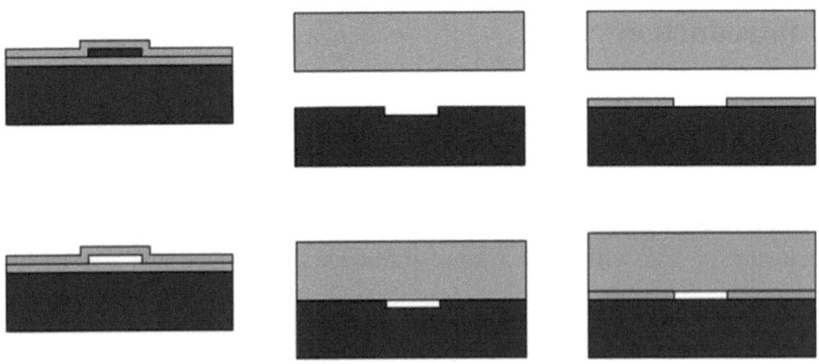

Fig. 3.1 The three main techniques for the creation of (1D-confined) nanochannels: Sacrificial layer etching (*left*), bulk etching (*middle*), and spacer layer technique (*right*)

[1]This section is based on: van Honschoten JW, Brunets N, Tas NR (2012) Capillary action in silicon based nanochannels In: Ondarcuhu T, Aime JP (eds) Nanoscale liquid interfaces: wetting, patterning and force microscopy at molecular scale. Pan Stanford Publishing, Singapore.

resulting in a tapered channel height (in the length direction). In the bulk etching technique [3] a trench is slowly and accurately etched into the silicon substrate. Wet anisotropic etching can produce the required smoothness and uniformity [4]. An enclosed channel is formed by bonding of a second wafer to the substrate wafer containing the trench. This second wafer can be a glass wafer producing an optically transparent channel capping. Using wet etching [4] uniform channel heights typically in the range of 20–100 nm can be produced, with an accuracy of a few nanometer. To define the channel height for even smaller channels, the spacer layer technique [5, 6] is particularly suitable. Uniform and well defined spacer layers of silicon oxide can be created on a silicon substrate by thermal oxidation. Their thickness can be accurately measured before wafer bonding by ellipsometry, creating channels as small as 5 nm in height with a precision better than 1 nm [7]. Note that in both the bulk etching and the spacer layer technique, in practice, at least a native oxide layer of 1– 2 nm silicon oxide is formed prior to wafer bonding, ensuring complete silica-based channel walls.

3.3 Static Capillarity in Nanochannels[2]

In sufficiently narrow lyophilic capillaries the negative Laplace pressure can exceed the atmospheric pressure (often 1 bar in absolute value), and a negative absolute pressure can develop in the liquid. Liquids can withstand a considerable tension due to the cohesive forces between the molecules. The tensile stress developed when the liquid is stretched is isotropic normal stress, therefore it is justified to refer to it as a negative pressure. Reports about liquids under tension date back as far as the late seventeenth and early eighteenth century, and originate from some of the greatest scientists of all time, including Huygens [8] and Newton [9]. They address the phenomenon that when a tube of a Torricelli barometer is completely filled with mercury and then perfectly purged of air and inverted, the mercury inside sometimes stayed top-full in the tube, suspended to heights as much as 75 in. [8]. This is much higher than the expected 29 inches caused by the weight of the atmosphere. In particular the explanation of Newton is remarkably accurate, referring to the cohesion of the liquid particles to each other and to the glass composing the tube [9]. Tension in water plays a central role in the cohesion-tension theory of sap ascent in trees as first proposed by Dixon and Joly [10], and has therefore drawn the interest of plant physiologists. Classical laboratory methods to induce negative pressures in liquids are the Berthelot tube method [11] based on isochoric cooling, and the rotating capillary technique introduced by Briggs [12].

Bulk liquid at negative absolute pressure is superheated, which is a metastable state. It is a metastable state because the system can move to a state of lower free energy by creating cavities filled by vapour, however to create a cavity (vapour bubble) an energy barrier has to be overcome [13]. If the liquid is at a pressure

---

[2]This section is based on: van Honschoten JW, Brunets N, Tas NR (2010) Chem Soc Rev 39: 1096.

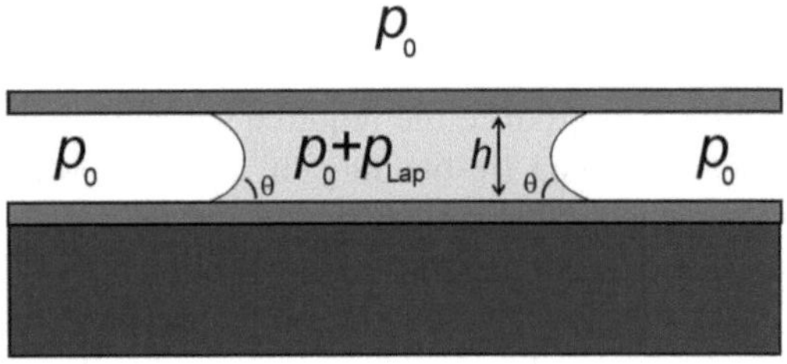

Fig. 3.2 Longitudinal section of a partially filled flat rectangular capillary. The pressure in the liquid is equal to the ambient pressure p_0 plus thc negative Laplace pressure p_{Lap}. The channel height is h the contact angle of the liquid to the channel walls is represented by θ

p, then the work required to create a bubble of volume V and surface area A, filled with vapour at a pressure p_v is given by [13]:

$$W = \gamma A + (p - p_V)\,V = 4\pi r^2\gamma + \frac{4}{3}\pi r^3\,(p - p_v) \tag{3.1}$$

where r is the radius of the bubble, and γ is the surface tension of the liquid. Assuming that the magnitude of p is much larger than p_v, the work required to create the bubble has a maximum $16\pi\gamma^3/3p^2$ for a critical radius [13]:

$$r_{crit} = -2\gamma/p \tag{3.2}$$

The capillary action in narrow lyophilic tubes, in open communication with the surrounding atmosphere, can easily be stronger in magnitude than the surrounding absolute atmospheric pressure. Figure 3.2 shows a longitudinal section of such a capillary, partially filled with liquid. The absolute pressure in the liquid is given by the sum of the ambient pressure p_0 and the negative Laplace pressure p_{Lap}.

Assume in Fig. 3.2 that we have a thin rectangular capillary (height $h <<$ width). Neglecting the meniscus curvature in the lateral plane, the pressure of the liquid phase p_{liq} is given by:

$$p_{liq} = p_0 - (2\gamma\cos\theta)\,/\,h \tag{3.3}$$

where h is the channel height, γ the surface tension of the liquid, and θ the contact angle of the liquid to the channel walls. If the liquid partially filling the hydrophilic channel is water, capillarity induced negative pressure will occur when the channel height is below approximately 1 μm. The first notion that capillary action in narrow tubes can induce tension in the liquid was probably by Kelvin when he derived what is now known as the Kelvin equation [14]. Experimental proof of the existence of capillary negative pressure was provided by Wiig and Juhola [15] through the measurement of a reduced density of water condensed in nano-porous activated charcoal, though quantitative interpretation was hampered by the unknown compressibility of water in the negative pressure regime as well as lack of exact knowledge of the pore diameter.

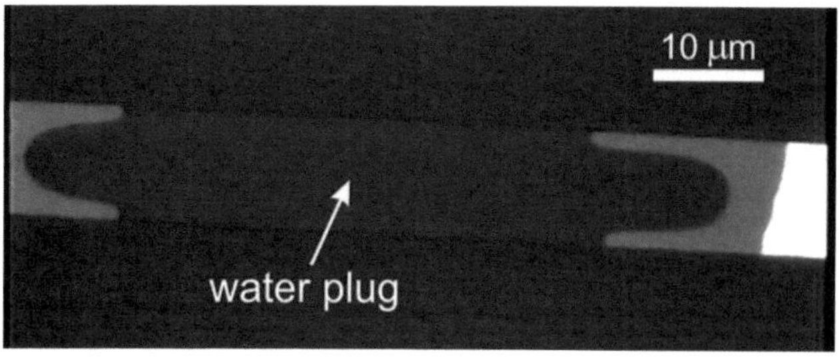

Fig. 3.3 Top view of a silicon oxide nanochannel (flat rectangular cross section) partially filled by water [17]. The channel capping is a silicon oxide thin film which deforms under the negative pressure load. This deformation can explain the peculiar shape of the meniscus [17, 18]

Through pull-off force measurements in a surface forces apparatus [16] consistency with Young-Laplace equation could be confirmed in nanometric gaps, however in this method the conversion from forces to pressures requires application of the Kelvin equation. A direct indication for the existence of capillary negative pressures was recently found by measuring the elasto-capillary deformation of silicon based nanochannels [17]. Figure 3.3 shows a microscope image of the typical shape of the meniscus (top view of a flat rectangular channel). The peculiar curvature of the meniscus in the lateral plane can be explained by a downward bending of the flexible channel cover. The downward bending can be quantified by detailed analysis of the meniscus shape [17, 18]. Together with the known stiffness of the channel cover it was possible to estimate the negative pressure of the liquid to be, depending on the method to analyse the meniscus curvature, -17 ± 10 bar [17] or -14 ± 8 bar [18]. Taking into account the measurement accuracy, this is in accordance with Eq. (3.3) (-12 bar).

Liquid plugs like the one shown in Fig. 3.3 are remarkable stable and can (depending on their volume and the speed of evaporation) exist for several minutes up to hours. The explanation for this observation [17] is that when we substitute Eq. (3.3) in the expression for the critical bubble size (Eq. (3.2)) we find (neglecting p_0 with respect to the negative Laplace pressure) $r_{crit} = h/\cos(\theta)$. This shows that for capillary negative pressure the critical bubble size for cavitation is larger than the size of the channel, so the liquid at negative pressure is in a stable state. Note also that we have implicitly considered channels of uniform cross section.

A related technique to measure capillary negative pressure is to look at the critical width of total collapse for nanochannels capped by an elastic cover plate [19]. Using this technique we measured capillary negative pressure in 80 nm channels for different liquids (as a function of their surface tension). Figure 3.4 shows the measured mechanical deformation pressure needed for total collapse (vertical) as a function of the applied capillary pressure (horizontal). Although limited in accuracy these results are indicative for the validity of the Young-Laplace equation in the sub-100 nm regime.

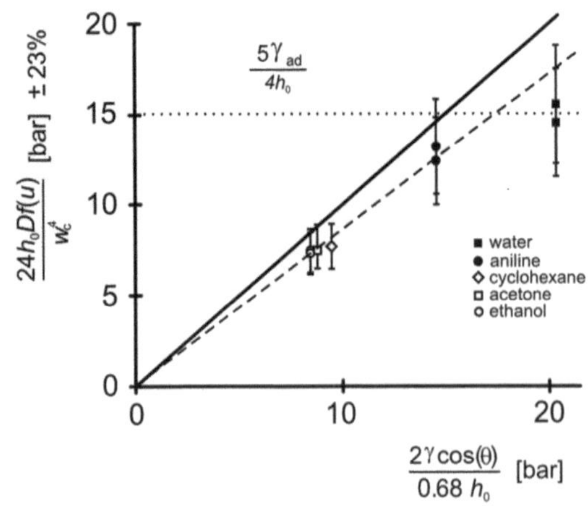

Fig. 3.4 Measured and expected mechanical deformation pressure versus the capillary pressure [19]. The expected relation is indicated by the *solid line* with slope one, the *dashed line* indicates the trend line in the experimental data. Where h_0 is the initial channel height, $D{\cdot}f(u)$ an elasticity parameter, and w_c is the critical width for full collapse [19]

3.4 Dynamic Capillarity in Nanochannels[3]

This section addresses the dynamics of filling of capillaries of arbitrary, though uniform cross section, typically in the 10–100 nm size range. For simple liquids at this length scale the continuum framework of hydrodynamics based on application of the Navier-Stokes equations is still valid, as is discussed in detail in the review by Bocquet and Charlaix [20]. Also the macroscopic contact angle and Young-Laplace equation can be applied without modifications. For smaller cross sections (hydrodynamic diameters less than roughly 10 nm) the influence of the disjoining pressure on the microscopic contact angle and the meniscus curvature can be become noticeable (in particular for water where the electrostatic contribution to the disjoining pressures can be relative long ranging) and care should be taken by applying the macroscopic Young and Young-Laplace equations The systematic study of capillary filling of sub-10 nm channels is still in its infancy and will most likely be subject of study of several groups in upcoming years.

3.4.1 Lucas-Washburn Equation for Infinite Stiff Channels

The filling dynamics of capillaries driven by the negative Laplace pressure was first modeled by Washburn [21]. The filling speed is determined by the surface tension γ of the liquid filling the capillary, its (advancing) contact angle θ to the channel walls, the channel diameter d (or in the case of a rectangular cross section by the width w and the height h) and finally by the viscosity η of the liquid. For sufficiently long and narrow capillaries the flow of the liquid entering the capillary may be

[3] This section is based on: van Honschoten JW, Brunets N, Tas NR (2010) Chem Soc Rev 39: 1096.

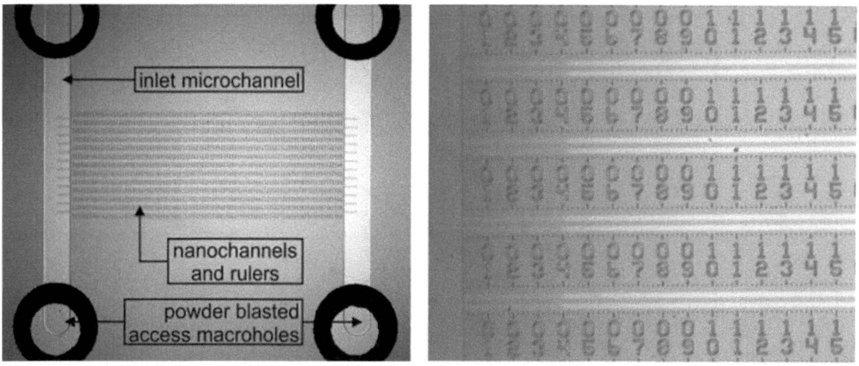

Fig. 3.5 *Top* view of a chip (appr. 1 × 1 cm^2) used in capillary filling experiments (*left*) [23]. Zoom in showing a still image of the filling of 30 nm channels (*right*) [7]

considered laminar and fully developed and the hydraulic resistance $R =$ (pressure drop Δp)/(volume flow ϕ) follows from the Poiseuille flow model. Taking the negative of the Laplace pressure for the pressure drop and realizing that the change of the position x of the moving meniscus is related to the volume flow through $dx/dt = \phi/A$ (where A is the cross sectional area of the channel) the following general expression for the position of the moving meniscus as a function of time results:

$$xdx = \frac{-p_{Lap}}{R^* \cdot A} dt \tag{3.4}$$

where R^* is the hydraulic resistance per unit length of the liquid plug ($R = xR^*$). As the hydraulic resistance is proportional to the liquid plug length x, it follows that the filling velocity decreases in time. More precisely, after integration the result is $x^2 \sim t$ or $x \sim t^{1/2}$. For example for a flat rectangular channel ($w >> h$) by substitution of R^* and p_{Lap} we find:

$$x = \sqrt{\frac{\gamma \cos(\theta)\, h}{3\eta} \cdot t} \tag{3.5}$$

In recent years this expression has been experimentally verified in silicon based nanochannels with a perfectly defined rectangular cross section, and down to channel heights of 5 nm [6, 7, 22–25] the qualitative $x = at^{1/2}$ dependence has been confirmed for water and aqueous solutions as well as for some organic liquids. Figure 3.5 shows a photograph of a chip used in such experiments. Liquid can be introduced in the access holes and fills the chip by capillary action, first through the inlet microchannel, then through the nanochannels. Rulers have been etched next to the nanochannels to follow the filling process in time.

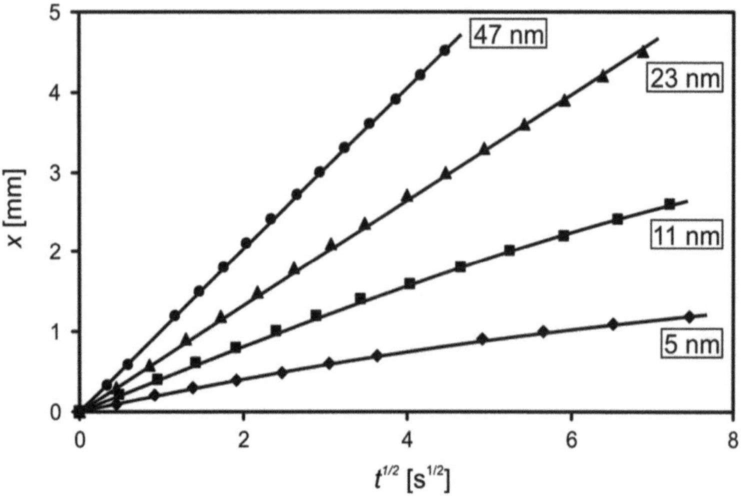

Fig. 3.6 Measured position of the moving meniscus as a function of $t^{1/2}$ for filling of 5, 11, 23, and 47 nm deep nanochannels with DI water at temperatures of 22.7, 22.2, 21.6, and 20.8 ± 0.5 °C, respectively [7]

3.4.2 Results of Capillary Filling Experiments in Nanochannels

Figure 3.6 shows an example of the measured position of the moving meniscus vs. time in capillary filling of hydrophilic nanochannels by de-ionized water (silicon-dioxide channel bottom and boro-silicate glass channel roof, rectangular cross section $h << w$). Note that down to the level of $h = 5$ nm the moving meniscus can be observed by optical microscopy, for the smallest channel heights using differential interference contrast (DIC) [7] or an embedded interferometer [25] to enhance the contrast. In the analysis of the filling dynamics usually macroscopic contact angles and surface tension values are used, which for small molecules should be correct when channel hydraulic diameters are in the order of 10 nm or more.

While the qualitative agreement with the Washburn model is good, several significant quantitative deviations have been observed. As they all concern slower than expected filling they are usually described by a dimensionless apparent viscosity η_{app}/η larger than 1 (where $\eta_{app}/\eta = 1$ refers to the situation described by Eq. (3.5)). Figure 3.7 shows the results as reported by Tas et al. [23] for filling of silica channels (20 nm SiO_2, boroslicate glass channel cover), by de-ionized water (DI water) and aqueous 0.1 M NaCl solution. A significant slower than expected filling for DI-water was attributed to the electroviscous effect, as the values of η_{app}/η returned to one when filling the channels by aqueous NaCl solutions. The electroviscous effect is related to the electrical double layer that is formed when the capillary walls in contact with the liquid are charged (see Schoch et al. [26]

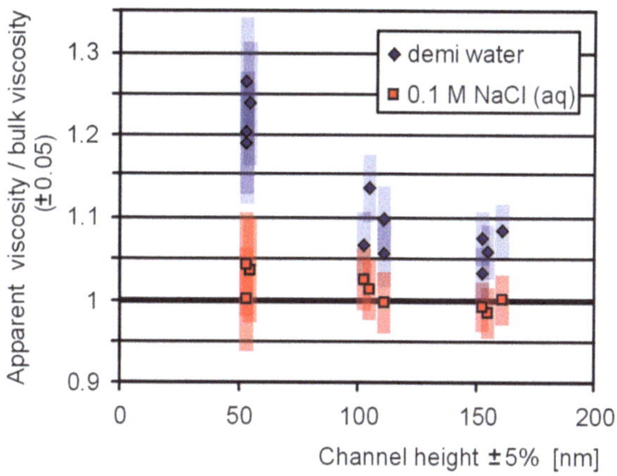

Fig. 3.7 Ratio of the apparent viscosity over the bulk viscosity of de-ionized water (*blue*) and aqueous NaCl solution (*red*) as derived from capillary filling experiments [23]

for an extensive review of electrokinetic effects in nanochannels). Due to the flow a net displacement of charge and an induced streaming potential results, which results into a counter electro-osmotic flow in a part of the electrical double layer that is closest to the channel walls [27]. A comparison [28] of the observed value of η_{app}/η with the theory of Levine, however, is not easy because of deprotonation of the hydroxylated channel walls during filling [29], leading to a modified and possibly axially non-uniform ionic composition of the solution filling the channels. In addition, it is likely that the constant potential assumption should be replaced by some form of charge regulation.

An important aspect of capillary filling, especially in channels having sharp inner corners, is the formation of bubbles [30]. In sharp inner corners a strong capillary suction develops, which leads to capillary "fingers" possibly proceeding in front of the main meniscus. The length of these fingers depends on the filling time and the "consumption" at the back by the proceeding main meniscus. As the main meniscus slows down in time the length and volume of the fingers tends to increase and eventually can lead to bubble formation through enclosure of air in front of the main meniscus. It has been suggested [30] that the presence of bubbles can be responsible for the sometimes observed slower than expected filling as shown in Fig. 3.7. However, the Washburn model is derived for filling without bubbles and should therefore only be applied up to the length where the first bubbles are formed [23].

3.4.3 *The Viscosity of Water in nm-Confinement*

Since the early 1970s there is a notion that water near polar or charged surfaces is somehow ordered or structured, leading to a so called structural component of the disjoining pressure. The first indications came from interpretation of the disjoining pressure isotherms of wetting films of water on hydrophilic substrates

[31], while unambiguous experimental proof came through application of the so called Surface Forces Apparatus (SFA) [32–36] showing short range repulsion forces between charged surfaces brought in close proximity (<5 nm), in addition to the double layer repulsion and van der Waals attraction as described by the DLVO theory. Despite the enormous importance of water as a liquid in nature as well as engineering systems, the exact nature of this hydration force as it is called for water has still not been conclusively determined. The main mechanisms that have been discussed in literature are related to the compression of hydrated ions (ionic surface groups, adsorbed ions, or counter-ions in the electrical double layer), orientation of water dipoles in the strong electric field near a charged wall, or the hydrogen bonding nature of water. An important question that emerges is if and how the supposed interfacial "structuring" of water affects its local viscosity. In the early 1970s measurements of the flow velocity of liquid plugs in quartz micro-capillarities driven by a disbalance between the capillary pressure and an applied external pressure revealed an elevated viscosity of water up to 40 % in capillaries of 0.04 μm [37]. These results could not be reproduced in nanometer thin films in the SFA. A dynamic SFA technique revealed a viscosity of water equal to the bulk viscosity, even in 2 nm thin films and under high salt concentrations necessary to induce a strong repulsive hydration force between the mica surfaces [38]. A similar conclusion was reached by SFA squeeze out experiments of thin water films between silica surfaces [36]. More recently, in a SFA where also shear force can be induced and measured, Raviv et al. [39] found that the viscosity of demineralized water when confined to films of 3.5 ± 1 to 0.0 ± 0.4 nm between mica surfaces remains within a factor three of its bulk value. Similar experiments with aqueous salt solution (in the presence of repulsive hydration forces) show that the bounded water molecules in the hydration layers retain the shear fluidity characteristics of bulk water, with an upper limit estimated for the viscosity of 0.08 Pa · s (maximum of 80 times the bulk viscosity) [40]. Remarkably, recent AFM experiments, measuring directly and simultaneously the normal and lateral forces of a nano-sized silicon tip approach a solid hydrophilic surface, reveals orders of magnitude increase in the viscosity of water with respect to bulk water in sub-nanometer confinement [41]. Capillary filling of silica nanochannels by water and aqueous salt solutions, reveals a significant slower than expected filling in channels with heights in the range of 10–50 nm [6, 7]. If interpreted as a viscosity effect, the averaged effective viscosity is up 30 % higher for 11 nm channels [7] (Fig. 3.8). The red line is a fit to the experimental data assuming a highly viscous layer next to the hydrophilic channel walls. The best fit is found for a thickness of this immobile layer of 0.9 nm at each interface. Filling experiments under different conditions are currently carried out in our lab to test the hypothesis that it is indeed the enhanced viscosity of hydration layers that causes the observed slower than expected filling. Based on all experimental data available, one must conclude that the question if and how hydration forces are connected to viscosity changes of interfacial water has still not been completely answered.

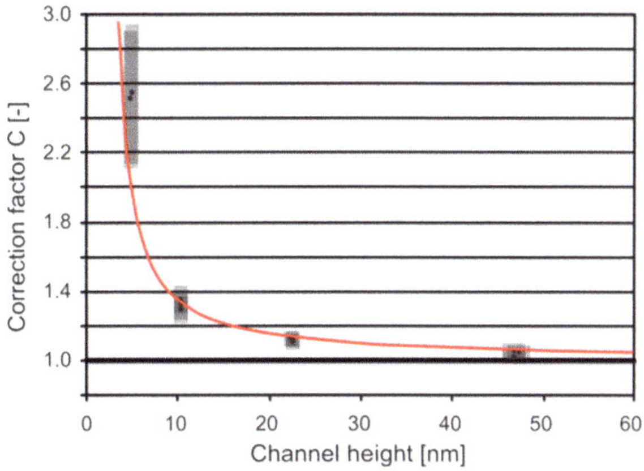

Fig. 3.8 Dimensionless correction factor (pre-factor of the denominator in Eq. (3.5)) indicating a slower than expected filling of water in flat rectangular silica nanochannels [7], as a function of channel height. The *red line* indicates the effect of a solidified layer next to the channel walls with a *thickness* of 0.9 nm or four monolayers

3.5 Conclusions

Planar silicon based nanochannels offer a unique platform for experimental studies of capillarity in extreme confinement. In many cases processes inside the nanochannels can be observed by standard optical microscopy. In lyophilic nanochannels the extreme confinement can easily lead to large negative pressures in the liquid. These can be measured by quantifying the elastic deformation of the channels. Dynamic capillarity studies using water or aqueous salt solutions revealed an increased apparent viscosity during capillary filling of nanochannels. The increased apparent viscosity is attributed to an electroviscous effect in the case of insulating channels. For channels well below 50 nm there is an additional effect leading to a strong increase of the apparent viscosity. It is unclear if this is a real increase in the viscosity of water strongly confined in hydrophilic channels. Alternatively, it could be caused by the combination of the properties of water and immobile adsorbed ions.

References

1. Stern MB, Geis MW, Curtin JE (1997) Nanochannel fabrication for chemical sensors. J Vac Sci Technol B 15:2887–2891
2. Turner SW, Perez AM, Lopez A, Craighead HG (1998) Monolithic nanofluid sieving structures for DNA manipulation. J Vac Sci Technol B 16:3835–3840
3. Han J, Craighead HG (1999) Entropic trapping and sieving of long DNA molecules in a nanofluidic channel. J Vac Sci Technol A 17:2142–2147
4. Haneveld J, Jansen H, Berenschot E, Tas N, Elwenspoek M (2003) Wet anisotropic etching for fluidic 1D nanochannels. J Micromech Microeng 13:S62–S66
5. Haneveld J (2006) Nanochannel fabrication and characterization using bond micromachining. Ph.D. Thesis, University of Twente

6. Persson F, Thamdrup LH, Mikkelsen MBL, Jaarlgard SE, Skafte-Pedersen P, Bruus H, Kristensen A (2007) Double thermal oxidation scheme for the fabrication of SiO_2 nanochannels. Nanotechnology 18(245301):1–4
7. Haneveld J, Tas NR, Brunets N, Jansen HV, Elwenspoek M (2008) Capillary filling of sub-10 nm nanochannels. J Appl Phys 104(014309):1–6
8. Huygens C (1672) An extract of a letter of M. Hugens to the author of the Journal des Scavans of July 25. 1672. Attempting to render the cause of that odd phenomenon of the quicksilver remaining suspended far above the usual height in Torricellian experiment. Philos Trans 7:5027–5030
9. Sir Isaac Newton (1721) Opticks: or a treatise of the reflections, refraction, inflections and colours of light, 3rd Book, 3rd edn, pp 365–366
10. Dixon HH, Joly J (1895) On the ascent of sap. Philos Trans R Soc Lond B 186:563–576
11. Berthelot M (1850) Sur quelques phénomènes de dilatation forcée des liquides. Annales de chimie et de Physique 30:232–237
12. Briggs LJ (1950) Limiting negative pressure of water. J Appl Phys 21:721–722
13. Fisher JC (1948) The fracture of liquids. J Appl Phys 19:1062–1067
14. Kelvin L (1870) On the equilibrium vapour at a curved surface of liquid. Proc R Soc Edinb 7:63–68
15. Wiig EO, Juhola AJ (1949) The adsorption of water vapor on activated charcoal. J Am Chem Soc 71:561–568
16. Fisher LR, Israelachvili JN (1980) Determination of the capillary pressure in menisci of molecular dimensions. Chem Phys Lett 76:325–328
17. Tas NR, Mela P, Kramer T, Berenschot JW, van den Berg A (2003) Capillarity induced negative pressure of water plugs in nanochannels. Nano Lett 3:1537–1540
18. van Honschoten JW, Escalante M, Tas NR, Elwenspoek M (2009) Formation of liquid menisci in flexible nanochannels. J Colloid Interface Sci 329:133–139
19. Tas NR, Escalante M, van Honschoten JW, Jansen HV, Elwenspoek M (2010) Capillary negative pressure measured by nanochannel collapse. Langmuir 26:1473–1476
20. Bocquet L, Charlaix E (2010) Nanofluidics, from bulk to interfaces. Chem Soc Rev 39:1073–1095
21. Washburn EW (1921) The dynamics of capillary flow. Phys Rev 17:273–283
22. Hibara A, Saito T, Kim H-B, Tokeshi M, Ooi T, Nakao M, Kitamori T (2002) Nanochannels on a fused-silica microchip and liquid properties investigated by time-resolved fluorescence measurements. Anal Chem 74:6170–6176
23. Tas NR, Haneveld J, Jansen HV, Elwenspoek M, van den Berg A (2004) Capillary filling speed of water in nanochannels. Appl Phys Lett 85:3274–3276
24. Han A, Mondin G, Hegelbach NG, de Rooij NF, Staufer U (2006) Filling kinetics of liquids in nanochannels as narrow as 27 nm by capillary force. J Colloid Interface Sci 293:151–157
25. van Delft KM, Eijkel JCT, Mijatovic D, Druzhinina TS, Rathgen H, Tas NR, van den Berg A, Mugele F (2007) Micromachined Fabry-Perot interferometer with embedded nanochannels for nanoscale fluid dynamics. Nano Lett 7:345–350
26. Schoch RB, Han J, Renaud P (2008) Transport phenomena in nanofluidics. Rev Mod Phys 80:839–883
27. Levine S, Marriott JR, Robinson K (1975) Theory of Electrokinetic flow in a narrow parallel-plate channel. J Chem Soc Faraday Trans 2(71):1–11
28. Mortensen NA, Kristensen A (2008) Electroviscous effects in capillary filling of nanochannels. Appl Phys Lett 92(063110):1–3
29. Jansen KGH, Hoang HT, Floris J, de Vries J, Tas NR, Eijkel JCT, Hankemeier T (2008) Anal Chem 80:8095–8101
30. Thamdrup LH, Persson F, Bruus H, Kristensen A, Flyvbjerg H (2007) Experimental investigation of bubble formation during capillary filling of SiO_2 nanoslits. Appl Phys Lett 91(163505):1–3
31. Derjaguin BV, Churaev NV (1974) Structural component of disjoining pressure. J Colloid Interface Sci 49:249–255

32. Israelachvili JN, Adams GE (1978) Measurement of forces between two mica surfaces in aqueous electrolyte solutions in the range 0–100 nm. J Chem Soc Faraday Trans 74:975–1001
33. Pashley RM (1981) Hydration forces between mica surfaces in aqueous electrolyte solutions. J Colloid Interface Sci 80:153–162
34. Pashley RM (1981) DLVO and hydration forces between mica surfaces in Li^{+}, Na^{+}, K^{+} and Cs^{+} electrolyte solutions: a correlation of double-layer and hydration forces with surface cation exchange properties. J Colloid Interface Sci 83:531–546
35. Pashley RM, Israelachvili JN (1984) Molecular layering of water in thin films between mica surfaces and its relation to hydration forces. J Colloid Interface Sci 101:511–523
36. Horn RG, Smith DT, Haller W (1989) Surface forces and viscosity of water measured between silica sheets. Chem Phys Lett 162:404–408
37. Churaev NV, Sobolev VD, Zorin ZM (1971) Special discussion in thin liquid films and boundary layers. Academic, New York, pp 213–220
38. Israelachvili JN (1986) Measurement of the viscosity of liquids in very thin films. J Colloid Interface Sci 110:263–271
39. Raviv U, Laurat P, Klein J (2001) Fluidity of water confined to subnanometre films. Nature 413:51–54
40. Raviv U, Klein J (2002) Fluidity of bound hydration layers. Science 297:1540–1543
41. Li T-D, Gao J, Szoszkiewicz R, Landman U, Riedo E (2007) Structures and viscous water in subnanometer gaps. Phys Rev B 75(115415):1–5

Chapter 4
Infrared-Thermodynamics Conversion as a Function of Temperature: Towards Confined Water

Isabelle Bergonzi and Lionel Mercury

Abstract An experimental method has been developed to calculate the thermodynamic properties of water from its vibrational properties, relevant to study (in near future) the properties of adsorbed or confined water. The infrared absorption of the intra-molecular OH stretching mode of liquid water has been measured over a wide range of temperature (from−10 to 90 °C). The corresponding large band has been decomposed into three Gaussian components standing for three different water connectivities (percolation model) that feature the liquid state as a function of temperature: network, intermediate, and multimer water. Measurements evidenced that the components are differently shifted with temperature, giving a quantitative insights into the internal energy change of liquid. A vibrational partition function has been used to calculate the corresponding thermodynamic properties, neglecting all energy components except the present intra-molecular vibrational mode. Interestingly, the vibrational free enthalpy thus computed differs of the total free enthalpy only by a multiplicative constant all along the thermal range.

4.1 Introduction

Liquid water presents a remarkable ability to fill in any cavity or channel, even of nanometric dimensions (confinement), and to stay as thin films onto almost any solid surfaces (adsorption). The term confinement relates to the influence of the close solids on the sandwiched liquid, while adsorption is a confinement between solid and air. Interestingly, confinement and adsorption are states that allow liquid water to stay inside a pore network under arid or semi-arid conditions, affording a minimum water content in soils, with the thin films hydraulically connecting

I. Bergonzi (✉) • L. Mercury
Institute of Earth Sciences, University of Orléans, Orléans, France
e-mail: isabelle.bergonzi@etu.univ-orleans.fr; lionel.mercury@univ-orleans.fr

L. Mercury et al. (eds.), *Transport and Reactivity of Solutions in Confined Hydrosystems*, NATO Science for Peace and Security Series C: Environmental Security,
DOI 10.1007/978-94-007-7534-3_4,

capillary clusters confined in wedges, crevices or small channels. This minimum water content directly impacts the soil wettability, its ability to host drought-tolerant species, and to maintain residual microbial activity and mobility. In terms of weathering, these water types are intimately contacting the host solid and also stay immobile for long period of time, making them a probably important (and overlooked) actor in the global mass balance.

Confined and adsorbed waters have properties that deviate from the bulk water (e.g., [1–7]), which give them specific driving forces in terms of transport, weathering, and biological habitats. Confined water is found in saturated and non-saturated media, actually whenever liquid is occluded in host spaces smaller than 0.1 μm (e.g., [8–12]). Adsorbed water is encountered in non-saturated media, as thin films of nanometer to hundreds of nanometers thickness (e.g., [1, 13]). Despite this large versatility in natural settings, the properties of these water types and the characteristic thresholds describing their evolution with sizes/thicknesses are not well apprehended in the natural sciences community. Meanwhile, their characteristic properties are essentially measured at very high confinement around the nanometer, and their thermodynamic behaviour in the 5–100 nm range is poorly known.

Infrared (IR) spectroscopy have been used to probe the water liquid state with solutes addition [14], when adsorbed on surfaces [11, 15] or filling restricted spaces [11, 12], at the solid-water or liquid-air interfaces [16]. These studies demonstrated the sensitivity of the IR signal to record slight changes of the properties of liquid, even for small amount of matter. This sensitivity indicates that vibrational energy changes correctly probe energetic changes with environment. Among all the IR absorption modes, the intra-molecular OH-stretching band revealed to be the most significant, showing shifts both in terms of global shape, intensity and location of the main peak. However, there is not a direct conversion of these IR shifts into macroscopic thermodynamic properties to date. Some studies proposed a statistical deconvolution of the IR and Raman spectra to obtain thermodynamic values of diluted solutions of HOD molecules in D_2O (or H_2O), but this approach cannot describe the IR spectra of normal water [17–19].

It is why infrared spectroscopic measurements have been targeted to record the change in the vibrational properties of water trapped into artificial rectangular channels having depths varying from 100 to 5 nm (1D nano-channels). However, this chapter lets aside the confinement topic to deal with a required prior step: devising a calculation method for the IR-thermodynamics conversion based on a partition function, and calibrating it on a well-known thermodynamic change, here the thermodynamic behavior of liquid water from −10 to 90 °C.

4.2 Experimental Section

IR measurements were recorded in the 400–4,000 cm^{-1} spectral region with a Fourier Transform Infra-Red (FTIR) micro-spectrometer (Nicolet Continuμm, Thermo scientific), equipped with a XT-KBr beam splitter and a liquid nitrogen cooled Mercury Cadmium Telluride (MCT) detector (250 μm). The microscope operates in confocal mode, using a 15× infinity corrected Schwarzchild objective and a matching 15× condenser.

To avoid saturating the IR absorption signal due to the enormous amount of water molecules per liquid volume, a thin film cell of varying thickness, down to micrometer, has been developed. This cell is composed of two steel parts compressing two BaF_2 windows (13 mm × 2 mm), sealed with a fluorocarbon O-ring. A liquid water drop is sandwiched between the two windows and the steel parts are tightened to form a bulk water film roughly 1 μm thick.

The thermal change is controlled by a heating-cooling stage (THMS-600, Linkam) with windows transparent to IR light. Briefly, the fluid cell is put on a silver block at controlled temperature (±0.1 °C), and the acquisition starts after 5 min, which is the time needed to afford thermal homogeneity between the two steel parts.

Each spectrum is recorded with a resolution of 4 cm^{-1} with 200 scans per spectrum and no mathematical correction. The infrared absorption coefficient is defined by:

$$A = -\log\left({}^{I_0}/_{I}\right) \tag{4.1}$$

where I_0 is the transmitted intensity of the empty cell and I the transmitted intensity of the cell filled with water. Spectra are recorded from −10 to 90 °C.

4.3 Experimental Results and Discussion

4.3.1 IR Absorption Band

The infrared spectrum of liquid water consists of four main bands, spreading from a few wavenumbers and 3,800 cm^{-1} [20]. The dominating feature of the IR water spectrum is the band located at 3,400 cm^{-1} which corresponds to the stretching motion of the covalent OH bonds. The HOH bending band (1,650 cm^{-1}) follows in the decreasing frequency order, then the libration band (675 cm^{-1}), and finally the "connectivity" band (200 cm^{-1}). The two former bands correspond to intra-molecular modes, either sensitive to the H-bonds environment (stretching) or not sensitive (bending). The two latter reflect inter-molecular modes. The libration mode, through small amplitude oscillations of the whole molecule, provides a mechanism of rupture and rearrangement of the hydrogen bonds at a sub-picosecond rate. The connectivity band arises from the longitudinal motion of the H-atom along the H-bond axis (H-bond stretching) and therefore characterizes the level of H-bonding between neighboring water molecules.

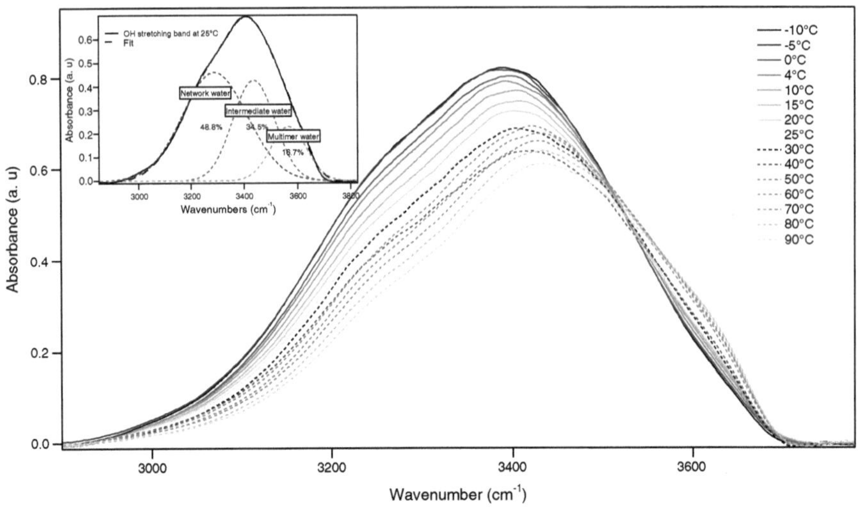

Fig. 4.1 OH stretching band of liquid water as a function of the temperature. Inset displays the three Gaussian components decomposition classically performed (see text)

4.3.2 OH Stretching Band

The OH stretching band centred at about 3,400 cm^{-1} is a sensitive probe of the H-bond: the wavenumber increases with the distance between O atoms of two molecules linked by H-bond (e.g., [21]). The absorption corresponds to intramolecular motions of the water molecules crossed by the IR beam, which depend on the molecular environment. As a result, this is a molecular signal averaged over macroscopic amount of matter. The width of the band echoes the energetic distribution of H-bonds along the beam pathway and then is an image of the molecular disorder, what justifies the intense study of this band (e.g., [22–24]).

The present measurements from −10 °C (supercooled water) to 90 °C (Fig. 4.1) are in agreement with previous studies [25–27]: upon heating, the main peak moves to higher wavenumbers (blue-shifted, from 3380.6 cm^{-1} at −10 °C up to 3436.5 cm^{-1} at 90 °C), while the intensity diminishes and the shape visually changes losing the lower wavenumbers and enriched on the higher ones. This shift is classically interpreted by the direct effect of heating that disrupts water intermolecular network and then promotes disorder (e.g., [22, 23]). This interpretation applies as well to account for the shape change under heating. The decreasing absorbance is readily accounted for, outlining that the resonance condition required for coupling the OH stretching motion is obtained at low temperature [28].

These visual observations cannot be turn easily into quantitative calculations because liquid water is a continuum of molecular configurations, each having its own behavior and influence on the change of the total energy. And the modeling of the whole band in terms of discrete molecular energies cannot be performed. The common strategy is to decompose the OH-stretching band into Gaussian components, each standing for a certain type of water that features the liquid at a given time. Decomposition into two (e.g., [29]), three (e.g., [20]) and four components (e.g., [14]) have been performed, most often related to the purposely existing water clusters. Actually, the vibrational mode causing the stretching motion of the intra-molecular OH bond is sensitive to the strength of H-bonding between molecules. The oscillator strength of a water molecule decrease when its local coordination increases and the corresponding OH frequency is downshifted. The notion of water cluster, and its relation to water structure, is problematic since it is well known that the relaxation time of water is around the picosecond, several order of magnitude quicker than the recording time of the IR absorption. It is why this classic approach has been recently reformulated in terms of the percolation model [30], each Gaussian component featuring an averaged coordination number, probabilistically ranging from 0 to 4 [12, 20]. The best and most relevant decomposition is done through three Gaussian components (inset in Fig. 4.1), ascribed to three different types of water molecules. The low-frequency Gaussian is associated to the fully coordinated water molecules, called Network Water (NW) also known as ice-like water since the Gaussian peak is close to the ice's (3,250 cm^{-1}, [24]). The high-frequency Gaussian is ascribed to the molecules with a low coordination number, called Multimer Water (MW) and close to the IR characteristics of vapour molecules. The Gaussian lying in-between has a coordination number close to two, is called Intermediate Water (MW) and refers to the incompletely coordinated liquid molecules. At the scale of one molecule, it is clear that each molecule experiences the three coordination states over the experimental time (minute). The final band represents how the molecular network promotes the different types of coordination related to the energy available in the system.

4.3.3 Band Decomposition

The experimental bands were fitted with the software Igor and its Multipeak fitting 2.0 package. Brubach et al. [20] suggested that the three Gaussian components were linked to the five species described by the percolation model [30]. This model calculates the variation of the coordination number of one water molecules with its immediate neighbors, with temperature. The number of intact H-bonds per molecule is given by the following probability:

$$p_B = {}^{n_{HB}}/_{z} \tag{4.2}$$

with n_{HB} the number of established H-bonds and $z = 4$ the maximum number of hydrogen atoms bonded to one oxygen atom (by covalent or intermolecular bond).

Molecules are thus divided into five species, by the possible number of intact H-bonds they form with their neighbors, from 0 to 4. The fraction f_j of water molecule establishing j bonds is given by the relation:

$$f_j = \frac{z!}{j!\,(z-j)!} {p_B}^j (1 - p_B)^{z-j} \tag{4.3}$$

with j ranging from 0 to 4. The fraction of species j is a function of the temperature T via the temperature dependence of p_B written $p_B = 1.8 - 0.004T(K)$.

Brubach et al. [20] proposed to ascribe NW to species $j = 4$ and $j = 3$, IW to species $j = 2$ and MW to species $j = 1$ and $j = 0$. The interesting point of such interpretation is the implicit use of the ergodic principle to describe the thermodynamic features of liquid water: the IR signal averaged over time echoes the averaged energetic states making up this liquid.

Practically, fits were constrained to make the full width at half maximum of the calculated Gaussian component to match the fraction of water molecules "types" calculated by Eq. (4.3) with temperature (Fig. 4.2). Over −10 to 90 °C, NW fraction dominates up to 40 °C, from which temperature IW starts predominating. Interestingly, IW is the fraction that varies the less over the thermal range, while the two others are anti-correlated in the expected way: NW decreases with temperature while the MW increases, in agreement with the chaotropic trend afforded by heating individual molecules that tend toward a vapour-like network organization. This variation is in good agreement with the shape evolution of the OH band with temperature: decreasing absorbance at low wavenumbers echoes the diminution of NW molecular fraction while increasing absorbance at high wavenumbers is naturally linked to the increasing MW molecular fraction. Gaussian maximum are shifted toward lower frequencies with temperature (Fig. 4.2): NW main peak moves from 3,315 cm^{-1} at −10 °C to 3,219 cm^{-1} at 90 °C, IW peak is displaced from 3,455 cm^{-1} at −10 °C to 3,384 cm^{-1} at 90 °C and MW peak slides from 3,583 cm^{-1} at −10 °C to 3,528 cm^{-1} at 90 °C. This IR downshifting evidences that the characteristic H-bond lengths of each water "types" decreases, what means that the energy to maintain the corresponding network organization must increase due to the additional thermal energy in the system. Otherwise, to keep its connectivity level, each water "types" uses more energy.

As a consequence, the OH-stretching band highlights that the IR shift is a direct measure of the internal energy change of the probed liquid. Summing up the three contributions, should result in deducing the whole thermodynamic change from a given reference state. In all that follows, the reference state is the pure liquid at 25 °C and 0.1 MPa.

4.3.4 Thermodynamic Treatment

The conversion of the IR absorption band into thermodynamic terms requires a description of the way the characteristics of each energetic state combine into

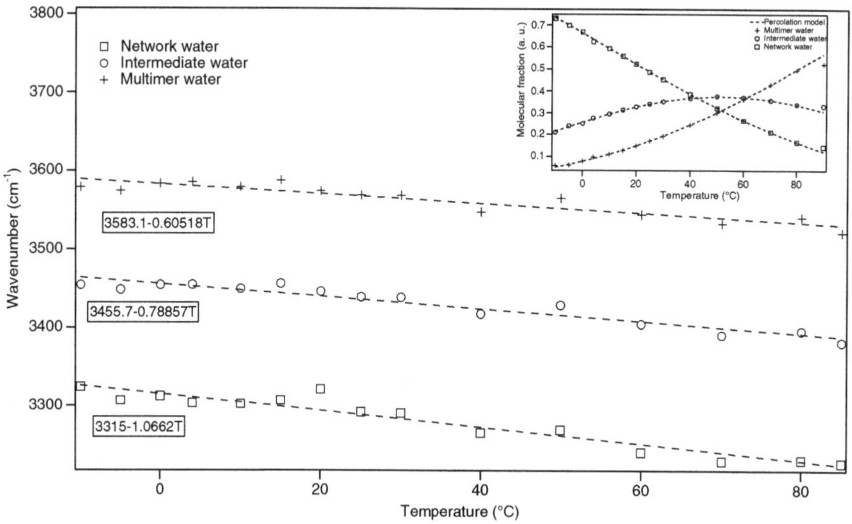

Fig. 4.2 Location of the main peak of the Gaussian components with temperature. The inset shows the evolution of the molecular fraction of the three water populations: NW, IW, MW (see text and Eq. (4.3))

the macroscopic value of the thermodynamic property. The relevant tool is the partition function, here that proposed by Vand and Senior [31]. It is based on the Némethy and Scheraga model [32], which is a continuum model based on the concept of "flickering clusters". The H-bonds formation in liquid water is described as a cooperative process, i.e. H-bonds are established or broken simultaneously depending on the local energy fluctuations. Short-lived regions of highly H-bonded clusters thus take place surrounded by non H-bonded molecules. As a consequence, the model features liquid with two main structures, clusters and non H-bonded molecules. The water clusters can be monomer (one bond), dimer (two bonds), trimer (three bonds), or tetramers (four bonds). Despite this "water structure" background, the partition function is built through five species distinguished by their coordination number (four water clusters and the non H-bonded molecules) and an energy level is ascribed to each species, in an analytical approach completely compatible with the percolation model. The molecules distribution between the energy levels is determined by the corresponding Boltzmann factors, $\exp(E_i/kT)$ where E_i is the energy of a given energy level and k is the Boltzmann constant, and by the degrees of freedom allowed for the motion of each species. The corresponding partition function Z writes as follows [32]:

$$Z = \sum_{[n_0, y_1, x_u]} g\left(n_0, y_1, x_u\right) \prod_{i=u}^{i=4} \left[f_i \exp\left(-\frac{E_i}{RT}\right)\right]^{N_0 x_i} \tag{4.4}$$

where g is a combinatorial factor written $g(n_0,y_1,x_u) = N_0\,!/(N_4\,!\,N_3\,!\,N_2\,!\,N_1\,!\,N_u\,!)$, $N_i = N_0 xi$ is the number of molecules of each species, N_0 is the Avogadro's number,

x_i are mole fractions of each species in the bulk liquid, E_i are the energies of each species and f_i are the weighting factors describing the vibrational, rotational and translational freedom allowed to each species, ascribed from IR and Raman experimental data.

The present approach by IR-shift relative to a reference enables to neglect the modes that are not sensitive to the varying parameter (temperature or confinement, or any other). The intra-molecular IR changes are entirely featured by the OH-stretching band, since the bending mode has a negligible thermal coefficient (-0.1 cm^{-1}.K^{-1}; [33]). The inter-molecular modes (60–1,000 cm^{-1}; libration and connectivity bands) include as well the cooperativity between vibration and rotation modes, and both change with temperature (e.g., [20]). Nemethy and Scheraga [32] hypothesized that these intermolecular bands that appear in ice, express the contribution of the clustered water. With their partition function, and experimental values for the IR bands, they were able to accurately re-calculate the entropy, free energy and internal energy of liquid water but failed with the specific heat of water. Based on experimental measurements [34, 35], Vand and Senior [31] alternatively proposed to describe the partition function with three species, each having a certain distribution around a mean effective length $\overline{l}$ of H-bond in liquid water. As a consequence, the energy of molecules in an energy band (mean effective length $\overline{l}$) will have a similar regular distribution about a mean value E_i. Vand and Senior [31] selected arbitrarily a Gaussian distribution to describe the distribution of the three energy bands, and the partition function becomes:

$$Z = \sum_{i=1}^{3} \frac{g_i X^{m_i}}{\Delta_i} exp \left\{ \frac{1}{4} \left(\Delta_i{}^2 X^2 \right) - E_i \right\} \quad (4.5)$$

where $X = {}^{1}\!/_{RT}$ E_i is the energy of the ith species, $g_i X^{mi}$ is the power law expression for the weighting factor f_i that integrate thermal variation, Δ is the bandwidth of the energy distribution and B is a constant.

At first sight, the partition function with three energy bands fits well the decomposition into three Gaussian components done from the OH-stretching band. Using the Gaussian changes as a function of temperature, the Vand and Senior function offers a simple pathway to transform IR shifts into macroscopic water energy changes. In liquid water, the bending and OH-stretching bands are intra-molecular modes and only the latter moves with the temperature. In the meantime, there are frustrated (with respect to vapor) rotational motions (librations, around 675 cm^{-1}) due to the networking that changes the inter-molecular vibrational mode of H-atoms. A second inter-molecular mode is featured by the hindered translational (or connectivity) band (around 200 cm^{-1}), which correspond to the longitudinal motion of the H-atom along the H-bond axis (H-bond stretching) and therefore characterizes the level of H bonding between neighboring water molecules. Among these four bands, only two exhibit pronounced substructures that significantly evolve with temperature (and probably confinement) (e.g., [12, 20]): the OH-stretching band and the connectivity bands. These substructures characterize the changing levels of H bonding in the molecular network that are directly related

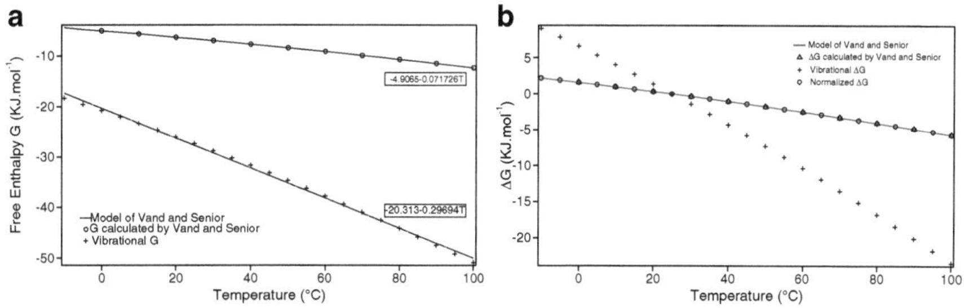

Fig. 4.3 (**a**) Evolution of the total and vibrational free enthalpies (or Gibbs free energies) as a function of temperature. *Dotted line* is only guide for eyes. (**b**) Evolution of vibrational and total free enthalpies relative to 25 °C. The *solid line* (calculated Gibbs free energy) is perfectly fitted by the vibrational Gibbs free energy (*open circles*) multiplied by the constant C (C = 0.241546, see text)

to the thermodynamic properties. As a consequence, the coming interpretation is weakened by the sole consideration of the stretching band, while (at least) the connectivity band contributes to the internal energy.

The IR-thermodynamics conversion corresponds to transform the total mean energies value in Eq. (4.5) [31] into the three vibrational mean energies given by the peak of the three Gaussian components, resulting in the "vibrational" thermodynamic functions of liquid water as a function of temperature (Fig. 4.3a). Consistently, the vibrational Gibbs free energy is smaller than the total Gibbs free energy ignoring the contribution of the inter-molecular modes.

This first absolute calculation can be easily normalized at 25 °C, and only the relative change to this reference is calculated (Fig. 4.3b), removing all the modes that are not sensitive to thermal change. Here again, ΔG_{vib} has a larger slope than total ΔG, outlining that the intra-molecular vibrational mode is compensated by another energetic contribution of opposite trend with temperature. Consistently, rotational and/or translational motions tend to increase with increasing temperature due to the decreasing of the coordination number of water molecules that favors non H-bond molecules. Assuming that the frustrated rotation (libration) does not change with temperature, and that any other rotation and/or translation modes impact the total energy, we may propose that a simple subtraction lead to predict the change of the inter-molecular connectivity mode. At this stage, obviously, an experimental control of such statement must be done.

For the time being, the multiplicative constant C that enables to retrieve the total ΔG (i.e. $\Delta G_{total} = C \times \Delta G_{vib}$;) all along the thermal range demonstrates that the thermodynamic changes under temperature, are as linear as the intra-molecular OH-stretching mode. Thus, the decomposition of the OH stretching band into three Gaussian components, coupled with the Vand and Senior partition function, allows us to obtain a thermodynamic description, though incomplete, of water. Technically, it is interesting to outline the compatibility between the percolation model and the original Nemethy and Scheraga approach. The probabilistic H-bonds networking can be assumed in the same continuum framework as that afforded by the flickering water-structured clusters.

4.4 Conclusion

The IR-to-bulk thermodynamics conversion has been explored following the OH-stretching band behaviour with temperature. The method, based on the percolation model and using mechanistic partition function, appears reliable and gives the correct trend. However, it fails to become strictly quantitative, most probably by overlooking the other significant contribution, mainly the connectivity band. These effects need now to be evaluated with temperature, to decipher the missing link between ΔG_{vib} and ΔG_{total}.

These results give confidence about the ability of the FTIR spectroscopy to probe the thermodynamic properties of liquid water under given conditions. Thus, the protocols to record IR absorption under confinement have been developed as well as the design of the artificial channels with varying depth (1D nano-confinement).

Acknowledgments This work has received support from the French Agency for Research (Agence Nationale de la Recherche, ANR) through the grants CONGE BLAN-61001 and Labex Voltaire ANR-10-LABX-100-01.

References

1. Derjaguin BV, Churaev NV, Muller VM (1987) Surface forces. Plenum Publishing Corporation, New York
2. Handa YP, Zakrzewski M, Fairbridge C (1992) Effect of restricted geometries on the structure and thermodynamic properties of ice. J Phys Chem 96:8594–8599
3. Teixeira J, Zanotti J-M, Bellissent-Funel M-C, Chen S-H (1997) Water in confined geometries. Physica B 234:370–374
4. Dore J (2000) Structural studies of water in confined geometry by neutron diffraction. Chem Phys 258:327–347
5. Christenson HK (2001) Confinement effects on freezing and melting. J Phys Condens Matter 13:95–133
6. Johari GP (2005) Water's size-dependent freezing to cubic ice. J Chem Phys 122:art. 194504, 5p
7. Tombari E, Salvetti G, Ferrari C, Johari GP (2005) Thermodynamic functions of water and ice confined to 2 nm radius pores. J Chem Phys 122:art. 104712, 9p
8. Shelton DP (2000) Collective molecular rotation in water and other simple liquids. Chem Phys Lett 325:513–516
9. Putnis A, Mauthe G (2001) The effect of pore size on cementation in porous rocks. Geofluids 1:37–41
10. Rijniers LA, Huinink HP, Pel L, Kopinga K (2005) Experimental evidence of crystallization pressure inside porous media. Phys Rev Lett 94:075503
11. Musat R, Renault JP, Candelaresi M, Palmer DJ, Le Caer S, Righini P, Pommeret S (2008) Finite size effects on hydrogen bonds in confined water. Angew Chem Int Ed Engl 47:8033–8035
12. Le Caër S, Pin S, Esnouf S, Raffy Q, Renault JP, Brubach JB, Greff G, Roy P (2011) A trapped water network in nanoporous material: the role of interfaces. Phys Chem Chem Phys 13:17658–17666

13. Pashley RM, Kitchener JA (1979) Surface forces in adsorbed multilayers of water on quartz. J Colloid Interface Sci 71(3):491–500
14. Masuda K, Haramaki T, Nakashima S, Habert B, Martinez I, Kashiwabara S (2003) Structural change of water with solutes and temperature up to 100°C in aqueous solutions as revealed by attenuated total reflectance infrared spectroscopy. Appl Spectrosc 57(3):274–281
15. Richard T, Mercury L, Poulet F, d'Hendecourt L (2006) Diffuse reflectance infrared Fourier transform spectroscopy as a tool to characterize water in adsorption/confinement situations. J Colloid Interface Sci 304:125–136
16. Mercury L, Jamme F, Dumas P (2012) Infrared imaging of bulk water and water-solid interface under stable and metastable conditions. Phys Chem Chem Phys 14:2864–2874
17. Efimov YY, Naberukhin YI (2003) Fluctuation theory of hydrogen bonding applied to vibration spectra of HOD molecules in liquid water. I. Raman spectra. Mol Phys 101:459–468
18. Efimov YY, Naberukhin YI (2004) Fluctuation theory of hydrogen bonding applied to vibration spectra of HOD molecules in liquid water. II. Infrared spectra: contour shape, integrated intensity, temperature dependence. Mol Phys 102:1407–1414
19. Efimov YY, Naberukhin YI (2004) Thermodynamic functions of liquid water calculated from the temperature evolution of the vibration spectrum contour. Spectrochim Acta A 61:1789–1794
20. Brubach J-B, Mermet A, Filabozzi A, Gershel A, Roy P (2005) Signatures of the hydrogen bonding in the infrared bands of water. J Chem Phys 122:184509
21. Bratos S, Leicknam J-C, Pommeret S (2009) Relation between the OH-stretching frequency and the OO distance in time-resolved infrared spectroscopy of hydrogen bonding. Chem Phys 359:53–57
22. Nucci NV, Vanderkooi JM (2005) Temperature dependence of hydrogen bonding and freezing behavior of water in reverse micelles. J Phys Chem B 109:18301–18309
23. Crupi V, Longo F, Majolino D, Venuti V (2007) Raman spectroscopy: probing dynamics of water molecules confined in nanoporous silica glasses. Eur Phys J Spec Top 141:61–64
24. Maréchal Y (2007) The hydrogen bond and the water molecule: the physics and chemistry of water, aqueous, and bio media. Elsevier, Amsterdam, The Netherlands
25. Maréchal Y (1991) Infrared spectra of water. I. Effect of temperature and of H/D isotopic dilution. J Chem Phys 95:5565–5573
26. Libnau FO, Kvalheim OM, Christy AA, Toft J (1994) Spectra of water in the near- and mid-infrared region. Vib Spectrosc 7:243–254
27. Czranik-Matusewicz B, Pilorz S, Hawranek JP (2005) Temperature-dependent water structural transitions examined by near-IR and mid-IR spectra analyzed by multivariate curve resolution and two-dimensional correlation spectroscopy. Anal Chim Acta 544:15–25
28. Green JL, Lacey AR, Sceats MG (1986) Spectroscopic evidence for spatial correlations of hydrogen bonds in liquid water. J Phys Chem 90:3958–3964
29. Libnau FO, Toft J, Christy AA, Kvalheim OM (1994) Structure of liquid water determined from infrared temperature profiling and evolutionary curve resolution. J Am Chem Soc 116:8311
30. Stanley HE, Teixeira J (1980) Interpretation of the unusual behavior of H_2O and D_2O at low temperatures: tests of a percolation model. J Chem Phys 73:3404–3422
31. Vand V, Senior WA (1965) Structure and partition function of liquid water. III. Development of the partition function for a band model of water. J Chem Phys 43:1869–1884
32. Némethy G, Scheraga HA (1962) Structure of water and hydrophobic bonding in proteins. I. A model for the thermodynamic properties of liquid water. J Chem Phys 36:3382–3400
33. Eisenberg D, Kauzmann W (1969) The structure and properties of water. Clarendon Press, Oxford
34. Buijs K, Chopin GR (1963) Near infrared studies of the structure of water. I. Pure water. J Chem Phys 39:2035–2045
35. Buijs K, Chopin GR (1964) Assignment of the near-infrared bands of water and ionic solutions. J Chem Phys 40:3120–3121

Chapter 5
Interchange of Infiltrating and Resident Water in Partially Saturated Media

Brian Berkowitz

Abstract The interplay between resident water ("old water") and infiltrating water ("new water") in porous media is examined through experiments in an idealized 2D glass micromodel. The analysis is motivated by the recognition that two complementary processes occur as water migrates through the vadose zone: infiltration, when water advances downward through this zone, and drainage, which follows infiltration events (precipitation and irrigation) when air displaces water as the water migrates deeper into the subsurface. Residual water saturation after a cycle of drainage is controlled by pores and pore clusters that retain "old water". During an infiltration cycle, "new water" flowing through the porous medium will intermittently connect to these clusters and then will leave a new pattern of disconnected clusters containing, presumably, both old and new water. In the micromodel experiments, image analysis is used to quantify the miscible interplay between old and new water over a cycle of imbibition by new water and drainage by air in a domain that is partially saturated with old water. We demonstrate that some old water remains in the system at long times within stable water pockets; these pockets may remain stable even after a second cycle of infiltration. An implication of this finding is that water scarcity under dry climate conditions may increase the influence of mixing of infiltrating and resident water on water quality, particularly in the uppermost aquifer layers.

B. Berkowitz (✉)
Department of Environmental Sciences and Energy Research, Weizmann Institute of Science, Rehovot 76100, Israel
e-mail: Brian.Berkowitz@weizmann.ac.il

L. Mercury et al. (eds.), *Transport and Reactivity of Solutions in Confined Hydrosystems*, NATO Science for Peace and Security Series C: Environmental Security,
DOI 10.1007/978-94-007-7534-3_5,

5.1 Introduction

The region situated from the soil surface to the water table can be regarded as a porous medium partially saturated by water. During an infiltration event, either from precipitation, irrigation or leakage, infiltrating, "new water" encounters the resident, "old water", and the two waters interact with each other [e.g., 1–4]. Using water isotopes and/or conservative chemical tracers as proxies, several field studies have demonstrated that the old water remaining in the vadose zone can be a significant component of outflow into, e.g., streams or deeper aquifer layers for several hours [e.g., 3, 5, 6], days [e.g., 7] and even years [e.g., 8] after infiltration of new water. A few studies have attempted to investigate this phenomenon at the catchment scale from a theoretical perspective [9]. From a phenomenological point of view, the old water affects the flow pattern of the new water at the pore scale [10], controlling the availability of pore space and solid surface to the new water. Moreover, this region is a major environmental interface, with solid-liquid, solid-air, liquid-air and liquid-liquid possible interactions. Because the old water might have had time to undergo chemical interactions with the soil environment [11], and the new water may carry various chemicals with it (e.g., fertilizers, pesticides, contaminants), the old and the new water may widely differ in composition.

In these contexts, infiltration and drainage processes are relevant. Infiltration refers to the invasion of new water into a pore space partially or fully saturated with air, while drainage entails air entering a pore space partially or fully saturated by water. During both infiltration and drainage, water and chemicals may move through specific pore channels, leading to a highly non-uniform infiltration pattern; this well-established phenomenon is often referred to as "preferential flow" [e.g., 4, 12, 13]. Moreover, the new water flowing through the porous medium will intermittently connect to old water pockets inherited from the previous infiltration event. The drainage that follows infiltration then leaves a new pattern of disconnected clusters containing, presumably, both old and new water, which might in turn mediate physical and biological processes. Depending on the nature of the flow pattern, the amount of water actually leaving the domain may differ from the amount that entered if smaller or larger amounts of residual water are trapped in the new clusters. Three mechanisms evidently contribute to the old-new water interplay: (1) displacement and flow of old water clusters as they become connected to new channels; (2) partial or full mixing of old water with new water, possibly with redistribution of the residual water clusters; and (3) diffusive transfer of water molecules from old water to new water as new water flows through a channel.

A common technique to study infiltration and drainage mechanisms at the pore scale is the use of chemically-etched 2D glass micromodels. Fabrication of micromodels by chemical etching is described widely in the literature [e.g., 14]. The technique was pioneered by [15, 16] for visualizing immiscible displacements, focusing on drainage of an oil-saturated medium by air; drainage of an oil-, hexane-, or air-saturated medium by mercury; drainage of an oil-saturated medium by a glucose aqueous solution; and infiltration of oil in an air-saturated medium.

A similar approach was used by [17] to examine the effect of gravity on drainage of a glycerol-saturated porous medium; [18] and [19] focused on pressure-saturation relations, while [20] and [21] investigated transport of colloids in unsaturated, fracture-matrix type micromodels. These experiments were in part motivated by interest in the mobilization of residual oil by water and/or gas flooding [22], and thus focused on immiscible displacements.

However, pore scale dynamics of old-new water mixing have received little attention. Relative to the experiments mentioned above, characterization of the dynamics of the old-new water interplay is further complicated by the variably-saturated nature of the system and the need for an effective means to differentiate between the old and the new water. In this study, a glass-etched 2D micromodel was used to visualize the flow and pattern of the old water, in a partially saturated medium subjected to alternative infiltration and drainage. Three dyes were successively used to color the injected water, enabling the distinction between the old and the new water and providing a basis for image analysis of the patterns in time.

5.2 Experimental Section

The glass micromodel consisted of a regular lattice 73.3 mm (length) $\times$ 42.7 mm (width), containing 4,500 regular diamond grains of size 660 μm on each side, separated by eye-shaped channels 100 μm wide and 50 μm deep. The upper and lower boundaries of the micromodel were designed to allow injection and drainage along the entire inlet and outlet faces of the domain, respectively. The lattice inlet consisted of a fine comb-like structure to ensure development of a uniform, homogeneous front parallel to the inlet boundary; the outlet of the lattice was a drain connected to two outflow channels, to ensure that air could enter the system even after the initial water breakthrough. The outflow channels were open to air (i.e., free drainage to atmospheric pressure), with the discharge point at the same height as the bottom of the micromodel. The capillary potential of the lattice was estimated from the pore dimensions of the micromodel [23, 24], with a value in the range 5,000–6,000 Pa. Flow through the micromodel was controlled by a syringe pump (Kd Scientific 100) to inject dyed water and air intermittently at controlled rates. Experiments were conducted at 20 °C under atmospheric pressure.

Water colored by molecular dyes was used as a visual tracer for the old-new water interplay. Air was injected into the micromodel between each dyed-water injection to achieve partial desaturation. Three dyed waters were injected sequentially: aqueous solutions of blue erioglaucine at 0.01 wt. % (Fluka standard, increasing the density of water by 0.01 %), yellow tartrazine at 0.1 wt. % (Sigma, dye content $\geq$ 85 %, increasing the density of water by 0.1 %), and red carmoisine at 0.1 wt. % (Fluka, dye content $\geq$ 98 %, increasing the density of water by 0.1 %). These dyes are used commonly in the context of vadose zone hydrology for field and/or laboratory studies [25]. All of the solutions were prepared using deionized water (18.2 M$\Omega \cdot$cm), and no air bubbles appeared during the experiments. Contact angles of the three solutions were measured using the static sessile drop method; on clean glass, all of them were below 5°. The dyes were confirmed to be conservative by two preliminary experiments. First, small glass beads were plunged into baths of, respectively, erioglaucine, tartrazine and carmoisine, and their mixes at various ratios, for 2 h. The beads were used to represent a large surface area akin to the surface area of the micromodel. Absorption spectra of the baths before and after the experiment were compared for possible adsorption of the dyes or their mix onto the glass beads; the spectra were identical. This experiment also controlled the stability of the dye solutions and their mixes by sampling the surface and the bottom of the baths separately. The duration of the experiment was extended to 24 h with the same results. Second, each dye solution and mix was injected into the empty micromodel. Analysis along the micromodel showed that the dyes were not retarded under the experimental conditions.

At the beginning of a run, an erioglaucine solution was injected into the dry micromodel, followed by injection of air. The resulting configuration, partially saturated with old water (erioglaucine solution), was the starting point of a first old-new water experiment. A tartrazine solution was then injected to represent the new water, and interacted with the old water. After the tartrazine solution injection, air was again injected into the system. Then, the system was partially saturated with erioglaucine and tartrazine solutions. This new configuration was the starting point of a second old-new water experiment, where the old water consisted of the erioglaucine and tartrazine solutions. Note that the initial degree of saturation in the micromodel for this second old-new water experiment was different than that at the beginning of the first experiment with the tartrazine solution injection. In this second old-new water experiment, a carmoisine solution was injected to represent the new water, followed by air. Thus, every run corresponded to two successive old-new water experiments, and the same analysis of the old-new water dynamics was carried out on the two experiments. The change of syringes was a critical moment of the runs as it could create fluid pulses. To minimize error due to alternating syringes, the following methodology was applied: at the end of an injection, a clamp was applied to the micromodel inlet tube, the syringe was changed, the pump was reconnected to the micromodel, and the clamp was removed. A picture was then taken; if this picture was nearly identical to the last picture of the previous injection, the change of syringes was validated and the run proceeded; otherwise, the run was discarded. Each run was performed twice; although the specific patterns differed between replicates because of the stochastic nature of this type of phenomenon [e.g., 15].

For every injection (erioglaucine, tartrazine, carmoisine, and air), the same flux and flux duration were used: 6.3 cm/h during 54 min (i.e., 0.14 mL/h or about 1.5 saturated pore volumes per hour, 0.13 mL total) for each injection of experiments 1 and 2 (letters g and h in the figures, respectively), 4.4 cm/h during 97 min (i.e., 0.09 mL/h or about 1.1 saturated pore volumes per hour, 0.16 mL total) for each injection of experiments 3 and 4 (b and c, respectively), 3.1 cm/h during 111 min (i.e., 0.07 mL/h or about 0.74 saturated pore volumes per hour, 0.13 mL total) for each injection of experiments 5 and 6 (a and f, respectively), and 1.2 cm/h during 220 min (i.e., 0.03 mL/h or about 0.32 saturated pore volumes per hour, 0.10 mL total) for each injection of experiments 7 and 8 (e and d, respectively). The fluxes and flux durations for all of the experiments, and corresponding experiments and letters, are summarized in Table 5.1.

These fluxes were chosen because they cover a range of flows relative to the scale of the micromodel (0.03–0.14 mL/h corresponding to 0.32–1.5 saturated pore volumes per hour); they are also comparable to those used in a field study [5]. At the end of each injection, the syringe pump was stopped briefly, and the syringe quickly replaced to inject the following phase. In all of the experiments, the capillary number, *Ca*, the ratio of the effect of viscous forces to surface tension ($Ca = \mu v/\sigma$, with μ the viscosity of water (Pa · s), v a characteristic velocity (m/s), and σ the surface tension between water and air (N/m)), was between 0.4×10^{-5} and 2×10^{-5}, which is relevant to flow in porous media under a capillary-dominated regime [26]. Between runs, the micromodel was rinsed with (1) deionized water, (2) laboratory-grade ethanol, (3) deionized water, (4) 5 % hydrochloric acid, and (5) deionized water, then left to dry at 50 °C with desiccants under a light vacuum. This methodology was tested previously on glass slides, dipped successively in the three solutions for 2 h each, to ensure that the original contact angles were recovered.

The glass micromodel was placed vertically and lit uniformly from behind using a transparency viewer. During the runs, photographs were taken at regular time intervals using a digital single lens reflex camera (Olympus E-330, 13.5 × 18.0 mm NMOS sensor, 7.5 megapixels) mounted with a Zuiko Digital 35 mm 1:3.5 macro lens. The photographs were captured at the base value of ISO 100 to minimize luminance and chrominance noise. To cover the size of the micromodel, the optical magnification was approximately 1:4, corresponding to a spatial resolution of about 670 μm^2 per pixel (average over all of the experiments). Depending on the flux, the interval between two snapshots was 1 min (experiments 1–2), 2 min (experiments 3–6) or 3 min (experiments 7–8). To obtain statistics on the dynamics of the old and new water distributions at the pore scale, image analysis of the snapshots was performed. The raw files were processed using Olympus STUDIO 2.3, and then cropped to isolate the central 85 % of the pore space to minimize wall effects. Using The GIMP 2.2, the pictures were decomposed according to CMY, RGB and LAB color-encoding systems, and a mask of the micromodel was superimposed to isolate the pore space. The C layer, the G layer, and the Y layer minus the L layer were found to be the best suited to identify the pure erioglaucine, tartrazine, and carmoisine solution components, respectively. These best layers and adequate thresholds were identified by trial and error on drops of different sizes on the micromodel, under the same lighting conditions as the experiments; the goal was to isolate all of the drops of a given dye while including none of the others. The pixels containing a mixture of dyes were identified by addition/subtraction of other layers. The air distribution was recovered by a simple subtraction of the pore space minus the dyed water. Applying the thresholds also converted each layer into its binary counterpart, which was then further analyzed using The Mathworks MATLAB R2007b-R2009b. Processing and analyzing the snapshots took about 3 days for each experiment on a desktop computer (Intel PENTIUM4 CPU, 3.00 GHz, 2.5 GB RAM, 50 GB temporary memory). To validate the image analysis procedure,

Table 5.1 Summary of the experimental conditions and some of the results. The flux and duration are for each successive injection of water and air, respectively

Figure letter	a	b	c	d	e	f	g	h
Experiment No.	5	3	4	8	7	6	1	2
New water	Tartrazine	Tartrazine	Carmoisine	Carmoisine	Tartrazine	Carmoisine	Tartrazine	Carmoisine
Initial water content, wc_i (%)	4.6	11.0	15.2	15.7	16.0	17.9	20.2	23.1
Flux (cm/h)	3.1	4.4	4.4	1.2	1.2	3.1	6.3	6.3
Duration (min)	111	97	97	220	220	111	54	54
Remaining old water (%)	10	7	5	18	12	26	23	29
Remaining first old water (%)	N/A	N/A	7	1.2	N/A	10	N/A	23
t_t	0.16	0.14	0.23	0.23	0.19	0.27	0.30	0.32

The remaining old water (%) is given relative to the initial amount of old water in the system (see Fig. 5.1 for its evolution in time). The remaining first old water refers the remaining erioglaucine solution after injection of air, tartrazine, air, and carmoisine. t_t is the transition time between the two regimes discussed in the text known amounts of dye were injected into the micromodel; image analysis was then conducted, and the resulting amount of dye compared with the known value. The relative error was approximately ±5 %.

5.3 Experimental Results

Figure 5.1 shows the amount of old water remaining unmixed in the system relative to the initial old water content (wc_i).

In the figure, time is normalized by the experiment duration to allow direct comparison among the experiments. Thus, $t = 0$ corresponds to the beginning of new water injection, $t = 1$ corresponds to the change from new water injection to air injection at the inlet, and $t = 2$ corresponds to the end of the air injection, i.e., the end of each old and new water experiment. Note that the normalization is linear, i.e., a time difference of 0.1 represents the same duration from the beginning to the end of a given experiment. The degree of saturation was calculated by dividing the amount of water in the lattice by the pore volume. Here, it was used to obtain the initial (old) water content (wc_i) at the beginning of each old-new water experiment. A flow rate of 6.3 cm/h (experiments 1–2) gave initial saturations, wc_i, of 20.2 and 23.1 % (which correspond to g and h). A flow rate of 4.4 cm/h (experiments 3–4) gave wc_i of 11.0 and 15.2 % (b and c). A flow rate of 3.1 cm/h (experiments 5–6) gave wc_i of 4.6 and 17.9 % (a and f). A flow rate of 1.2 cm/h (experiments 7–8) gave wc_i of 16.0 and 15.7 % (e and d).

All experiments (Fig. 5.1) display an evolution in two steps: first, an initial rapid drop in old water content; second, a slow decrease to a constant value. The amount

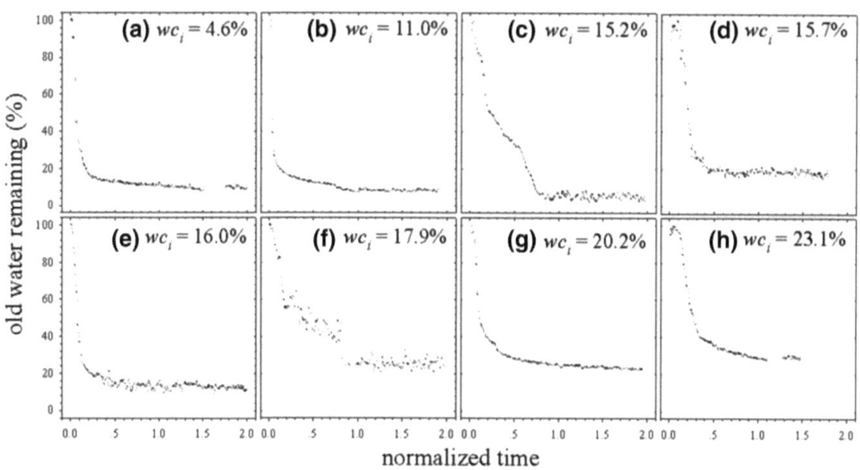

Fig. 5.1 Evolution in time of the old water remaining in the lattice as a percentage of the initial old water content (wc_i)

of initial old water remaining in the system at the end of each experiment, relative to the initial amount of old water, tended to increase with wc_i (Table 5.1). The lowest wc_i (4.6 %) is above this trend (10 % of the initial old water remained in the system), likely due to the small amount of old water in the lattice to begin with and the resulting poor old water connectivity. In general, the first step took place for $t < 0.5$, with the transition time t_t between the two steps increasing with wc_i (from $t_t = 0.16$ for $wc_i = 4.6$ % to $t_t = 0.32$ for $wc_i = 23.1$ %; see Table 5.1). The change from the first to the second step corresponds to the new water breakthrough, wherein the new water found its way through the pore space along a preferential pathway. Because the second step represents a slow decrease, the amount of initial old water remaining was governed mainly by the first step, and may remain in the system even at relatively long times.

Because of the two successive old-new water experiments in each of the runs, it was also possible to compute the amount of the very first old water (i.e., erioglaucine solution) remaining unmixed as the system went through two successive cycles of infiltration of new water and drainage (relative to the erioglaucine solution, the new water was first tartrazine solution, then carmoisine solution). The amount of pure erioglaucine solution remaining in the system at the end of the runs was 23 % for experiments $1+2$, 7 % for experiments $3+4$, 10 % for experiments $5+6$, and 1.2 % for experiments $7+8$, relative to the initial amount of erioglaucine solution at the beginning of experiments 1, 3, 5, and 7, respectively. The same amounts of pure erioglaucine solution remained in the system at the end of the corresponding experiments 1, 3 and 5; thus, the second infiltration cycle did not displace or mix with the pockets of old water inherited from the first infiltration experiment. For experiment 7, 12 % of the pure erioglaucine solution was found, while only 1.2 % remained after experiments $7+8$; note that these experiments employed the lowest of the four flow rates.

Conservative tracer experiments were used to quantify the interplay between old and new water in micromodel experiments. At the beginning of each experiment, the new water occupied the pore space and established the main flow paths (Fig. 5.2a, b). Subsequent air injection had little effect on the new water surface area but it caused a decrease in the volumetric fraction of the new water, with air displacing the new water away from the main pathways (Fig. 5.2c). In contrast, the distribution of old water was not affected by the air injection: old water pockets connected to the flow during the initial new water infiltration phase, but then remained relatively stable in shape and location, and air injection (drainage) did not fragment them significantly. The amount of old water that remained in the system at the end of each experiment was generally higher for higher initial old water contents.

The time regime $0 < t < 0.5$ ended when the new water reached the outlet (i.e., the inlet and outlet became connected, and the preferential pathways were established, see Fig. 5.2a). The second zone ($0.5 < t < 1$) corresponds to entry of new water into a pore space with already established preferential pathways. The new water then flowed into additional, secondary channels, but generally followed the established pathways (see Fig. 5.2b), which translated into a slow increase of the new water volumetric fraction. The third zone ($1 < t < 1.4$) represents the replacement of new

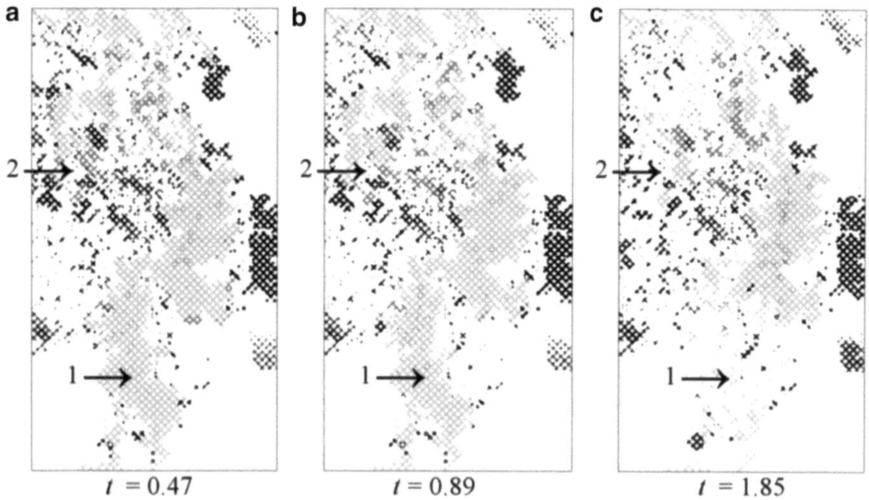

Fig. 5.2 Infiltration of new water in the lattice micromodel partially occupied by old water: (**a**) a preferential pathway has been established through Area 1; (**b**) the new water occupies more of the pore space (e.g., around Area 2) while the previous preferential pathway remains dominant; (**c**) air has replaced new water at the inlet and the new water pattern is disconnected and less dense (in particular, the new water receded from Area 1). These images were taken from one of the replicates of Exp. 2 (h in Fig. 5.1), with a flux of 6.3 cm/h for each injection. *Black* denotes the old water, *light gray* the new water, and *dark gray* the mixing of old and new water; the air and the solid grains are in *white*. The remaining pockets can be noted, even close to the outlet, and the water does not connect to the outlet

water injection by air injection; this led to a rapid decrease in the volumetric fraction of the new water, with the distribution pattern of new water becoming disconnected and less dense. This zone ends when the air at the inlet connected with outlet, and the volumetric fraction for both the new and the old water did not significantly evolve after that (fourth zone, $1.4 < t < 2$; see Fig. 5.1c for an example of the third and fourth zone pattern). Note that for the first and second zones, evolution of the volumetric fractions for the old and the new water were approximately mirror images of each other. This suggests that the new water flowed mostly into pathways that were previously occupied, fully or partially, by the old water. Once air replaced new water at the inlet, evolution of the volumetric fractions for the old and the new water stabilized.

5.4 Discussion and Concluding Remarks

Two successive regimes were identified throughout the experiments. The initial regime corresponded to exploration of the pore space by the new water, while the second regime involved the establishment of stable old water pockets not connected to the main water or air pathways. The amount of old water remaining in the system was significant, and some old water pockets resisted invasion by new water even after a second infiltration-drainage cycle. In some cases, nearly one quarter of the old water remained in the system after a second cycle.

While relating laboratory experiments to real field-scale systems poses clear difficulties in terms of scale and setup [e.g., 27], consideration of mass balance data available from some field-scale experiments can be instructive for qualitative comparison of the old-new water dynamics to the present laboratory results. In particular, a two-step, two-regime behavior similar to the present results was observed in field experiments, where contaminants were found to initially flush rapidly, followed by a long-time low-level contamination of streams [28]. A study of the response of a catchment in the Tawhai State Forest (Westland, New Zealand) to a storm [2] found that the storm runoff is generated by rapid transmission of new water to the stream, a phenomenon analogous to the initial regime described in this study.

In another experiment, a conservative tracer was applied to an undisturbed 2.05×6.3 m soil block, 0.64 m deep isolated in the Ouachita National Forest (Arkansas, USA) [5]. Bromide was added to the first precipitation, and successive precipitations containing no Br^- were then applied; Br^- measured in subsurface flow during the second and third precipitations (no Br^-) would therefore have come from water stored in the soil (i.e., old water). Bromide concentrations in subsurface flow were measured at various depths (14, 26, 44 and 64 cm); the upper two thirds of the depth were a gravelly loam, and the lower third was a clay loam. Old water contributions were calculated by mass balance. Various inflow intensities were investigated, but each displayed a similar pattern of old water concentration changes with time. The old water percentage in the outflow was seen to decrease in two stages, which can be explained by the two regimes observed in the present experiments (Fig. 5.1).

The combination of several dyes to separate individual water phases and quantify their interplay at the pore scale constitutes a novel experimental effort to analyze the fate of old water through successive infiltration events. It is interesting to note that results of the present study exhibit traits similar to those reported in field studies. In most cases, a significant amount of old water remains in the system. In particular, the persistence of a significant part of the old water leads to a soil solution with a different composition, as it has more time to react and to equilibrate with the local soil components. This old water can act as a sink, and later as a source, of contaminants [29]. Moreover, these microenvironments can play an important role for flow and transport of water and solutes from land surface to the water table, and more generally, in the coupled biological, geochemical and physical processes occurring in this region.

The present experiments were conducted in a highly idealized 2D system. Whereas previous studies at the catchment scale attributed the observed behavior to the complex geometry of natural soils [30, 31], the present results suggest that such behavior may also rise in regular, homogeneous, relatively high porosity media. Undisturbed soils, which exhibit lower porosities, a wider variety of grain sizes and a combination of macropores, aggregates, fractures and other structural features (bedding, lamination), likely increase this behavior with more significant air and water trapping [32]. The amount of old water left in the system may thus be underestimated relative to such systems. Full details and analysis of the experiments discussed here appears elsewhere [33].

Extrapolating to consider the effect of changes in rainfall patterns, the proportion of resident water can be expected to increase under dry climate conditions (at low average water content of the unsaturated zone). Moreover, this water may partially evaporate between successive rainwater inputs, increasing local concentrations of chemical species; some of these stored chemicals could be harmful for human health. Heavy rainfall, which occurs occasionally even in arid and semi-arid regions, could dissolve and flush these chemicals into groundwater reservoirs, causing abrupt deterioration of groundwater quality. For example, after the anomalously rainy winter of 1991–1992 in Israel, with about 200 % of the average rainfall, salinity of groundwater in pumping wells increased suddenly by several hundred mg L^{-1} [34].

Acknowledgement Maxime Gouet-Kaplan is thanked for performing the experiments.

References

1. Pearce AJ, Stewart MK, Sklash MG (1986) Storm runoff generation in humid headwater catchments 1. Where does the water come from? Water Resour Res 22:1263–1272
2. Sklash MG, Stewart MK, Pearce AJ (1986) Storm runoff generation in humid headwater catchments 2. A case study of hillslope and low-order stream response. Water Resour Res 22:1273–1282
3. Luxmoore RJ, Ferrand LA (1993) Towards pore-scale analysis of preferential flow and chemical transport. Adv Ser Agric Sci 20:45–60. In Russo D, Dagan G (eds) Water flow and solute transport in soils. Springer, Berlin, pp 45–60
4. Hendrickx JMH, Flury M (2001) Uniform and preferential flow mechanisms in the vadose zone. In: Panel on Conceptual Models of Flow and Transport in the Fractured Vadose Zone, U.S. National Committee for Rock Mechanics, Board on Earth Sciences and Resources, National Research Council (eds) Conceptual models of flow and transport in the fractured vadose zone. National Academy Press, Washington, DC, pp 149–187
5. Turton DJ, Barnes DR, de Jesus NJ (1995) Old and new water in subsurface flow from a forest soil block. J Environ Qual 24:139–146
6. Collins R, Jenkins A, Harrow MA (2000) The contribution of old and new water to a storm hydrograph determined by tracer addition to a whole catchment. Hydrol Process 14:701–711
7. Sklash MG, Farvolden RN (1979) The role of groundwater in storm runoff. J Hydrol 43:45–65
8. Gvirtzman H, Paldor N, Magaritz M, Bachmat Y (1988) Mass exchange between mobile freshwater and immobile saline water in the unsaturated zone. Water Resour Res 24:1638–1644
9. Botter G, Bertuzzo E, Rinaldo A (2010) Transport in the hydrologic response: travel time distributions, soil moisture dynamics, and the old water paradox. Water Resour Res 46:W03514
10. Flury M, Flühler H, Jury WA, Leuenberger J (1994) Susceptibility of soils to preferential flow of water: a field study. Water Resour Res 30:1945–1954
11. Fisher R, Binkley D (2000) Ecology and management of forest soils, 3rd edn. Wiley, New York
12. Peck AJ (1965) Moisture profile development and air compression during water uptake by bounded porous bodies 3. Vertical columns. Soil Sci 100:44–51

13. Hill DE, Parlange JY (1972) Wetting front instability in layered soils. Soil Sci Soc Am J 36:697–702
14. Buckley JS (1990) Multiphase displacements in micromodels. In: Morrow NR (ed) Interfacial phenomena in petroleum recovery, Surfactant science series 36. Marcel Dekker, New York, pp 157–190
15. Lenormand R, Touboul E, Zarcone C (1988) Numerical models and experiments on immiscible displacement in porous media. J Fluid Mech 189:165–187
16. Lenormand R (1990) Liquids in porous media. J Phys Condens Matter 2(S):SA79–SA88
17. Birovljev A, Furuberg L, Feder J, Jossang T, Maloy KJ, Aharony A (1991) Gravity invasion percolation in two dimensions: experiment and simulation. Phys Rev Lett 67:584–587
18. Soll WE, Celia MA, Wilson JL (1993) Micromodel studies of three-fluid porous media systems: pore-scale processes relating to capillary pressure-saturation relationships. Water Resour Res 29:2963–2974
19. Cheng JT, Pyrak-Nolte LJ, Nolte DD, Giordano NJ (2004) Linking pressure and saturation through interfacial areas in porous media. Geophys Res Lett 31:L08502
20. Wan J, Tokunaga TK, Tsang CF, Bodvarsson GS (1996) Improved glass micromodel methods for studies of flow and transport in fractured porous media. Water Resour Res 32:1955–1964
21. Chen G, Flury M (2005) Retention of mineral colloids in unsaturated porous media as related to their surface properties. Coll Surf A 256:207–216
22. Øren PE, Billiotte J, Pinczewski WV (1992) Mobilization of waterflood residual oil by gas injection for water-wet conditions. SPE Formation Eval 7(1):70–78. doi:10.2118/20185-PA
23. Smith JD, Chatzis I, Ioannidis MA (2002) A new technique for measuring the breakthrough capillary pressure. In: Proceedings of the SCA 2002 annual international symposium, Monterey, 22–26 September, Paper no. 2002–40
24. Kirkham MB (2004) Principles of soil and plant water relations. Elsevier Academic Press, Burlington
25. Flury M, Wai NN (2003) Dyes as tracers for vadose zone hydrology. Rev Geophys 41:1002
26. Aker E (1996) A simulation model for two-phase flow in porous media. Ph.D. thesis, University of Oslo, Oslo
27. van Genabeek O, Rothman DH (1996) Macroscopic manifestations of microscopic flows through porous media. Annu Rev Earth Planet Sci 24:63–87
28. Kirchner JW, Feng X, Neal C (2000) Fractal stream chemistry and its implications for contaminant transport in catchments. Nature 403:524–527
29. Mayes MA, Jardine PM, Mehlhorn TL, Bjornstad BN, Ladd JL, Zachara JM (2003) Transport of multiple tracers in variably saturated humid region structured soils and semi-arid region laminated sediments. J Hydrol 275:141–161
30. Brammer DD, McDonnell JJ (1996) An evolving perceptual model of hillslope flow at the Maimai catchment. In: Anderson MG, Brooks SM (eds) Advances in hillslope processes, vol 1. Brit Geomor/Wiley, New York, pp 35–60
31. Kirchner JW, Feng X, Neal C (2001) Catchment-scale advection and dispersion as a mechanism for fractal scaling in stream tracer concentrations. J Hydrol 254:82–101
32. van Genuchten MT, Wierenga PJ (1976) Mass transfer studies in sorbing porous media I. Analytical solutions. Soil Sci Soc Am J 40:473–480
33. Gouet-Kaplan M, Berkowitz B (2011) Measurements of interactions between resident and infiltrating water in a lattice micromodel. Vadose Zone J 10:624–633
34. Development of Exploitation and State of Water Resources in Israel till the fall of 2011 (2012) Report to Israel Water Commission, Jerusalem, 480 pp

Chapter 6
Impact of Heterogeneity on Evaporation from Bare Soils

Shmuel Assouline, Kfir Narkis, Stéphanie Veran-Tissoires, Manuel Marcoux, and Marc Prat

Abstract Heterogeneity in soil hydraulic properties has a significant impact on evaporation, and could be harnessed to reduce water losses and improve soil water conservation. This is illustrated through the consideration of the effect of Darcy scale heterogeneities resulting from horizontal layering. The impact of permeability gradient and thickness of layers has been investigated from evaporation experiments performed from homogeneous as well as horizontally multi-layered soil columns. Two main cases are distinguished depending on the sign of the permeability gradient, the unstable case when the permeability increases with depth and the stable case when, on the contrary, the permeability decreases with depth. The results indicate an interesting competition between stabilizing gravity effects and destabilizing or stabilizing permeability gradient effects and lead to the emergence of the concept of two-scale evaporation process.

6.1 Introduction

Evaporation is a key process for water exchange between soil and atmosphere. Evaporation from bare soils depletes the water stored in the root zone during the rainfall season and limits the establishment and spreading of vegetation cover, thus initiating a vicious cycle leading to desertification. In cultivated fields, evaporation reduces the amount of water available for crop growth, determines the water and salinity regimes in the root zone, and consequently, affects yield and product quality.

S. Assouline (✉) • K. Narkis
Institute of Soil, Water and Environmental Sciences, Agricultural Research Organization (ARO), Bet Dagan, Israel
e-mail: vwshmuel@volcani.agri.gov.il

S. Veran-Tissoires • M. Marcoux • M. Prat
Institut National Polytechnique de Toulouse (INPT), Institut de Mécanique des Fluides de Toulouse (IMFT), UMR 5502 CNRS-UPS, Toulouse, France

L. Mercury et al. (eds.), *Transport and Reactivity of Solutions in Confined Hydrosystems*, NATO Science for Peace and Security Series C: Environmental Security,
DOI 10.1007/978-94-007-7534-3_6,

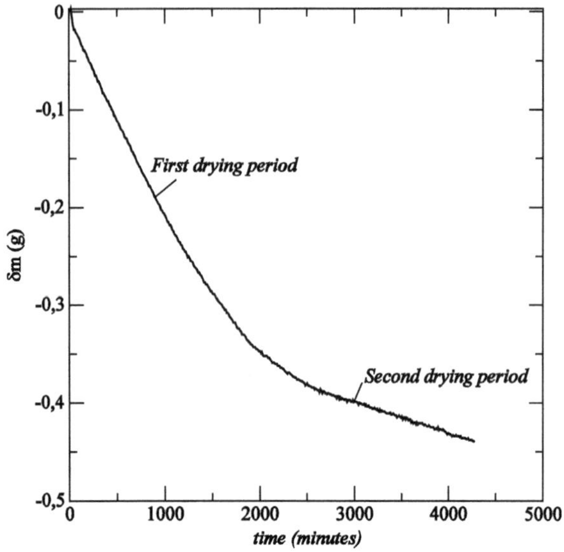

Fig. 6.1 Evaporated mass as a function of time: the two main regimes of evaporation in a classical evaporation situation

Drying processes are also of significant industrial and engineering importance including food processing, paper production, cosmetic industry, and construction activities. Evaporation from soils drew attention of researchers along decades [1–3]. Notwithstanding the prominent role of evaporation, prediction of drying rates from porous media remains a challenge. The evaporation rate is affected by both atmospheric demand (humidity, temperature, and velocity of ambient air) and by porous-medium pore space and transport properties (thermal and hydraulic conductivities and vapor diffusion). Consequently, complex and highly dynamic interactions between medium properties, transport processes, and boundary conditions result in a wide range of evaporation behaviors as discussed in [4, 5]. The understanding and the analysis of evaporation in homogeneous porous systems have been greatly improved since the early 1990s thanks to pore network modeling and invasion percolation (IP) theory, e.g. [6–14]. This type of approach is also well adapted to the study of heterogeneous systems [15] and therefore is one of the cornerstones of the present study.

In a typical evaporation situation (see Fig. 6.1), two main regimes are generally distinguished. As sketched in Fig. 6.1, the first regime is characterized by the presence of liquid water up to the surface of the porous medium. The pumping effect due to capillary forces is sufficient to bring the liquid water from the inside of porous medium up to the surface. The evaporation rate is the highest in this regime and very often (quasi-) constant. Explaining why it is so, that is why the evaporation rate remains constant whereas the overall liquid saturation decreases is still an open question. The main factors, referred to as the invasion percolation effect, the Suzuki and Maeda effect [16], and the films effect, are well known, e.g. [17] and references therein, but the details, that is the relative significance of the various factors for a specific situation, are generally not well understood. The second regime is characterized by the development of a dry zone adjacent to the surface of the porous medium. Because of the poor efficiency of the transport of the water vapor by diffusion across the dry zone, the evaporation rate drops dramatically in this regime. Notice that it is customary and more accurate to distinguish a cross-over regime, referred to as the falling rate regime [4], between the two regimes depicted in Fig. 6.1. Distinguishing only two main regimes is, however, sufficient in the context of the present chapter.

Evaporation from a sample fully saturated initially is thus characterized by the development of a two-phase zone where liquid and gas coexist. A key time in the analysis of evaporation from a porous sample is when the surface of the porous medium ceases to be fed by liquid water. This time marks the end of the first period of high evaporation and the beginning of the second period of much slower evaporation. Thus, this time corresponds to the time when the two-phase zone is not in contact anymore with the porous medium surface. This is why characterizing the extent of the two-phase zone has attracted a lot of attention in the studies on evaporation. Suppose we are able to estimate the maximum vertical length L_{max} of the two-phase zone, then the end of first regime is when the length of the two-phase zone reaches L_{max}. This length depends on several factors: the forces or the effects acting on the distribution of the fluid, namely the gravity, capillary and viscous forces, the temperature gradients, the species gradients (when species like ions, colloidal suspension, other molecules are present in the water), the structural heterogeneities (as discussed below), the wettability properties, etc. This length is easy to estimate from drainage theory in the simple cases where the temperature gradients can be neglected, the liquid is a pure liquid and is wetting, and the effect of liquid films can be also neglected. Liquid films refer to the capillary films that can be present in the corners of the pore space, e.g. [18–21]. As shown in [21] for the case of a model porous medium made of glass beads, the capillary films can be present over a large zone outside what we would be referred to as the two phase zone in drainage theory. This is sketched in Fig. 6.2. The presence of capillary films can dramatically extent the duration of the first regime of evaporation. Thus, when the films effect is significant, the length L_{max} must of course include the film region. The significance of the film effects depends on several factors, such as the contact angle (the lower the contact angle, the greater the probability to observe a significant film effect), the geometry of the pore space (corners or other geometrical singularities must be present), the evaporation rate and the pore size (viscous effects in the films limit the extent of films; thus smaller pore sizes for a given evaporation rate lead to a lower extent of film region).

In what follows we assume that the film effects can be neglected. Under these circumstances, the length L_{max} of the two phase zone can be determined thanks to the theory of invasion percolation in a gradient (IPG), e.g. [9, 22] and references therein. This shows that the scaling in 2D systems is quite different from the scaling in 3D. Whereas the length of the two-phase zone can be reasonably estimated using the classical continuum approach to porous media, e.g. [23], the situation is quite different in two-dimensional systems because the mean position of the drainage front corresponds in this case to the percolation threshold. The invasion

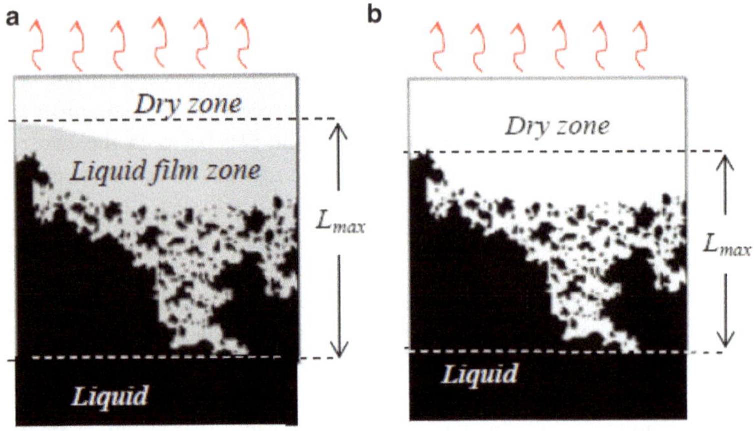

Fig. 6.2 Sketch of main zones during evaporation in a stabilizing gradient when capillary films effect is significant (**a**) or negligible (**b**); L_{max} is the size of the two-phase zone where liquid and gas coexist in the pore space

front (= the two-phase zone) is then an example of self organized criticality (SOC) and this leads to non-trivial scalings [9, 24]. This brings once in while some confusion when the fundamental differences between the 2D and the 3D cases are not understood.

The consideration of L_{max} is useful not only to determine the transition between the two evaporation regimes but also to delineate the factors controlling the evaporation process. As discussed in more details in [4], see also [23], one can estimate the length L_{max} corresponding to the competition between the capillary and gravity effect, denoted by L_g , as well as the length L_{vis} corresponding to a two-phase zone controlled by the competition between the capillary and viscous effects only. The smaller of these two scales indicates which factor (in addition to capillarity) is dominant. For example the extent of the phase zone is controlled by the gravity and capillary forces when $L_g << L_{vis}$ whereas gravity and viscous effects are both important when $L_g \sim L_{vis}$. When these lengths are much larger than the height H of the sample, then the process is dominated by the capillary forces and the two-phase zone can extent over the entire sample. This regime is referred to as the capillary regime [4].

The IPG theory is also useful to distinguish two main situations, referred to as invasion percolation in a stabilizing gradient (IPSG) and invasion percolation in a destabilizing gradient (IPDG) respectively. The gradient evoked here is the gradient in the pore occupation probability; see for example [9] and references therein for more details. An IPSG pattern corresponds to the one shown in Fig. 6.2 and is characterized by the presence of the two-phase zone (regardless of the liquid films) over a region of finite thickness L. This is observed for example when gravity or viscous effects are sufficiently important compared to capillary effects, $L_g < H$ or $L_{vis} < H$. A typical IPDG pattern can be obtained when the gas invasion resulting

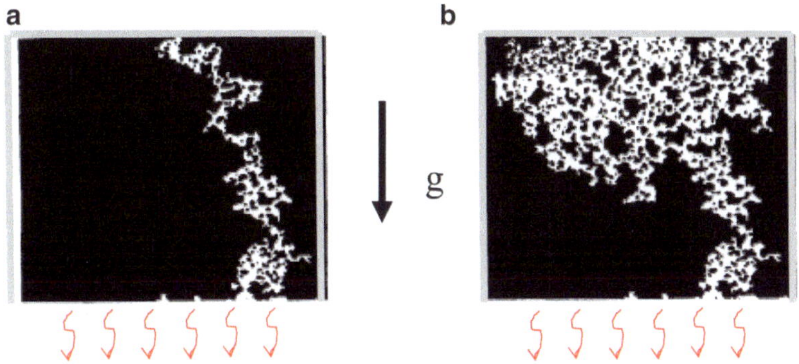

Fig. 6.3 Example of IPDG patterns at two stages of the invasion obtained in a two-dimensional model porous medium when gravity effects induce a destabilizing gradient in the occupation probability. Liquid phase in black and gas phase in white. Vapor escapes from the bottom of the micromodel only

from the evaporation process is from the bottom side of the system, and gravity and capillary effects control the phase distribution. A typical IPDG pattern at two stages of the invasion process is shown in Fig. 6.3. In contrast to the case where gas invasion is from the top and gravity effects are stabilizing (Fig. 6.2), gravity effects are destabilizing when evaporation is from the bottom. Thus, here, L_g is still lower than H but the occupation probability gradient is now negative. The occupation probability decreases with the distance from the sample top as can be seen from Fig. 6.3b, which shows a preferential invasion in the upper region of sample.

In the IPDG case the porous medium surface remains connected to the liquid phase over a much longer period than in the IPSG case. As a result, the duration of fast evaporation regime (first drying period) is much longer in the IPDG case. Thus IPDG patterns favor fast evaporation whereas IPSG situations favor the formation of a dry zone and thus a slower evaporation.

As discussed in the next section, the consideration of the above length scales and a clear distinction between IPSG and IPDG situations help analyze the effect of layering on evaporation.

6.2 Experimental Section

Laboratory scale experiments were carried out on columns. Columns of 50-cm height and of 8-cm in diameter were packed to create bi-layered or multi-layered systems using sands of different mean grain size. As a result, heterogeneous columns with successive horizontal layers presenting increasing or decreasing permeability ($\sim$ grain size) were prepared. Two thicknesses for the layers were considered: 2 and 7 cm thickness. The columns were gently saturated from below with tap water. The bottom was then sealed and the columns exposed to drying.

Fig. 6.4 Experimental set-up of multi-layer sand columns

The corresponding lower boundary condition is thus of zero flux at the bottom. A small individual fan was installed close to the open upper top of each column. The columns were weighted at very high frequency (every 10–15 min) using electronic scales connected to a data logger (Fig. 6.4). Water mass loss and evaporation rate were computed from the data. Reference evaporation rate was characterized by evaporation from columns filled with water and subjected to similar conditions.

This leads to the results depicted in Fig. 6.5, which clearly show a significant effect of the layering. For reasons that will be made clear below, the case where the permeability (i.e. the grain size) increases with the distance from the column top is referred to as the unstable case whereas the case where the permeability decreases with the distance from the column top is referred to as the stable case.

The main results can be summarized as follows:

1 Unstable arrangements lead to faster drying than stable arrangements
2 The 2 cm layers-stable arrangement leads to slowest drying
3 The 7 cm layers unstable arrangement leads to fastest drying
4 The 7 cm layers unstable arrangement is first slower than the 2 cm layers unstable arrangement then faster

To analyze and explain these results one must consider the effects influencing the phase distribution. The viscous effects can be discarded because even the finer layer is made of a relatively coarse material (mean grain size ~ 100 μm). Thus, gravity and capillary effects have to be considered (because estimate of length L_g is lower than column height H). The fact that the duration of first drying period is greater with the fine sand in the case of the homogenous column is thus easily explained by the fact that $L_{gfine} > L_{gcoarse}$ ($L_g \propto d_p{}^{-2}$ where d_p is the grain size [9]). As discussed

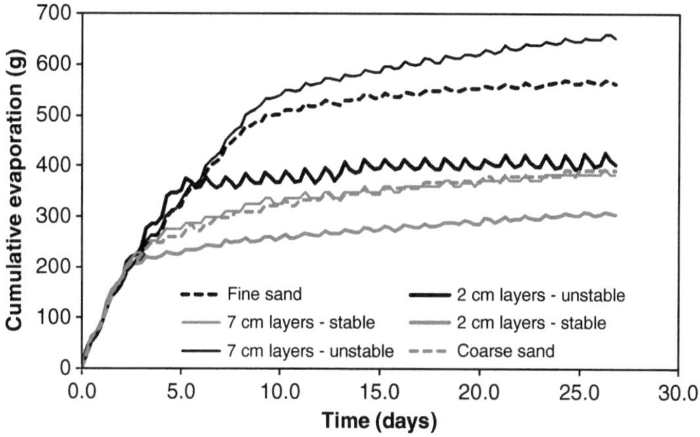

Fig. 6.5 Variation of evaporated mass as a function of time for the different layers arrangement

for example in [25], layering in the absence of gravity and/or viscous forces can lead to IPSG or IPDG invasion patterns depending on the sign of the permeability gradient. In the case of a permeability gradient, the length scale L_k characterizing the extent of the two-phase in the IPSG case depends on the grain size gradient, $L_k \sim \left(-\frac{d\sqrt{k}}{dz}\right)^{-1}$, where k is the permeability. In a system made of grains of size d_p, $k \propto {d_p}^2$ and thus $L_k \sim \left(-\frac{d(d_p)}{dz}\right)^{-1}$. As reported in [25], a IPSG pattern is obtained when $\frac{d\sqrt{k}}{dz} < 0$ whereas a IPDG pattern is obtained when $\frac{d\sqrt{k}}{dz} > 0$ (with z directed from the top to the bottom). In the case of our column experiments, the increase or decrease in pore size is not continuous but discrete since the column is made of a limited number of homogeneous layers. This does not change the qualitative behaviour. Thus an IPDG pattern due to layering is expected when the grain size increases with the distance from the column top whereas an IPSG pattern is expected when the mean grain size in each layer decreases from the top. This is why the arrangement where the grain size increases (decreases respectively) with depth is referred to unstable (stable respectively). Thus point # 1 is fully consistent with the fact that IPDG patterns lead to fastest evaporation. The discrete gradient $\frac{d\sqrt{k}}{dz}$ is smaller when the layer thickness is 7 cm than for the 2 cm layer since this gradient is roughly equal to difference in grain size between two successive layers divided by the thickness of layers (= distance between the centres of two successive layers). This implies $L_{k7cm} > L_{k2cm}$, which is fully consistent with point # 2. This is also fully consistent with the first part of point # 4, that is the fact that the 7 cm layers unstable arrangement first leads to a slower evaporation than for the 2 cm layers unstable arrangement

To explain the second part of point #4, the fact that the 7 cm layers unstable arrangement eventually leads to a faster evaporation than for the 2 cm layers unstable arrangement, one must realize that both the layering effect and gravity effect affect the phase distribution during evaporation, which means that L_k and L_g are of the same order of magnitude, at least at some moments during the evaporation process. Such a situation, in which more than one length scale affects the evaporation process, is referred to as a multiscale evaporation process [26]. Here gravity effects are stabilizing regardless of the layer arrangement since evaporation is from the top. Since $L_g \propto {d_p}^{-2}$, L_g decreases with the distance from the top when the arrangement is unstable and decreases faster when the thickness of layers is 2 cm. It turns out that L_g becomes smaller than L_k faster with the 2 cm layers and also reaches a smaller value since the grains forming the bottom layer in the 2 cm layer unstable arrangement are bigger than for the 7 cm layer unstable arrangement. Thus gravity effects eventually stabilize the invasion, which of course leads to a slower evaporation than for a purely IPDG case (i.e. layering without gravity effects), and this stabilizing effects appears sooner and is stronger with the 2 cm layers. This explains the second part of point #4 and has been confirmed from pore network simulations where it is of course easy to run the simulations with and without gravity [26]. A more detailed analysis of the multi-layered column experiments is presented in [27].

6.3 Conclusion

Evaporation from bare soil is greatly dependent on the factors affecting the liquid and gas distribution during the process. In particular, it is important to predict when the surface of the porous ceases to be hydraulically connected to the liquid still present in the porous medium. The influence of the various factors can be assessed from the consideration of the length scale characterizing the maximum extent of the two-phase zone in the system. As other factors, such as the viscous effects or the gravity effects, large scale heterogeneities such as horizontal layering can have a great impact on the evaporation kinetics. This is a factor which one can play with to enhance or, on the contrary, limit the evaporation.

To large extent, evaporation process in a porous medium can be conveniently analyzed within the theoretical framework of invasion percolation in a gradient. Although predicting the evaporation curve accurately remains a challenge, the drying theory can thus be considered as well advanced, at least as regards the microporous systems which have been the focus of most recent studies. By contrast the nanoporous materials have been much less studied and specific phenomena not important or present in microporous systems are expected. The emergence of nanofluidics and associated investigation tools is clearly an opportunity to develop or at least revisit the theory of evaporation in nanoporous materials, a widely open and exciting topic.

Acknowledgement Financial support from the Israeli French High Council for Scientific and Technological Research (Project ISRAEL 10 F 8/EAU) is gratefully acknowledged.

References

1. Gardner WR, Hillel DI (1962) The relation of external evaporative conditions to the drying of soils. J Geophys Res 67:4319
2. Hanks RJ, Gardner WR, Fairbour MI (1967) Evaporation of water from soils as influenced by drying with wind or radiation. Soil Sci Soc Am Proc 31:593–598
3. Aydin M, Yang SL, Kart N, Yano T (2005) Test of a simple model for estimating evaporation from bare soils in different environments. Ecol Model 182:91–105
4. Prat M (2002) Recent advances in pore-scale models for drying of porous media. Chem Eng J 86:153–164
5. Van Brakel J (1980) Mass transfer in convective drying. Adv Dry 1:217–267
6. Prat M (1993) Percolation model of drying under isothermal conditions in porous media. Int J Multiph Flow 19:691–704
7. Laurindo JB, Prat M (1996) Numerical and experimental network study of evaporation in capillary porous media. Phase distributions. Chem Eng Sci 51:5171–5185
8. Laurindo JB, Prat M (1998) Numerical and experimental network study of evaporation in capillary porous media: drying rates. Chem Eng Sci 53:2257–2269
9. Prat M, Bouleux F (1999) Drying of capillary porous media with stabilized front in two dimensions. Phys Rev E 60:5647–5656
10. Huinink HP, Pel L, Michels MAJ, Prat M (2002) Drying processes in the presence of temperature gradients. Pore-scale modeling. Eur Phys J E 9:487–498
11. Plourde F, Prat M (2003) Pore network simulations of drying of capillary media. Influence of thermal gradients. Int J Heat and Mass Transf 46:1293–1307
12. Prat M (2007) On the influence of pore shape, contact angle and film flows on drying of capillary porous media. Int J Heat and Mass Transf 50:1455–1468
13. Chapuis O, Prat M (2007) Influence of wettability conditions on slow evaporation in two-dimensional porous media. Phys Rev E 75:046311
14. Chraibi H, Prat M, Chapuis O (2009) Influence of contact angle on slow evaporation in two dimensional porous media. Phys Rev E 79:026313
15. Pillai KM, Prat M, Marcoux M (2009) A study on slow evaporation of liquids in a dual-porosity porous medium using square network model. Int J Heat and Mass Transfr 52:1643–1656
16. Suzuki M, Maeda S (1968) On the mechanism of drying of granular beds. J Chem Eng Jpn 1(1):26–31
17. Prat M (2011) Pore network models of drying, contact angle and films flows. Chem Eng Technol 34(7):1029–1038
18. Eijkel JCT, Dan B, Reemeijer HW, Hermes DC, Bomer JG, van den Berg A (2005) Strongly accelerated and humidity-independent drying of nanochannels induced by sharp corners. Phys Rev Lett 95:256107
19. Chauvet F, Duru P, Geoffroy S, Prat M (2009) Three periods of drying of a single square capillary tube. Phys Rev Lett 103:124502
20. Yiotis AG, Salin D, Tajer ES, Yortsos YC (2012) Analytical solutions of drying in porous media for gravity-stabilized fronts. Phys Rev E 85:046308
21. Yiotis AG, Salin D, Tajer ES, Yortsos YC (2012) Drying in porous media with gravity-stabilized fronts: experimental results. Phys Rev E 86:026310
22. Tsimpanogiannis IN, Yortsos YC, Poulou S, Kanellopoulos N, Stubos AK (1999) Scaling theory of drying in porous media. Phys Rev E 59(4):4353–4365
23. Lehmann P, Assouline S, Or D (2008) Characteristic lengths affecting evaporative drying of porous media. Phys Rev E 77:056309
24. Shaw TM (1987) Drying as an immiscible displacement process with fluid counterflow. Phys Rev Lett 59(15):1671–1674

25. Chaouche M, Rakotomalala N, Salin D, Xu B, Yortsos YC (1994) Invasion percolation in a hydrostatic or permeability gradient: experiments and simulations. Phys Rev E 49: 4133–4139
26. Prat M, Assouline S, Marcoux M, Veran-Tissoires S Multi-scale drying in porous media. (to be submitted)
27. Assouline S, Narkis K, Gherabli R, Lefort P, Prat M (2013) Impact of surface layer properties on evaporation from porous systems. (under review)

Part II
Ions, Hydration, and Transport

Aqueous solutions containing ions are ubiquitous in the world around us and their properties and behavior influence a variety of processes. In biology, regulation of ionic transport and pH is critical for the functioning of biological cells and tissues and for the transmission of nerve impulses. This regulation is accomplished by proteins called ion channels that transport or pump specific ions in and out of the cell. In geology, dissolution and deposition of minerals and their precipitation can sculpt rocks, shape landscapes, erode mountains, and generate intriguing features such as stalagmites and stalactites. The stability of colloidal solutions is influenced by ions in the solution into which they are dispersed; addition of alum that dissociates into multivalent ions is widely used to clean water by flocculation of colloidal impurities. The manner in which certain materials interact with ions is also critical for realization of technologies such as seawater desalination by reverse osmosis using semi-permeable membranes. While ions in aqueous solutions play such a critical role in natural as well as artificial systems, only 100 years ago the nature of ionic solutions was largely unknown.

The term "ion" was first coined by Michael Faraday in 1834, who observed that something (unknown to him at the time) carried electrical current through a solution in an electrolytic cell. "Ion" derives from Greek "ιoν" which means to go (from one electrode to another) [1]. The first major advance in understanding of the nature of ionic solutions was made by Svante Arrhenius [2]. In his doctoral thesis submitted in 1884, Arrhenius pointed out that although neither salts nor pure water are very good conductors of electricity, solutions of salts in water are. He was therefore the first person to posit that electrolytes separate into ions when in solution. This idea failed to impress his professors and earned him a low grade in his thesis, but the same thesis later won him the Nobel Prize in Chemistry in 1903! Rapid developments in the understanding of the nature and behavior of ionic solutions occurred in the first decade of the twentieth century, when Nernst and others probed the question as to whether ions are associated with water molecules, i.e. whether ions are hydrated [3]. The experiments carried out to probe ion hydration seem surprisingly crude in today's days of advanced instrumentation: The basic hypothesis was that if water molecules were associated with ions, then water molecules will move along with ions and get concentrated at the electrodes in an electrolysis cell! The concentration of water at electrode surfaces (i.e. depletion of a dissolved non-electrolyte) suggested that indeed ions "carry" water molecules.

As more evidence of the nature of electrolytes and hydration of ions was emerging, Born's theory of ion solvation published in 1920 was successful in qualitatively explaining the energies of dissolution of different salts; the theory explained how the dielectric shielding of charges allows ions to be dissolved in water [4]. A major conceptual advance occurred in the same year, when Latimer and Rodebush published a paper that pointed out the existence of the hydrogen bond [5]. They wrote "... a free pair of electrons on one water molecule might be able to exert sufficient force on a hydrogen held by a pair of electrons on another water molecule to bind the two molecules together... Such combination need not be limited to the formation of double or triple molecules. Indeed the liquid may be made up of large aggregates of molecules, continually breaking up and reforming under the influence of thermal agitation." It is now well-appreciated how hydrogen bonding in water plays a critical role in governing the structure and behavior of water and biomolecules, but it was only half a century later that sufficient progress was made to fully appreciate Latimer's conjecture. Three years after Latimer and Born's papers, Debye and Hückel proposed their famous theory that accounted for electrostatic interactions between ions in solution, which explained the deviation of electrolyte behavior from ideality [6]. In parallel, the nature of the interface between water and electrolyte solutions was also being investigated. Louis Guoy in 1910 and David Chapman in 1913 had developed the diffuse electrical double layer model, which described the distribution of ions in solution next to a charged solid surface, forming a "double layer" of charges (one in solution and the other on the solid surface). In 1924, Otto Stern proposed a new model for the electric double layer, which comprised a layer of surface-adsorbed ions first proposed by Helmholtz in addition to the diffuse double layer concepts developed by Guoy and Chapman. Thus, we can say that the foundations of today's understanding of the structure of water and ionic solutions, as well as the field of electrokinetics, were laid in the first quarter of the twentieth century.

Later significant developments in understanding of the nature of ionic solutions and their behavior at interfaces included the DLVO theory of colloidal stability in the 1940s, which explained how the balance between van der Waals attraction and electrostatic repulsion determined the stability of colloids. Even with these developments, much debate remained about the structure of water and hydration of ions well into the 1970s [7]. During this period several theories of the structure of water were proposed, with debates in the early years ranging from whether water molecule was (H_2O), $(H_2O)_2$, $(H_2O)_3$ or a mixture of these forms [8], to the "polywater" controversy in the 1960s. By the 1970s, new tools such as NMR, vibrational spectroscopy, and surface force apparatus, as well as computer simulation techniques were being employed to study the structure of water and ionic solutions. The realization occurred that the structure of water was dynamic, and it made sense to talk about correlations between the relative positions of water molecules or atoms with respect to an ion or water molecule ("radial distribution function"), rather than talk about a fixed number of water molecules bound to an ion [9]. The same decade saw the first measurements of ionic currents through single ion channels in cell membranes by Sakmann and Neher [10], (which won them the Nobel Prize in 1991), and direct measurements of forces between surfaces that quantified the effects of van der Waals, electrostatic, and hydration [11]. The 1990s saw the discovery of the aquaporin water transporter channels by Peter Agre, which also led to a Nobel Prize in 2003 shared with MacKinnon who elucidated the mechanism of potassium ion channels. Electron microscopy and crystallography helped shed light on the structure and function of ion and water channels in cell membranes, with the interesting mechanism of selective water transport in aquaporin being revealed only in 2001 [12].

These historical developments have led to our current understanding of ions and hydration, in which two important length scales that affect ionic transport have emerged [13]. The first length scale concerns the size of the hydrated ions, most effectively described using the radial distribution function of water molecules around the ion. This length scale is on the order of 1 nm. The second important length scale concerns the range of electrostatic interactions in solution. This length scale is characterized by the Debye length, which depends on the ionic strength of the solution and ranges from about 1 μm in deionized water to less than 1 nm in concentrated salt solutions. Transport of ions and water in confined systems changes dramatically as either of these two length scales is approached. For example, if channel geometries approach the nanometer length scale, hydration effects can significantly alter transport properties. Such systems include carbon nanotubes, small nanochannels or nanopores, and porous materials such as clays, zeolites, graphene, and polymers that have nanometer-scale conduits or spaces. In contrast to hydration effects, electrostatic and electrokinetic effects can play a role in larger channels or conduits approaching the millimeter length scale. The layer of ions next to a charged surface moves in response to electric fields, leading to electroosmotic flows and other electrokinetic effects in microchannels and porous media. In smaller channels, the surface charge can govern ionic transport and co-ions are excluded [13]. Transport of ions and water in confined systems is thus intimately connected to electrokinetic and hydration effects that occur at small length scales.

In perspective, it is only in the last two to three decades that sufficient computational resources and advanced experimental techniques have been employed for understanding the structure and interactions of ions and water. On the experimental side, characterization, control, and interrogation of interfacial phenomena are difficult. Going beyond the continuum double-layer theories to account for the behavior of ionic solutions at interfaces involves complexities that often necessitate simulations. At the same time, the first decade of the twenty-first century has seen a burst of research on nanoscale and microscale systems that often operate in aqueous solutions containing ions where interfaces play a major role. As a result, tools and techniques for controlling the structure of matter at the nanoscale have become ubiquitous. Research has also shifted much of its focus from basic to applied, with interdisciplinary methods and collaborations becoming the norm rather than exceptions. Examples of problems being tackled in this new era range from how to design a drug molecule to bind a target protein in physiological solution, to flow of water in carbon nanotubes for filtration, to control of surfaces of biosensors. The molecular structures and behaviors of these systems are varied, and there is no simple analytical theory that enables quantitative prediction of their behavior. Simulations have become important in advancing our understanding of the behavior of ionic solutions and interfaces, complementing and helping the interpretation of experimental results.

The following chapters in this section describe various studies concerning the behavior of ions and water molecules in systems ranging from nanoscale devices to geological length scales, where confinement at the length scales of electrostatic or hydration effects dominates the observed transport behavior. The chapter by Duan describes deviations of ionic mobilities from bulk values observed in 2-nm slit-like nanochannels. This behavior is attributed to the fact that the channel height is comparable to the hydration length scales, possibly affecting the structure of water and ionic hydration. The chapter by Karnik outlines transport of ions through nanoscale defects in graphene, a recently discovered and much-studied 2-dimensional material that holds promise for applications in electronics, energy storage, membranes, and other areas. The chapter by Kalinichev describes advances in computational modeling of the structure and dynamics of water at crystal surfaces. The final chapter in this section by Shilov provides an analytical treatment of the siesmoelectric effect, where vibrations in the earth result in corresponding electric potential oscillations due to the interactions between the charged surfaces in the soil or rock and the ionic solutions contained therein. These chapters illustrate the diverse range of ionic and interfacial phenomena in confined systems being investigated in the twenty-first century using tools, resources, and conceptual frameworks that were probably unimaginable just a century ago.

Rohit Karnik

References

1. Online Etymology Dictionary (2013) http://www.etymonline.com/index.php?term=ion
2. Arrhenius G, Caldwell K, Wold S (2008) A tribute to the memory of Svante Arrhenius (1859–1927) – a scientist ahead of his time. http://www.iva.se/upload/Verksamhet/H%C3%B6gtidssammankomst/Minnesskrift%202008.pdf
3. Washburn EW (1909) The hydration of ions determined by transference experiments in the presence of a non-electrolyte. J Am Chem Soc 31:322–355
4. Born M (1920) Volumes and hydration warmth of ions. Z Phys 1:45–48
5. Latimer WM, Rodebush WH (1920) Polarity and ionization from the standpoint of the Lewis theory of valence. J Am Chem Soc 42:1419–1433
6. Debye P, Huckel E (1923) The theory of electrolytes I. The lowering of the freezing point and related occurrences. Phys Z 24:185–206
7. Hinton JF, Amis ES (1971) Solvation numbers of ions. Chem Rev 71:627–674
8. Chadwell HM (1927) The molecular structure of water. Chem Rev 4:375–398
9. Enderby JE (1995) Ion solvation via neutron-scattering. Chem Soc Rev 24:159–168

10. Neher E, Sakmann B (1976) Single-channel currents recorded from membrane of denervated frog muscle-fibers. Nature 260:799–802
11. Israelachvili JN, Adams GE (1978) Measurement of forces between two mica surfaces in aqueous electrolyte solutions in the range 0–100 nm. J Chem Soc Faraday Trans 1 74:975–1001
12. de Groot BL, Grubmuller H (2001) Water permeation across biological membranes: mechanism and dynamics of aquaporin-1 and GlpF. Science 294:2353–2357
13. Sparreboom W, van den Berg A, Eijkel JC (2009) Principles and applications of nanofluidic transport. Nat Nanotechnol 4:713–720

Chapter 7
Enhanced Ion Transport in 2-nm Silica Nanochannels

Chuanhua Duan

Abstract Fluidic nanochannels with 1–2 nm in size are functional mimics of protein channels, and have recently attracted significant attention for exploring the transport of ions and molecules in confined liquids. Here we report ion transport in 2 nm deep nanochannels fabricated by standard semiconductor manufacturing processes. Ion transport in these nanochannels is dominated by surface charge until the ion concentration exceeds 100 mM. At low concentrations, proton mobility increases by a factor of four over its bulk value, possibly due to overlap of the two hydration layers adjacent to hydrophilic surfaces. The mobility of K+/Na + ions also increases as the bulk concentration decreases, although the reasons are not completely understood.

7.1 Introduction

Protein channels have inspired the study of a new regime of fluid mechanics under nanoscale confinement in artificial structures, which is now called nanofluidics [1]. One of the unique aspects of nanofluidics is that the relevant length scale is comparable with the range of various surface or interfacial forces in liquids. These forces include steric interactions (0.1–2 nm in range), van der Waals forces (0.1–50 nm in range) and/or electrostatic forces (1–100 nm in range). Among them, electrostatic interaction is the most studied one. The changes in ionic concentration [2] due to overlap of the electric double layers in channels that are 10–100 nm in size have been extensively studied, leading to various discoveries such as local ion depletion/enrichment [3, 4], active gated control of ion/molecules [5–8], rectified ion transport [9, 10], and energy conversion from streaming current [11, 12].

C. Duan (✉)
Department of Mechanical Engineering, Boston University, 110 Cummington Street, Boston, MA 02215, USA
e-mail: duan@bu.edu

L. Mercury et al. (eds.), *Transport and Reactivity of Solutions in Confined Hydrosystems*, NATO Science for Peace and Security Series C: Environmental Security, DOI 10.1007/978-94-007-7534-3_7,

The size of the electric double layer is characterized by the Debye screening length, which follows the relation $l_D \propto 1/\sqrt{\eta}$ where η is the ion concentration. For channels 10–100 nm in size, the effects of double layer overlap can only be observed for $\eta < 10$ mM, which is one order of magnitude lower than the concentration of realistic physiological solutions. Since the Debye screening length at 100 mM is about 1 nm, it is necessary to reduce the size of nanochannels to ~2 nm to manipulate practical physiological media. In addition to electrostatic interactions, steric forces, such as hydration forces in water, as well as van der Waals forces can become very important at these length scales. Hence, the interplay between these multiple interactions in this size regime could give rise to phenomena that are both rich in science and with multiple applications in biology [13], energy [14] and environment [15]. However, the fluid mechanics in this regime have not been fully explored, since it is difficult to make individual artificial structures of this size [13, 16]. Water and ion flows through single-wall carbon nanotubes have already shown anomalous enhancements due to hydrophobic interactions [13, 15], which warrant much deeper study. Similar flows in hydrophilic channels have yet to be studied. Herein, we present a method to fabricate nanochannels that are 2 nm in height and report results of ion transport in this unexplored regime.

7.2 Experimental Section

Figure 7.1 shows a 1-D confined 2-nm silica nanochannel device, which includes two microchannels and four reservoirs. These reservoirs are located at the ends of the microchannels for solution introduction/change and conductance measurements. There are ten nanochannels bridging two microchannels in the middle of the device. Each of them is 140 μm long, 2 μm wide and 2 nm deep. The height of these

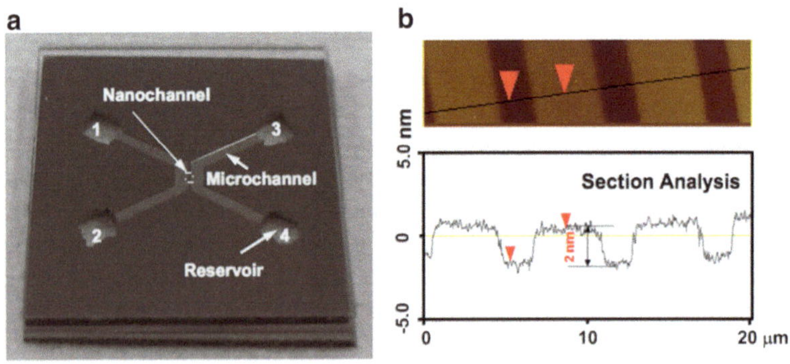

Fig. 7.1 2-nm nanochannel device. (**a**) A bonded 2-nm nanochannel device with Pyrex substrate on *top*. Microchannels are 1 cm long, 500 μm wide and 60 μm deep. Reservoirs are 2 by 2 mm^2 through-holes. (**b**) AFM scan image of three nanochannels after surface oxidization

nanochannels was confirmed by AFM measurements, which also indicated that the channel wall has similar surface roughness (2 Å) as the original wafer surface.

This 2-D nanochannel device was fabricated based on an "etching and bonding" scheme [17, 18]. Briefly, photolithography was first used to define the width and length of the nanochannels. A delicate reactive ion etching (RIE) step was then employed to create open 2-nm nanochannels. Afterwards, two microchannels were patterned and etched using deep reactive ion etching. The top surface was then deposited a 1-μm-thick oxide layer to protect both micro- and nanochannels during the subsequent reservoir etching step. Reservoirs were patterned from the back side of the wafer and DRIE was again used to form straight etch-through holes. After this step, the oxide layer was removed by HF solution and the wafer was sent to a 1,000 °C furnace to uniformly re-grown 500-nm-thick thermal oxide on all exposed surfaces. The wafer was then diced into 1 cm^2 dies and bonded with the same size clean Pyrex die to seal the nano- and microchannels. This bonding step included a 1-min anodic bonding process (performed at 400 °C under 600 V), and a 10-h fusing bonding process (performed at 400 °C) [18].

To test whether the nanochannels survived after bonding, we measured the ionic conductance of pure deionized (DI) water, 1:1 DI water and ethanol, and pure ethanol filled nanochannels. Ethanol filled nanochannels yielded almost zero conductance (10.42 ± 5.04 pS) whereas water filled nanochannels provide a relatively high value (147.80 ± 2.01 pS). The conductance of 1:1 mixture of water and ethanol was almost half of conductance of DI water (74.65 ± 2.61 pS). These data indicate that there were indeed nanochannels bridging two microchannels. Further test of channel leakage were carried out by filling a solution containing fluorescent dyes (100 μM Rhodamine 6G in 1 M KCl) into the microchannels. Rhodamine 6G is a positively charged fluorescent molecule with diameter around 1.3 nm. We failed to observe any fluorescent signal from 2-nm-deep nanochannels although our system has the capability to detect the fluorescent signal from a monolayer of R6G [18]. Since these fluorescent molecules have a comparable size with the channel height, they may not be able to enter the channel due to steric repulsion, despite the fact that negatively charged surface of the glass provide an attractive interaction for these positively charged molecules. Similar tests were also run in 30-nm nanochannel devices fabricated by the same bonding process. Fluorescent signals were only observed from the designed channel area but not any other place, providing an indirect but strong evidence for our leakage test.

Ionic conductance of various electrolyte solutions was measured by a Keithley 6430 sourcemeter. A matlab program was used to control the sourcemeter and measurements were carried out at low voltage bias ($\pm$50 mV) using a single spiral out sweep sequence 0, +10, −10... +50, −50 mV. Ag/AgCl electrodes were used to make contact with the solutions. For each measurement, we measured the conductance from reservoir 1 and 3 (total conductance) as well as reservoir 1 and 2 (microchannel conductance). The conductance from the nanochannels was then extracted from these two measurements.

7.3 Experimental Results and Discussion

7.3.1 Time-Dependent Nanochannel Conductance

In the present investigation, we studied ion transport by experimentally measuring the time and concentration dependence on ionic conductance along nanochannels. Figure 7.2a shows the time dependence of KCl conductance at 10 μM, 1 mM and 1 M. Since the nanochannel is 140 μm long, K^+ and Cl^- should diffuse from the microchannels to the nanochannels in a really short time (within 5 s based on bulk diffusivity). It was not expected to see any conductance change with time after we completely rinsed the microchannels for 5 min. However, ionic conductance indeed changed with time. At 10 μM and 1 mM, it took about 10 h to reach a saturation point. Although 1 M concentration took less time, the duration is still much longer than the bulk diffusion time. Similar time dependence was also observed in NaCl solutions. A possible explanation for such time dependence is that ion diffusivity decreases significantly in such small channels. However, it is found that the diffusivity remain at least within the same order of magnitude according to the final conductance of 1 M KCl/NaCl solutions. Therefore, there must be some other reason.

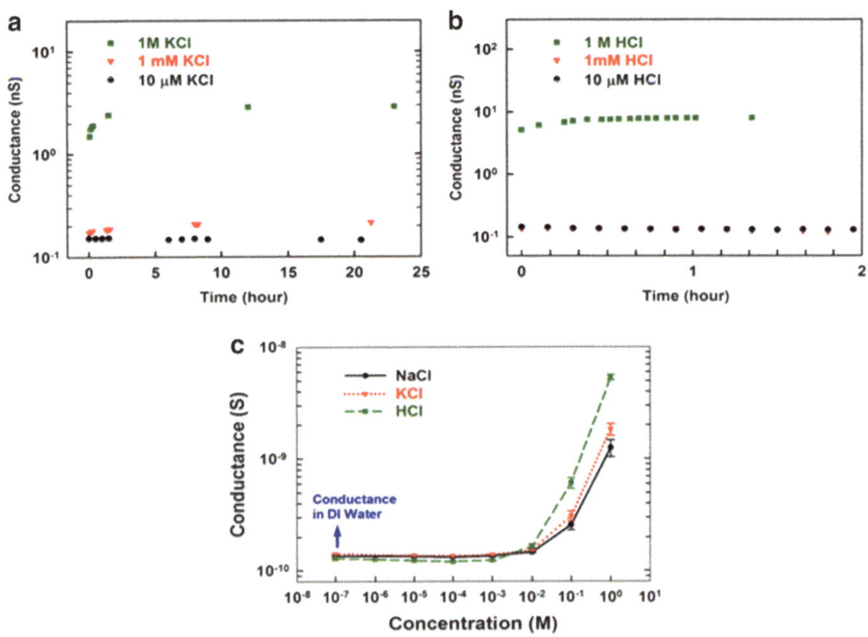

Fig. 7.2 Time dependent conductance in 2-nm nanochannels. (**a**) Time-resolved conductance of KCl solutions. (**b**) Time-resolved conductance of HCl solutions. (**c**) Initial conductance at t = 5 min of HCl, KCl and NaCl

It is well known that silanol groups on silicon dioxide surfaces would dissociate when it is brought into contact with aqueous solutions, giving protons to the solution and resulting in negatively charged surface. These protons nearby the surface would be slowly replaced by other cations in solution, which in turn can break the local dissociation equilibrium and give rise to slow surface dissociation. This proton-cation exchange process thus can be a slow process, especially when local proton concentration is high [19].

Although proton concentration in KCl/NaCl bulk solutions is low (pH ~ 5 due to CO_2 adsorption), it could be relatively high inside 2-nm nanochannels due to electrostatic attraction of the negative surface charge. For example, given a surface charge density $\sigma = 1$ mC/m^2, the proton concentration could be as high as 10 mM ($n_{H^+} \approx 2\sigma/eh$) in a unipolar solution [2, 5], where h is the channel height and e is the electron charge. Such a high proton concentration could significantly delay the proton-cation exchange process and thus result in time-dependent conductance.

To prove this argument, the time dependence of HCl conductance at various concentrations was also investigated. As protons are the only cations in HCl solution, the proposed proton-cation exchange process would not occur and the conductance of HCl should exhibit a much weaker time dependence. This was exactly what we observed: the conductance of HCl did not change with time at concentration of 10 μM and 1 mM and reach saturation at 1 M only after 10 min (Fig. 7.2b). We also compared the initial conductance of KCl/NaCl/HCl solutions at various concentrations. As shown in Fig. 7.2c, although the conductance of these three solutions is different at high concentrations, at concentration below 10 mM, it is found to be the conductance of DI water in which protons are the only cations. These data suggest that protons likely occupy all available cation sites first. The fact that initial conductance differs at high concentration suggests that corresponding cations (K^+/Na^+) become more dominant at those concentrations. Consequently, the cation exchange rate increases and the time duration to reach equilibration decreases, being consistent with the time-dependent conductance in Fig. 7.2a. These results add strong evidence for the hypothesis of the protons-cations exchange mechanism.

7.3.2 Enhanced Ion Transport in 2-nm Nanochannels

The measured final conductance after 12 h incubation and the bulk predictions for KCl/NaCl/HCl solutions are plotted in Fig. 7.3a. To better understand the conductance characteristics, we also plot theoretical predictions based on the classic PNP model and bulk mobility in Fig. 7.3b, c. This model considers change of surface charge density based on the dissociation reaction and has quantitatively predicted conductance of nanochannels with a feature height of 25 nm and above [18].

It is well known that conductance of nanochannels would generally fall into two separate regimes: the surface-charge-governed regime and the bulk-behavior regime. At low ionic concentrations, the conductance is governed by surface charge,

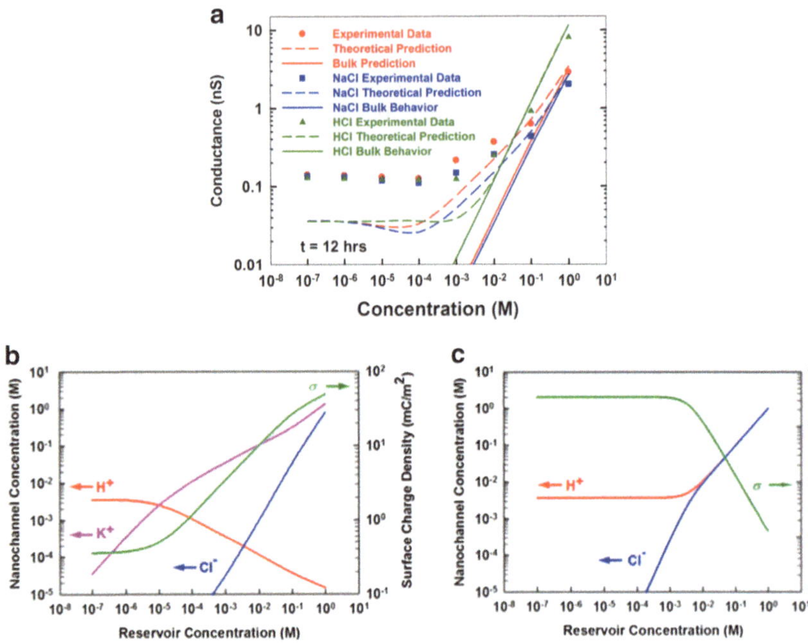

Fig. 7.3 Concentration dependence on conductance in 2-nm nanochannels. (**a**) Measured final conductance of HCl/KCl/NaCl and corresponding theoretical predictions. (**b**) Estimated nanochannel ionic concentration and surface charge density for KCl/NaCl solutions. (**c**) Estimated nanochannel ionic concentration and surface charge density for HCl solutions. All predications were made based on the classic PNP model where surface charge change due to surface dissociation reaction and bulk mobility change due to overall ionic concentration have been considered [18]. This model has quantitatively predicted conductance of nanochannels with a feature height of 25 nm and above [18]

exhibiting a plateau with small but gradually increase with increasing concentration; At high ionic concentrations, the conductance is determined by nanochannel geometry and bulk ionic concentration [2], showing a linear increase with the increasing concentration. For KCl/NaCl solutions, the measured final conductance approaches bulk behavior only at concentrations close to 1 M (Fig. 7.3a). In other words, the surface-charge-governed regime extends beyond 100 mM in 2-nm nanochannels, making it relevant for physiological conditions. Different from these salt solutions, the final conductance of HCl solutions starts deviating from bulk behavior around 10 mM, which means the surface-charge-governed regime of HCl solutions is one order of magnitude smaller than those of KCl/NaCl solutions. Such two different ranges of the surface-charge-governed regime are well predicted by the theoretical calculations. Consequently, we can use the calculated ion concentration and surface charge to explain this conductance characteristic. As shown in Fig. 7.3b, for KCl/NaCl solutions, K^+/Na^+ concentration in 2-nm nanochannels is at least one order of magnitude higher than Cl^- concentration until bulk concentration reaches 100 mM as a result of surface charge and channel confinement. Even at bulk concentration of 1 M, K^+/Na^+ concentration (1.33 M) is still much higher than Cl^- concentration (0.79 M), indicating that 1 M is a transition concentration between surface-charge governed regime and bulk behavior regime. In contrast, there is no surface charge anymore when HCl concentration is above 10 mM due to full protonation of the silica surface (Fig. 7.3c, also see ref. [7]), leading to a smaller surface-charge-governed regime for HCl solutions in 2 nm nanochannels.

Besides the extension of the surface-charge-governed regime, another interesting observation of the final conductance is the presence of a minimum in the conductance vs. concentration curve. It was found that conductance in the surface-charge-governed regime does not show a monotonic increase as the concentration increases. Instead, it first slowly decreases and then increases with the increasing concentration, showing a minimum between 10^{-5} and 10^{-4} M. The appearance of this valley was also captured by the theoretical prediction and can be explained by proton to salt-dominating conductance [20]. As evident in Fig. 7.3b, protons actually contribute a large portion of the conductance at low concentrations. Although the overall surface charge density and the total cation concentration increase with the increasing concentration in the surface-charge-governed regime, the proton concentration inside the nanochannel continuously decreases. Since proton mobility is much larger than K^+/Na^+ mobilites, the conductance increase due to concentration increase eventually cannot compensate the conductance decrease induced by mobility change, leading to the presence of the minimum, which essentially provide another strong evidence of the effects of surface charge and geometrical confinement on channel conductance.

Although the theoretical predications can qualitatively reveal same conductance characteristics as the experimental results show, these two groups of data do not quantitatively match each other in the surface-charge-governed regime. There are only two factors that can affect nanochannel conductance for a given geometry: ionic concentration and ionic mobility. Since the PNP model are known to be able to precisely estimate ionic concentrations inside the nanochannels, differences between the experimental data and the theoretical predictions indicate that ionic mobilities must have been changed in 2-nm nanochannels. We investigated such mobility change of each individual cations in certain areas of the surface-charge-governed regimes where only one type of cation is dominant. In these regimes, cation mobility $\mu_{\rm nc}$ in nanochannels was extracted from the measured conductance G and surface charge density σ as $\mu_{nc} = Gl/2w\sigma$, where l and w are the channel length and width, respectively. The corresponding bulk mobility $\mu_{\rm bulk}$ at this ionic concentration $(2\sigma/eh)$ was calculated based on ref. [21], where the dependence of ionic strength on ionic mobility had been considered. The ionic mobility at concentrations beyond the surface-charge-governed regime is difficult to estimate due to the presence of multiple ions and thus is not discussed in this work. Figure 7.4a plots the ratio of $\mu_{\rm nc}:\mu_{\rm bulk}$ for 2-nm nanochannels as a function of bulk concentration. Proton mobility was investigated in HCl solutions of concentrations from 10^{-7} to 10^{-3} M and also in KCl/NaCl solutions at 10^{-7} and 10^{-6} M because protons were the dominant ions in these solutions (Fig. 7.3b, c). The extracted proton mobility was

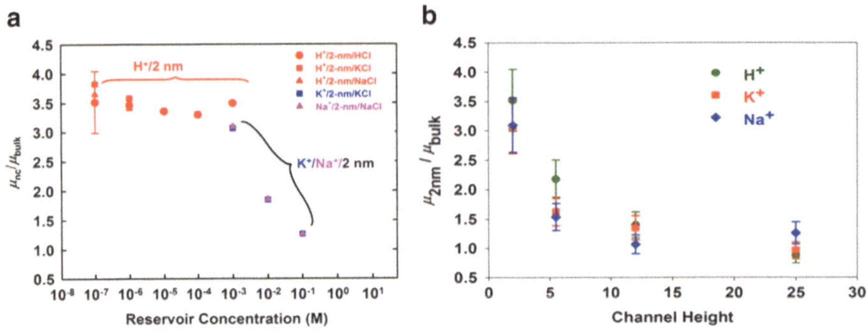

Fig. 7.4 Enhanced cation mobility in 2-nm nanochannels. (**a**) Ratio of ionic mobility in nanochannel (μ_{nc}) to the corresponding bulk mobility (μ_{bulk}) as a function of reservoir concentration. (**b**) Mobility as a function of channel height

found to be around 3.5 ~ 4 times the bulk proton mobility. K^+/Na^+ mobility could only be obtained in KCl/NaCl solutions from concentration 10^{-3} to 10^{-1} M where K^+/Na^+ became the dominant ions as shown in Fig. 7.4a. The calculated $\mu_{2nm}:\mu_{bulk}$ for K^+ and Na^+ overlaps at each concentration, indicating that there is no ion specificity for these two monovalent ions in such 2-nm nanochannels although they are indeed different in size [22]. It is found that the mobility of K^+/Na^+ in this regime shows a decreasing trend as the bulk concentration increases; whereas the ratio of $\mu_{2nm}:\mu_{bulk}$ is around 3 at 10^{-3} M, it approaches to unity at 10^{-1} M.

We believe that the increase of proton mobility comes from the geometrical confinement and the structure of water near the surface. It is well known that proton transport in water is based on a Grotthuss mechanism, where protons hop from one water molecule to the next, resulting in transport of charge defects rather than individual protons [23, 24]. This relay mechanism benefits from the organized hydrogen bond network, yielding an increase of proton mobility in ice [25] and 1-D ordered water chain [26, 27]. Adjacent to a smooth mica/silica surface is a 1-nm hydration region where water molecules are more organized than bulk water [28–30], i.e. during random fluctuations of water molecules, the timescale associated with correlation in orientation is longer in the hydration layer than in the bulk, which allows for better streamlining of proton hopping. For a 2-nm nanochannel, the two hydration regions near the channel surface are likely to overlap. Consequently, water in 2-nm nanochannels is much more organized and thus could exhibit higher proton mobility than bulk water. The increase of K^+/Na^+ mobility may also be attributed to the organized water structure since monovalent ions could stay between two organized water layers [30]. Although these ions still share the full hydration layers, they could experience much less viscous force since water density is low between two organized layers [31]. Another possible reason is that monovalent cations could be dehydrated close to the charged silicon/mica surface in 1–3 nanochannels [32, 33]. These dehydrated ions are smaller than the fully hydrated ones, and thus could move more rapidly.

To further confirm our hypotheses, we investigated the height dependence of $H^+/K^+/Na^+$ mobilities. As the hydration layer is only around 1 nm thick, the portion of "organized" water in nanochannel should decrease as the channel height increases. Consequently, a monotonic decreasing trend of the cation mobilities is expected. This is exactly what we observed. Figure 7.4b plots cation mobility in nanochannels with a height ranging from 2 to 25 nm. Mobilities of K^+ and Na^+ ions were calculated based on conductances of 1 mM electrolyte solutions, while proton mobility was obtained from conductance of DI water. There is a clear trend that cation mobilities decrease as the channel height increases from 2 to 25 nm. In 25 nm nanochannels, the mobility enhancement is almost negligible and all cation mobilities are close to their bulk values. These results provide strong support for the hydration force induced cation mobility enhancement. It is worth noting that although the above hypothesis explains mobility enhancement in the surface-charge-governed regime, it cannot explain the decrease of K^+/Na^+ mobilities as the bulk concentration rises from 10^{-3} to 10^{-1} M. It is clear from the above discussion that, while surface charge governed transport can be well explained, a complete quantitative understanding about ionic mobility in 2-nm nanochannels will require further study.

7.4 Conclusion

In summary, 2 nm nanochannels have been successfully fabricated using traditional lithography approaches. We studied ion transport in these nanochannels by investigating time and concentration dependence of the ionic conductance. Our results show that the ion transport characteristics of 2-nm nanochannels are not only different from bulk behavior, but also different from characteristics of larger nanochannels. 2-nm nanochannels showed a much broader surface charge governed regime (up to 100 mM), which could be used to study biological phenomena in physiological solution. They also present a relatively slow kinetics for cation adsorption. Furthermore, they exhibit up to 4-fold increase in ionic mobility at low electrolyte concentrations. For protons, we hypothesize that this is due to overlap of hydration layer of the two surfaces and thereby providing a more correlated hydrogen bonding network. The reason for the observed increase in K^+ and Na^+ mobility is not understood. These enhancements of ion transport in 2-nm nanochannels, in terms of both ion concentration and ion mobility, may find wide range of applications in single molecular detection, fuel cells and batteries.

Note: This work is a revision of a previous paper titled "Anomalous ion transport in 2-nm hydrophilic nanochannels", which has been published in Nature Nanotechnology 5, 848–852, 2010.

References

1. Eijkel JCT, van den Berg A (2005) Nanofluidics: what is it and what can we expect from it? Microfluid Nanofluid 1:249–267
2. Stein D, Kruithof M, Dekker C (2004) Surface-charge-governed ion transport in nanofluidic channels. Phys Rev Lett 93:035901
3. Pu QS, Yun JS, Temkin H, Liu SR (2004) Ion-enrichment and ion-depletion effect of nanochannel structures. Nano Lett 4:1099–1103
4. Kim SJ, Wang YC, Lee JH, Jang H, Han J (2007) Concentration polarization and nonlinear electrokinetic flow near a nanofluidic channel. Phys Rev Lett 99:044501
5. Karnik R, Fan R, Yue M, Li DY, Yang PD, Majumdar A (2005) Electrostatic control of ions and molecules in nanofluidic transistors. Nano Lett 5:943–948
6. Fan R, Yue M, Karnik R, Majumdar A, Yang PD (2005) Polarity switching and transient responses in single nanotube nanofluidic transistors. Phys Rev Lett 95:086607
7. Fan R, Huh S, Yan R, Arnold J, Yang PD (2008) Gated proton transport in aligned mesoporous silica films. Nat Mater 7:303–307
8. Karnik R, Castelino K, Majumdar A (2006) Field-effect control of protein transport in a nanofluidic transistor circuit. Appl Phys Lett 88:123114
9. Karnik R, Duan C, Castelino K, Daiguji H, Majumdar A (2007) Rectification of ionic current in a nanofluidic diode. Nano Lett 7:547–551
10. Vlassiouk I, Siwy ZS (2007) Nanofluidic diode. Nano Lett 7:552–556
11. Lu MC, Satyanarayana S, Karnik R, Majumdar A, Wang CC (2006) A mechanical-electrokinetic battery using a nano-porous membrane. J Micromech Microeng 16:667–675
12. van der Heyden FHJ, Bonthuis DJ, Stein D, Meyer C, Dekker C (2006) Electrokinetic energy conversion efficiency in nanofluidic channels. Nano Lett 6:2232–2237
13. Liu HT, He J, Tang JY, Liu H, Pang P, Cao D, Krstic P, Joseph S, Lindsay S, Nuckolls C (2010) Translocation of single-stranded DNA through single-walled carbon nanotubes. Science 327:64–67
14. Chmiola J, Yushin G, Gogotsi Y, Portet C, Simon P, Taberna PL (2006) Anomalous increase in carbon capacitance at pore sizes less than 1 nanometer. Science 313:1760–1763
15. Holt JK, Park HG, Wang YM, Stadermann M, Artyukhin AB, Grigoropoulos CP, Noy A, Bakajin O (2006) Fast mass transport through sub-2-nanometer carbon nanotubes. Science 312:1034–1037
16. Rhee M, Burns MA (2007) Nanopore sequencing technology: nanopore preparations. Trends Biotechnol 25:174–181
17. Mao P, Han JY (2005) Fabrication and characterization of 20 nm planar nanofluidic channels by glass-glass and glass-silicon bonding. Lab Chip 5:837–844
18. Duan C, Majumdar A (2010) Anomalous ion transport in 2-nm hydrophilic nanochannels. Nat Nanotechnol 5:848–852
19. Raider SI, Gregor LV, Flitsch R (1973) Transfer of mobile ions from aqueous-solutions to silicon dioxide surface. J Electrochem Soc 120:425–431
20. Jensen KL, Kristensen JT, Crumrine AM, Andersen MB, Bruus H, Pennathur S (2011) Hydronium-dominated ion transport in carbon-dioxide-saturated electrolytes at low salt concentrations in nanochannels. Phys Rev E 83:056307
21. Baldessari F (2008) Electrokinetics in nanochannels – Part I. Electric double layer overlap and channel-to-well equilibrium. J Colloid Interface Sci 325:526–538
22. Israelachvili J (2003) Intermolecular and surface forces, 2nd edn. Academic, London
23. von Grotthuss CJD (1806) Sur la décomposition de l'eau et des corps qu'elle tient en dissolution à l'aide de l'électricité galvanique. Ann Chim 58:54–73
24. Agmon N (1995) The Grotthuss mechanism. Chem Phys Lett 244:456–462
25. Kunst M, Warman JM (1980) Proton mobility in ice. Nature 288:465–467

26. Pomes R, Roux B (2002) Molecular mechanism of H^+ conduction in the single-file water chain of the gramicidin channel. Biophys J 82:2304–2316
27. Dellago C, Naor MM, Hummer G (2003) Proton transport through water-filled carbon nanotubes. Phys Rev Lett 90:105902
28. Du Q, Freysz E, Shen YR (1994) Vibrational-spectra of water-molecules at quartz water interfaces. Phys Rev Lett 72:238–241
29. Israelachvili JN, Pashley RM (1983) Molecular layering of water at surfaces and origin of repulsive hydration forces. Nature 306:249–250
30. Koplik J, Banavar JR, Willemsen JF (1989) Molecular-dynamics of fluid-flow at solid-surfaces. Phys Fluids A 1:781–794
31. Peter C, Hummer G (2005) Ion transport through membrane-spanning nanopores studied by molecular dynamics simulations and continuum electrostatics calculations. Biophys J 89:2222–2234
32. Leng YS, Cummings PT (2006) Hydration structure of water confined between mica surfaces. J Chem Phys 124:074711
33. Mattke T, Kecke HJ (1998) Molecular dynamic simulations of single, interacting, and sheared double layers 1. Configuration of a double layer. J Colloid Interface Sci 208:555–561

Chapter 8
Ionic and Molecular Transport Through Graphene Membranes

Rohit Karnik

Abstract New membrane materials have the potential to address some of the persistent challenges in water purification to improve the flux of water, selectivity to ions or contaminants, and fouling resistance. With its atomistic thickness and the ability to sustain nanometer-scale holes, graphene promises significant enhancement in the flux of water while offering potentially novel transport properties. In this work, the transport of ions and molecules through a single layer of graphene were measured as a first step towards realizing practical graphene membranes. Graphene grown by chemical vapor deposition (CVD) was transferred to a porous polycarbonate support membrane, and the diffusion of different salts and molecules was examined. While pressure-driven flow measurements revealed that the graphene covered the polycarbonate support membrane, diffusion experiments showed that it was permeable to salts, but not to larger molecules. This behavior was attributed to intrinsic defects in graphene in the 1–15 nm size range.

8.1 Introduction

Several regions of the world including the Middle East, North Africa, eastern Australia, parts of Central and South Asia, and the southwestern areas of North America, face increasing scarcity of water due to the rising human population and the consequent increase in the demand for water [1]. In addition to water scarcity, a large part of the human population does not have access to clean drinking water. Ensuring the security of water resources to meet the needs of the human

R. Karnik (✉)
Department of Mechanical Engineering, Massachusetts Institute of Technology, Cambridge, MA 02139, USA
e-mail: karnik@mit.edu

L. Mercury et al. (eds.), *Transport and Reactivity of Solutions in Confined Hydrosystems*, NATO Science for Peace and Security Series C: Environmental Security,
DOI 10.1007/978-94-007-7534-3_8,

society in future generations will require advances in conservation and efficient use of water, as well as technologies to purify contaminated water or to desalinate seawater and brackish water. Membrane-based technologies play a very significant role in water purification: more than 40 % of the global desalination capacity uses reverse osmosis through semi-permeable polymeric membranes [1], while membranes are also widely used in filtration and purification of contaminated water. Membrane-based processes are typically more energy-efficient and modular than other processes, but suffer from fouling and limited flux (flow rate per unit area). Furthermore, in the case of reverse osmosis membranes, the flux and selectivity are intimately coupled to the membrane material, making it difficult to increase flux while compromising selectivity. It is also difficult to create membranes that reject a wide range of species; rejection of contaminants such as boron and arsenic remains challenging. The performance of membranes for water purification and other applications depends critically on the membrane material and structure. In this context, new membrane materials may hold the key to further advancing membrane technologies.

Graphene with its atomistic thickness [2], remarkable mechanical strength [3], impermeability to even helium in its pristine state [4], and the ability to sustain nanometer-scale pores, is a recently discovered material that promises significant advances in membrane technology. While transport in polymeric membranes typically involves sorption and diffusion in the case of reverse osmosis and flow through relatively tortuous pores for nanofiltration and ultrafiltration, graphene promises an ideal thin membrane with non-tortuous pores for high-flux filtration where steric effects play a significant role. Recent simulations predict that graphene is promising for gas separations and water desalination [5–9], while experiments have demonstrated selective gas transport through graphene [10]. These studies are a preliminary indication of the promise of graphene to impact membrane technology for water purification. However, realizing practical working membranes is fairly challenging. It will require several breakthroughs in graphene membrane fabrication, understanding of how to create pores in graphene and the transport properties of the pores, development of defect-tolerant designs or strategies to eliminate defects, and methods for fabrication of membranes and membrane modules. While graphene can be obtained or synthesized in many different ways, chemical vapor deposition (CVD) is capable of growing large-area graphene and it may be a likely candidate for future membranes. Understanding the transport properties of CVD graphene is important to enable future design of graphene-based membranes. In this work, we report fabrication of macroscopic membranes comprising a single layer of CVD graphene on a porous support and investigate the transport properties of ions and molecules through the fabricated membrane. The work presented at the conference and discussed here is reported in detail in a recent paper by O'Hern et al. [11].

8.2 Experimental Section

8.2.1 Graphene Membrane Fabrication

Graphene grown by CVD on copper foil (ACS Materials Inc.) was transferred to a polycarbonate track-etched membrane support (Sterlitech Inc.) with 200 nm diameter pores. In the first step, the graphene on the back side of copper was removed by floating a piece of copper on ammonium persulfatecopper etchant (APS-100, Transcene Inc.). The graphene side of the copper piece was manually pressed against the polycarbonate membrane so that the two weakly adhered together. The assembly was then floated again on the copper etchant at a pressure of about 1.5 atm to completely remove the copper, leaving the layer of graphene on the polycarbonate support. The membrane was rinsed in water and in a water-ethanol mixture before being dried.

8.2.2 Transport Measurements

The graphene membrane was mounted in a stirred, custom-built side-bi-side diffusion cell (Permgear Inc.). Pressure-driven convective flow was used for measuring the coverage of graphene on the polycarbonate support. Both sides of cell were filled with a solution of a dye (Allura Red) in water. One side was connected to a larger water bath at the same water level by tubing, while the other side was filled to a higher level. The higher water level induced a flow into the opposite side, and the dye flowed into the larger water bath where its concentration was measured using *in situ* UV-V absorption using a Cary 60 Spectrophotometer with a dip-in probe. To measure the diffusion of salts or dye molecules, the corresponding solution was filled in one side of the cell while the other side was filled with water. A conductivity probe or UV-V probe was used to measure the increase in concentration of the salt or dye with time on the other side. Dye molecule diffusion experiments were performed in 0.5 M KCl to eliminate electrokinetic effects, and both sides were connected to an external water bath to eliminate pressure gradients.

8.2.3 Graphene Membrane Characterization

The fabricated membranes were imaged using scanning electron microscopy (SEM) using a Helios NanolabDualbeam 600 microscope at 1–5 kV. Images were captured in secondary electron mode at a sample tilt of 52° to enhance the contrast.

To characterize the nanometer-scale defects, graphene was transferred to a commercially available transmission electron microscopy (TEM) grid (Ted Pella Inc). Scanning Transmission Electron Microscopy (STEM) was performed using a NionUltraSTEM 100 at Oak Ridge National Laboratory at a voltage of 60 kV in the dark field mode. The images (including Fig. 8.4a) were adjusted for brightness, contrast, and gamma and spatially filtered to clearly reveal the graphene lattice and defects.

Further details of the materials and methods are given in reference [11].

8.3 Experimental Results and Discussion

8.3.1 *Graphene Membranes*

The overall process for fabrication of the graphene membranes is shown in Fig. 8.1a. Graphene transferred easily to the polycarbonate support membrane, and was visible as a darker patch on the polycarbonate (Fig. 8.1b). SEM images clearly showed graphene suspended over the 200 nm pores in the polycarbonate support (Fig. 8.1c). Nearly all polycarbonate pores in the transferred area (~25 mm^2) were covered with graphene, as shown by the absence of large defects in the zoomed-out image (Fig. 8.1d). A few defects were occasionally seen, most notably holes without graphene or tears in graphene. From the SEM images, it was estimated that about 90–98 % of the pores in polycarbonate were covered with graphene. Several factors were found to influence the quality of the transfers of graphene from copper to polycarbonate [11]. Roughness of the copper on which graphene was grown was detrimental to good quality membranes, presumably due to poor conformal contact with the polycarbonate. Use of ferric chloride based etchant was found to severely damage graphene due to formation of insoluble crystals during the etching process, and ammonium persulfate was therefore used in this work.

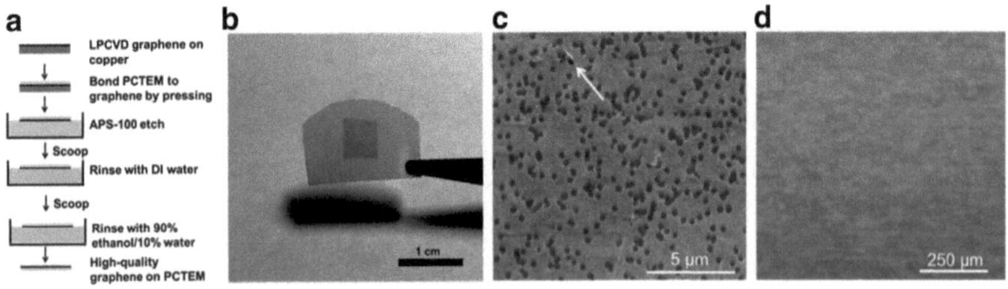

Fig. 8.1 Fabrication of graphene membranes on polycarbonate support. (**a**) Fabrication process. (**b**) Photograph of fabricated membrane. (**c**, **d**) SEM images of the membrane. *Arrow* indicates defect where a polycarbonate pore is not covered by graphene

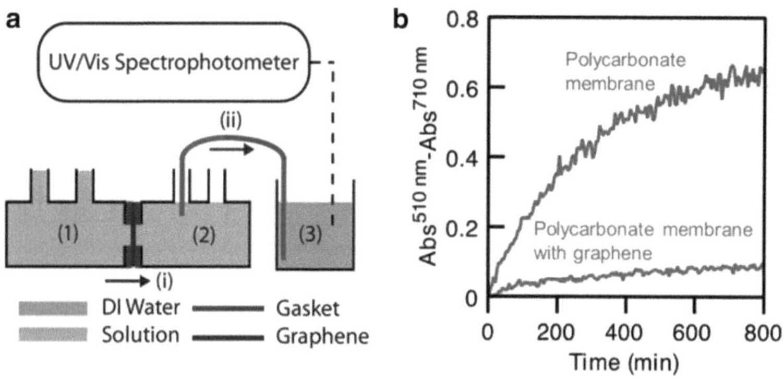

Fig. 8.2 Pressure-driven flow measurements across graphene membranes. (**a**) Schematic of the experimental setup. (*1*) and (*2*) are reservoirs on either side of the membrane, while (*3*) is the water bath in which absorbance of the dye is measured. Flow of water (*i*) through the membrane driven by the higher water level in reservoir (*1*) causes flow of dye into the bath (*3*). (**b**) UV-Vis absorption measured in bath (3) for a polycarbonate membrane and polycarbonate membrane with graphene. The leveling off of the curve in case of the polycarbonate membrane is due to the drop in the water level

8.3.2 Pressure-Driven Transport Across Graphene

Pressure-driven flow was first used to check the coverage of graphene on the polycarbonate (Fig. 8.2). If all polycarbonate pores were covered with graphene, no flow is expected. Hence, the flow rate through the graphene membrane normalized by the flow rate through just the polycarbonate support membrane indicates the extent of graphene coverage on the polycarbonate support. Due to the 4th power dependence of pressure-driven flow on pore diameter for cylindrical pores, any pores in graphene smaller than about 50 nm would allow negligible flow of water compared to the polycarbonate pore. Thus, this experiment measures the coverage of graphene on a large scale and is not sensitive to smaller pores or defects in graphene. As expected, transferring a single layer of graphene to the polycarbonate support reduced the flow of water by about 90 % (Fig. 8.2b), which is consistent with the coverage as observed under SEM (Fig. 8.1). This result verified that the graphene was indeed covering most of the surface of the polycarbonate support.

8.3.3 Diffusive Transport Across Graphene

Diffusion of different species- potassium chloride (KCl), tetramethylammonium chloride (TMAC), Allura Red dye, and tetramethylrhodamine-labeled dextran of MW 70 kDa (TMRD) was examined across the fabricated graphene membranes (Fig. 8.3). The approximate diameter of KCl and TMRD ions is 0.7 nm, while that of Allura Red and TMRD is 1.2 and 12 nm, respectively. If graphene were

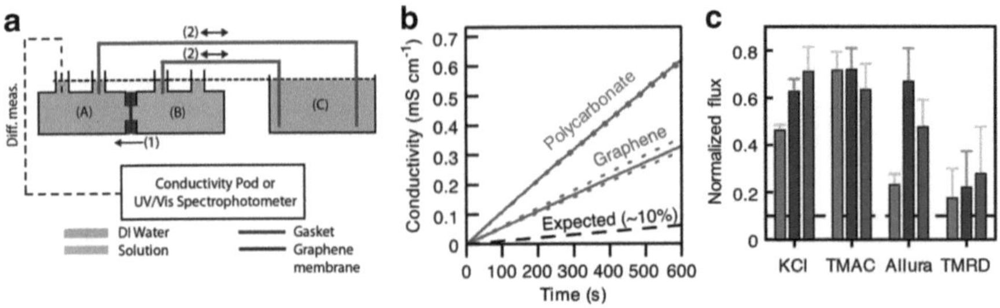

Fig. 8.3 Diffusive transport across graphene membranes. (**a**) Schematic of experimental setup. Diffusion of salt or dye from reservoir (*B*) across the membrane into reservoir (*A*) as indicated by (*1*) causes a change in solution conductivity or absorbance that is measured in (*A*). Bath (*C*) is used to equalize the pressure in the case of Allura Red and TMRD experiments. (**b**) Conductivity increase in reservoir (*A*) in the case of the polycarbonate membrane and polycarbonate with graphene. *Dashed lines* indicate range of data. The result expected if graphene were impermeable and covered 90 % of the polycarbonate pores is also indicated. (**c**) Diffusive flux through the graphene membrane normalized by flux through just the polycarbonate membrane. The three sets of bars denote results for three different membranes. Error bars indicate 95 % confidence. The horizontal *dashed line* denotes flux contribution expected from the polycarbonate pores that are not covered with graphene, assuming 90 % graphene coverage

defect-free, a decrease in the diffusion rate of about 90 % is expected compared to the polycarbonate membrane without graphene. However, diffusion measurements with KCl revealed that the graphene layer blocked less than 50 % of the transport, while the same membrane blocked about 90 % of the pressure-driven flow of water (Fig. 8.3b). This result clearly indicated that the graphene was permeable to KCl. In contrast to pressure-driven flow, the diffusive transport rate scales with the square of the diameter for the cylindrical polycarbonate pores and linearly with the diameter for isolated pores in thin membranes; for this reason, much smaller pores in graphene can have resistance to diffusive transport comparable to that through the polycarbonate pore over which the graphene is suspended, thereby producing a detectable change in the diffusive transport rate. These results imply the existence of smaller pores in graphene. The transport measurements of TMAC showed a similar behavior (Fig. 8.3c). However, the diffusive transport rate of Allura Red was lower for one of the three membranes measured, and that of TMRD was significantly lower and more consistent with the 90 % blockage of the pressure-driven flow rate (Fig. 8.3c). These results suggest that the graphene may have pores that are smaller than the ~12 nm size of the TMRD molecules.

8.3.4 Characterization of Intrinsic Pore Defects in Graphene

Direct visualization of the graphene lattice was performed using scanning transmission electron microscopy (STEM). The graphene lattice is clearly visible in the

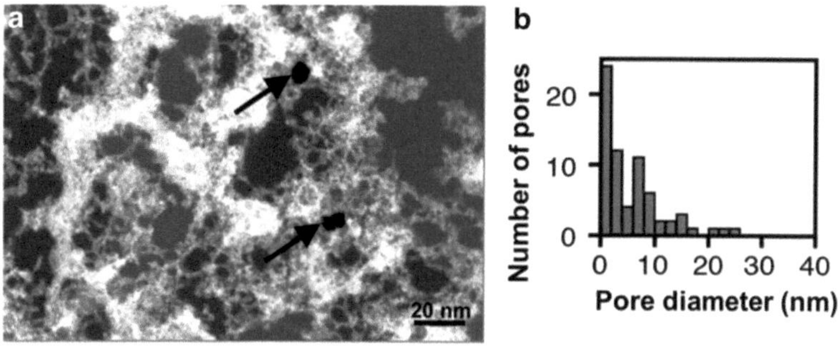

Fig. 8.4 STEM characterization of nanometer-scale pores in graphene. (**a**) Micrograph showing two pore defects indicated by *arrows*. (**b**) Pore size distribution

dark field mode, where the off-axis scattering intensity of electrons increases with increasing atomic number. Small pores primarily in the 1–15 nm size range were observed in the graphene (Fig. 8.4). These pores may have been introduced during the transfer process due to the etchant, or have formed during graphene synthesis. However, the APS-100 etchant used in the process did not significantly influence the transport through graphene even after 3 days of exposure, indicating that the defects must be intrinsic to the graphene. The existence of these pores can thus explain the measured transport behavior and lower transport rate of the TMRD molecules.

8.4 Conclusion

Macroscopic graphene membranes comprising a single layer of CVD graphene on a porous polycarbonate support were fabricated. Experimental setups were developed to measure transport through these membranes under pressure-driven flow and through diffusion. Pressure-driven flow measurements and SEM characterization revealed that the graphene covered about 90 % of the polycarbonate support. However, diffusion measurements showed that graphene was permeable to potassium chloride, but impermeable to the larger dextran molecules. This selective transport behavior was found to arise due to nanometer-scale pore defects that were intrinsic to the CVD graphene. These results demonstrate selective transport in graphene membranes, but also show that the effect of intrinsic defects has to be mitigated if graphene membranes selective to ions or small molecules are to be practically realized.

Acknowledgements The author thanks the organizers and NATO for the opportunity to attend the NATO Advanced Research Workshop on Alternative Water Resources in Arid Areas by Retrieving Water from Secondary Sources held at the Daniel Dead Sea Hotel, Israel, from May 7–11, 2012. The author would like to thank students and collaborators involved in the research, especially Sean O'Hern. The work discussed here was funded by King Fahd University of Petroleum and

Minerals in Dhahran, Saudi Arabia through the Center for Clean Water and Clean Energy at MIT and KFUPM under project number R10-CW-09. Part of the research was performed at ORNL's Shared Research Equipment (ShaRE) User Program sponsored by DOE, the Center for Nanoscale Systems (CNS) at Harvard, and the Center for Materials Science and Engineering at MIT.

References

1. Humplik T et al (2011) Nanostructured materials for water desalination. Nanotechnology 22(29):292001
2. Allen MJ, Tung VC, Kaner RB (2010) Honeycomb carbon: a review of graphene. Chem Rev 110(1):132–145
3. Lee C et al (2008) Measurement of the elastic properties and intrinsic strength of monolayer graphene. Science 321(5887):385–388
4. Bunch JS et al (2008) Impermeable atomic membranes from graphene sheets. Nano Lett 8(8):2458–2462
5. Sint K, Wang B, Kral P (2008) Selective ion passage through functionalized graphene nanopores. J Am Chem Soc 130(49):16448–16449
6. Jiang DE, Cooper VR, Dai S (2009) Porous graphene as the ultimate membrane for gas separation. Nano Lett 9(12):4019–4024
7. Schrier J (2010) Helium separation using porous graphene membranes. J Phys Chem Lett 1:2284–2287
8. Cohen-Tanugi D, Grossman JC (2012) Water desalination across nanoporous graphene. Nano Lett 12(7):3602–3608
9. Suk ME, Aluru NR (2010) Water transport through ultrathin graphene. J Phys Chem Lett 1(10):1590–1594
10. Koenig SP et al (2012) Selective molecular sieving through porous graphene. Nat Nanotechnol 7(11):728–732
11. O'Hern SC et al (2012) Selective molecular transport through intrinsic defects in a single layer of CVD graphene. ACS Nano 6(11):10130–10138

Chapter 9
Molecular Structure and Dynamics of Nano-Confined Water: Computer Simulations of Aqueous Species in Clay, Cement, and Polymer Membranes

Andrey G. Kalinichev

Abstract Molecular-level knowledge of the thermodynamic, structural, and transport properties of water confined by interfaces and nanopores of various materials is crucial for quantitative understanding and prediction of many natural and technological processes, including carbon sequestration, water desalination, nuclear waste storage, cement chemistry, fuel cell technology, etc. Computational molecular modeling is capable to significantly complement the experimental investigations of such systems by providing invaluable atomic-scale information leading to improved understanding of the specific effects of the substrate structure and composition on the structure, dynamics and reactivity of interfacial and nano-confined aqueous solutions. This paper offers a brief overview of recent efforts to quantify some of these effects for individual H_2O molecules and hydrated ions confined at the interfaces and in nanopores of several typical hydrophilic and hydrophobic materials. The first molecular layer of aqueous solution at all substrates is often highly ordered, indicating reduced translational and orientational mobility of the H_2O molecules. This ordering cannot be simply described as "ice-like", but rather resembles the behavior of supercooled water or amorphous ice, although with very significant substrate-specific variations.

9.1 Introduction

Most chemical reactions near the Earth's surface involve fluid (aqueous) phases and take place at fluid-solid interfaces or in confined spaces of mineral interlayers and nanopores. These reactions affect many important natural processes, including

A.G. Kalinichev (✉)
Laboratoire Subatech, UMR 6457, Ecole des Mines de Nantes, 44307 Nantes, France

Department of Chemistry, Michigan State University, East Lansing, MI 48824, USA
e-mail: Andrey.Kalinichev@subatech.in2p3.fr

L. Mercury et al. (eds.), *Transport and Reactivity of Solutions in Confined Hydrosystems*, NATO Science for Peace and Security Series C: Environmental Security,
DOI 10.1007/978-94-007-7534-3_9,

mineral weathering, adsorption or release of environmental contaminants in soil, drinking water quality, formation and behavior of ice nano-crystals and hydrated mineral nano-particles in the atmosphere, the fate of CO_2 in geologic carbon sequestration. The same interfacial phenomena are also at the core of many important technological processes related to water purification and desalination membranes, fuel cells, production and utilization of cement, etc. All these phenomena are inherently multi-scale in time and space, and addressing the scientific and technological problems they pose is multidisciplinary. Molecular-level knowledge of the chemistry and physics of interfacial and confined aqueous solutions interacting with mineral surfaces and other substrates is essential to their understanding. Important questions center on how water and dissolved species behave at the surfaces and in nano-confinement by different inorganic and organic substrates.

Current understanding of hydrated mineral-water interfaces is far from complete. Molecular-scale information is difficult to obtain experimentally due to the nature of the materials (e.g., interfacial and bulk structural disorder, the presence of molecular scale dynamical disorder with characteristic time scales that span many orders of magnitude, the physical limitations on the samples that can be effectively examined, and the difficulty of experiments under extreme conditions). Mineral surfaces themselves vary by the arrangements and reactivity of individual surface functional groups and the specific adsorption of inorganic and organic ions. Even the utility of macroscopic thermodynamic concepts, such as the dielectric constant and electric double-layer, remain open and are important questions driving the current research [1–3]. Modern experimental probes, including synchrotron light sources, high-resolution neutron scattering, time-resolved spectroscopy and diffraction, scanning force microscopy, and multi-dimensional NMR spectroscopy, are providing unprecedented, atomic-level and surface-specific information about mineral-fluid systems [3–5]. However, it is often difficult to interpret these experimental observations without parallel theoretical and computational studies [6, 7]. In addition, the interaction with experimentalists is imperative for verifying theoretical models and guiding larger scale simulation efforts. Connecting computational and experimental results is crucial in addressing these questions and in bridging the gap between the atomic- and nanometer-scales and the micron- and larger scales that is essential for environmental, geochemical and energy applications of practical importance [8].

The methods of atomistic computer simulations [9], often coupled to surface-specific experimental results, are already mature enough to provide otherwise unobtainable perspective into the structure, dynamics and energetics of fluid-mineral systems [10]. Computational approaches range from traditional quantum chemical methods to classical molecular dynamics (MD) and Monte Carlo (MC) simulations using semi-empirical interaction potentials to emerging methods including quantum molecular dynamics, metadynamics, and reactive force fields for molecular-scale modeling [11–35]. Critical questions currently revolve around the effect of solid surfaces on the properties of near-surface and nano-confined fluids, structural and chemical modification of the surface by the fluid, the pathways of chemical reactions at surfaces and in fluids, the connections between these reactions and fluid structure and dynamics, and the consequent effects on the pathways of dissolution and precipitation. This paper provided a brief overview of our group's recent efforts to quantify some of these properties and processes at the interfaces and in nano-confinement with several typical hydrophilic and hydrophobic materials by classical MD simulations using the CLAYFF force field [12].

9.2 Computational Models and Methods

We compare the molecular structure of several qualitatively different substrate-water interfaces. Crystals of metal hydroxides, like **brucite, $Mg(OH)_2$**, or **portlandite, $Ca(OH)_2$**, can be naturally cleaved parallel to their (001) crystallographic plane, resulting in an electrostatically neutral surface fully covered by OH groups oriented away from the surface. They can easily form hydrogen bonds with interfacial H_2O molecules. Therefore, we will use the surface of portlandite as a reference model of a hydrophilic interface [11, 20]. A similar hydrophilic surface can be created by cleaving the crystal structure of **quartz, SiO_2**, parallel to the (001) plane. In the presence of water and at relatively low pH, this surface will also be fully covered by OH groups [36]. In the structure of **hydrocalumite, $[Ca_2Al(OH)_6]Cl{\cdot}2H_2O$**, 1/3 of the divalent Ca cations are isomorphically substituted by trivalent Al, resulting in a positive structural charge compensated by the presence of Cl^-or some other anions in the interlayer space and on the surfaces of this material, an anionic clay [11, 20]. Similar to its neutral portlandite analog, its cleaved (001) surface is also fully covered by OH groups and can be considered hydrophilic.

The structure of **talc, $Mg_3Si_4O_{10}(OH)_2$**, consists of so-called TOT layer (an octahedral layer of $[MgO_4(OH)_2]$ sandwiched between two tetrahedral layers of $[SiO_4]$) stacked along the crystallographic *Z* direction [18, 21, 27]. All tetrahedral sites are occupied by Si forming $[SiO_4]$ structural units, and all octahedral sites are occupied by Mg forming $[MgO_4(OH)_2]$ units. The TOT layers are, thus, electrostatically neutral and interact with each other only by weak by van der Waals forces, reflecting the hydrophobic nature of the basal siloxane Si-O-Si (001) surface. The TOT layered structure is typical for many clay minerals, which can develop various degrees of negative structural charge due to the isomorphic substitutions possible for both tetrahedral (e.g., Al for Si) and octahedral (e.g., Li for Mg) layers. The emergence of negative layer charge is balanced by the presence of charge-balancing cations in the interlayer space and on the surfaces of these minerals and makes them more hydrophilic. Their interaction with water have already been extensively investigated [12, 25, 26, 29, 30, 32, 37] and we focus here only on the behavior of neutral talc, as a reference hydrophobic surface.

It is known experimentally (e.g., [38]) that mineral surfaces rarely exist in nature in the atomically smooth ideal form as the ones listed above. To illustrate the effects of nanoscale surface roughness of the substrate on the structural and dynamic properties of the interfacial aqueous solution, the surface of **tobermorite, $Ca_5Si_6O_{16}(OH)_2$**, is taken here for comparison. Tobermorite is also known as one

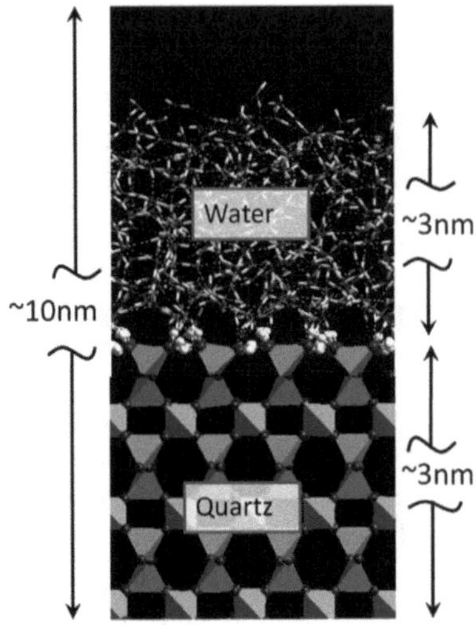

Fig. 9.1 A schematic view of the molecular dynamics periodic simulation cell for the modeling of a substrate-water interface. Bulk quartz SiO_2 crystal is shown as tetrahedra (Si) and *dark balls* (O). *White* (H) and *dark* (O) balls represent the O-H groups of the fully hydroxylated surface. Approximately half of the surface O-H groups are bent to make hydrogen bonds among themselves, while the other half – donate H-bonds to the interfacial H_2O molecules, which are shown as *white* (H) and *dark* (O) cylinders above the surface. H-bonds are indicated as *thin dashed lines*

of the so-called C-S-H (calcium silicate hydrate) phases in cement chemistry and its surface properties control the behavior of confined water in the nanopores of cement and concrete [11, 20].

The disordered interfaces of water filtration membranes are also typically characterized by the surface roughness on the nanoscale and above. In this case, the diversity of the surface adsorption sites can be very large and it is not always possible to identify and investigate all of them individually on a truly atomistic level. Nevertheless, computer simulations can still provide invaluable information about the molecular mechanisms of the processes of membrane filtration. The effects of various metal cations and their interactions with dissolved natural organic matter on the process of bio-fouling of **polyethersulfone (PES) desalination membrane** is taken here to illustrate the case [39].

All MD simulations described above were performed at ambient conditions in the *NVT* statistical ensemble (constant number of atoms, constant volume, and temperature $T = 300$ K) after a pre-equilibration in the *NPT* ensemble at $T = 300$ K and $P = 0.1$ MPa using standard MD algorithms [9]. The mineral surfaces were constructed by cleaving the bulk crystal structures parallel to the (001) plane. The irregular polymer surface was created in the course of preliminary MD runs of condensing a polyethersulfone polymer chain consisting of 60 monomer units [39]. Each complete interfacial MD simulation cell consisted of a substrate slab about 2–3 nm thick, and a layer of liquid water about 3–7 nm thick placed in contact with them, as illustrated in Fig. 9.1. The water was pre-equilibrated at

300 K and 0.1 MPa. Periodic boundary conditions [9] were then applied in all three dimensions to produce models of the interfaces formed by infinite (in x- and y directions) flat substrate layers interspersed with layers of water. In all cases, the thickness of the water layer and the size of the cell in the z-direction (normal to the surface) were sufficiently large to effectively exclude direct interactions between two different substrate/water interfaces created due to the periodicity of the system. The number of H_2O molecules in the layer was chosen to reproduce the density of bulk water under ambient conditions (~1 g/cm^3). The final x-, y-, and z-dimensions of the simulation supercells were approximately 2 nm × 2.5 nm × 10 nm with slight variations depending on the nature of the substrate.

The MD time step was 1.0 fs, and each model system was allowed to equilibrate for at least 500 ps of MD before an equilibrium MD trajectory for each model was accumulated at 10 fs intervals during additional 500–1,000 ps of MD simulation. The details of the MD simulation procedures and parameter settings for mineral-water systems are described in previous publications [11, 12, 18, 20, 27, 37, 39–41]. The structural analysis of the near-surface water films was undertaken using atomic density profiles in the direction perpendicular to the substrate surface and atomic density maps within defined slices of the water film parallel to the surface. These properties were all calculated by averaging over the last 500 ps equilibrium MD trajectory of each system. The position of the mineral surface ($z = 0$) was defined by the average position of the top-most surface oxygen atoms or of the corresponding OH groups, depending on the substrate.

9.3 Simulated Results and Discussion

9.3.1 Interfacial Water Structure

Figure 9.2 shows the time-averaged atomic density contour map of O_w in the first molecular layer of water on the (001) surface of quartz. It is important to note, that unlike the surfaces of brucite [18] and portlandite [11], only about 50 % of the surface hydroxyl groups on quartz are oriented away from the surface and easily donate H-bonds to the interfacial H_2O molecules. The other 50 % of the surface OH groups are oriented parallel to the surface of quartz, form H-bonds to other surface hydroxyls, and accept H-bonds from H_2O molecules (Figs. 9.1 and 9.2). The surface ordering of the H_2O molecules is clearly observed in the time-averaged patterns of the O_w distribution. Surface water molecules are significantly immobilized by each accepting an H-bond from one surface OH group and simultaneously donating one H-bond to another neighboring surface OH group. A similar pattern of H_2O ordering is also observed at the neutral hydroxylated surfaces of brucite [18] and portlandite [11]. However, the charged hydroxylated surface of hydrocalumite exhibits a completely different behavior where no H-bonds donation is possible for the interfacial water molecules [11].

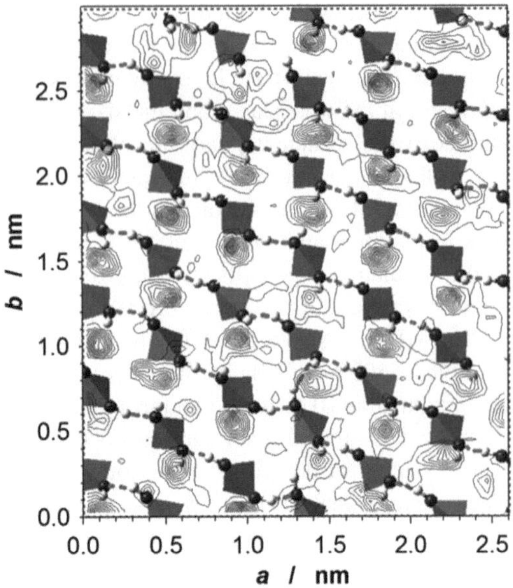

Fig. 9.2 Contour maps of the atomic density of oxygens of H_2O (O_w) within the first molecular layer of water at the fully hydroxylated (001) surface of quartz, which is shown below the contours as tetrahedra (Si) and *dark* (O) and *white* (H) balls. See Fig. 9.1 for the detailed notations

This is clearly visible from the comparison of the atomic density profiles for O_w and H_w of these surfaces, which provide important additional insight into the ordering of water molecules at the interface. Two orientations of H_2O molecules are energetically most favorable for the formation of stable hydrogen bonds to the hydroxylated surfaces of portlandite and quartz (Fig. 9.3a, b). In one of them, water molecules have one H-atom directed toward an O-atom of the surface hydroxyl, thus donating an H-bond to the surface. This is reflected in the presence of a strong peak at ~0.2 nm on the H_w density profiles. In this orientation, the second H-atom contributes to the peak at ~0.35 nm. However, a stronger contribution to the 0.35 nm peak in the H_w distribution arises from the other favorable H_2O orientation, in which it accepts an H-bond donated by the surface hydroxyl groups. In this case, both H_{H2O} of the surface water molecule contribute to the density profile peak at 0.35 nm. In both preferred orientations of the surface water molecules, their O-atoms are located approximately at the same distance from the surface (~0.25–0.28 nm). The ability of the surface water molecules to both donate and accept H-bonds to the substrate surface creates a very well developed H-bonding network across the interface with the structure strongly resembling that of bulk liquid water. O-O and O-H radial distribution functions of liquid water are shown as thick dashed lines in Fig. 9.3 to illustrate this similarity. The $g_{OH}(r)$ peak at ~0.2 nm is a definitive signature of donated hydrogen bonds in the system, while the $g_{OO}(r)$ peak at ~0.45 nm clearly indicates that the local ordering of the H-bonding network is close to tetrahedral – a typical characteristic of bulk liquid water structure [42].

For hydrocalumite (Fig. 9.3c) the computed near-surface water structure is strikingly different. Due to the positive structural charge of this substrate, the H-bond donation to the surface by H_2O molecules is prevented and the well-

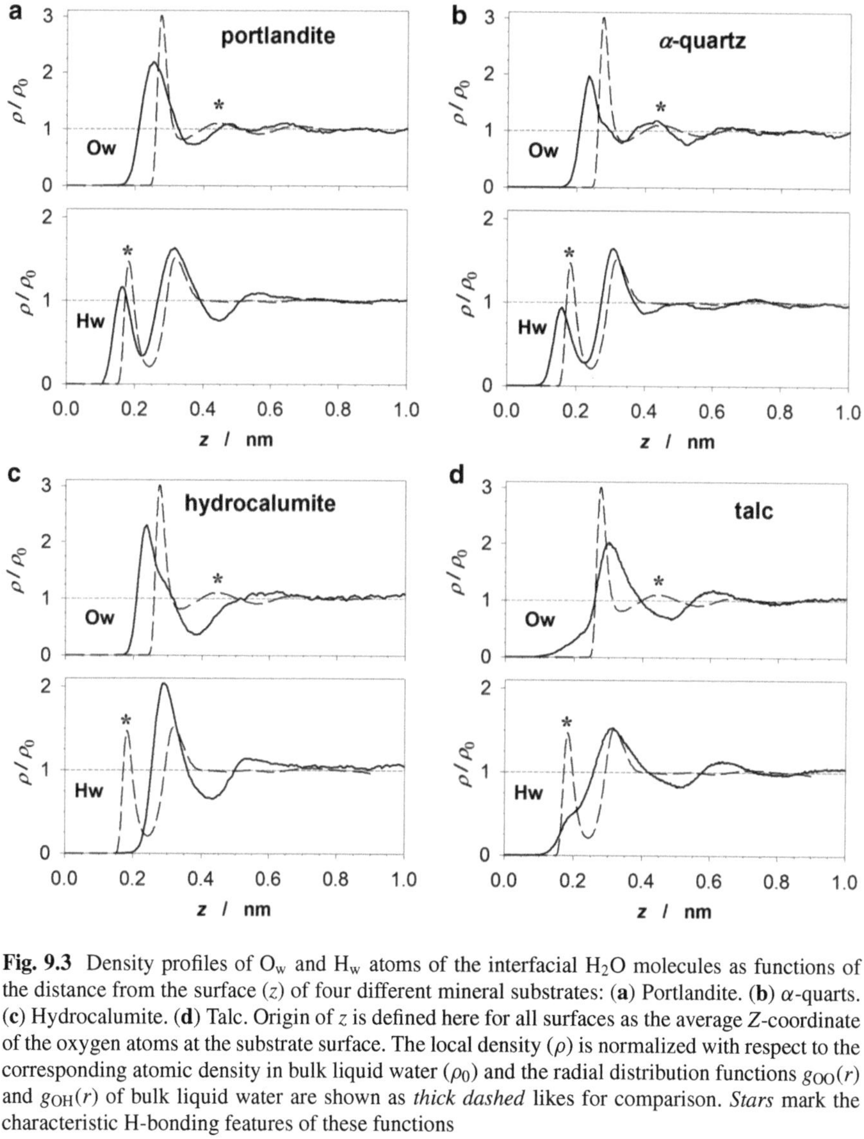

Fig. 9.3 Density profiles of O_w and H_w atoms of the interfacial H_2O molecules as functions of the distance from the surface (z) of four different mineral substrates: (**a**) Portlandite. (**b**) α-quarts. (**c**) Hydrocalumite. (**d**) Talc. Origin of z is defined here for all surfaces as the average Z-coordinate of the oxygen atoms at the substrate surface. The local density (ρ) is normalized with respect to the corresponding atomic density in bulk liquid water (ρ_0) and the radial distribution functions $g_{OO}(r)$ and $g_{OH}(r)$ of bulk liquid water are shown as *thick dashed* likes for comparison. *Stars* mark the characteristic H-bonding features of these functions

interconnected H-bonding network is not formed in the interfacial region [11], despite the fact that the H_2O molecules are strongly attracted to this surface and it is clearly hydrophilic.

The atomic density profiles of the surface water on talc (Fig. 9.3d) show that H_2O molecules are located somewhat farther away than from the previously discussed hydrophilic surfaces (Fig. 9.3a–c) and the H_2O molecular orientation is predominately parallel to the surface (the peaks of O_w and H_w are approximately at the same distance of 0.3–0.32 nm), in accordance with the hydrophobic character of talc. However, donation of weak H-bonds is still possible to this surface, as is clearly evident from the shoulders of the O_w and H_w distributions around 0.2 nm. Thus, talc surface exhibits a weak degree of hydrophilicity under ambient conditions, which can be most pronounced at very low relative humidity [33], or can even be turned into a strongly hydrophilic surface by the application of very high pressures and temperatures typical for the Earth's mantle conditions [40, 41].

9.3.2 *Interfacial Water Dynamics*

The simulated diffusional dynamics of surface water is in agreement with these observations: the H_2O molecules have noticeably higher mobility at the hydrophobic surface of talc and significantly lower mobility at all hydrophilic surfaces [18, 20]. Longer-time-scale MD simulations provide an additional opportunity to quantify these relatively slow diffusional motions of interfacial by means of the Van Hove self-correlation function (VHSCF) [9],

$$G_S\left(r,t\right) = \frac{1}{N}\left\langle \sum_{i=1}^{N} \delta\left[r + r_i(0) - r_i(t)\right] \right\rangle, \tag{9.1}$$

where r is the coordinate of the molecule, t is time, and N is the total number of molecules. This relationship describes the correlation in the positions of the same atom at different times, such that $4\pi r^2 G_S(r,t)\mathrm{d}r$ is the probability of finding an atom at distance r after a time t if the position of this atom was at the origin $r = 0$ at the initial time $t = 0$. The Fourier transform of the VHSCF represents the incoherent or self-intermediate scattering function (SISF), which can be directly measured in incoherent quasielastic neutron scattering experiments and contains detailed information concerning the single-molecule dynamics both in time and space domain.

The calculation of this self-correlation function for the water molecules strongly bound to the surface of tobermorite, which exhibits nanoscale surface roughness strongly affecting the first two monolayers of H_2O, shows a dramatic difference in the dynamic behavior of adsorbed H_2O (Fig. 9.4a), as compared with the diffusional dynamics in bulk liquid water (Fig. 9.4c), and yields a characteristic time scale of the diffusional process of the order of $\tau_m \sim 0.8$ ns, in excellent agreement with NMR measurements, indicating a mean time between diffusional jumps on the surface of ~0.8–1.0 ns [43]. The calculated long-time-scale VHSCF also points to a hopping diffusional mechanism with a length scale of $l_m \sim 5.5$ Å, i.e., close the characteristic lattice dimension of crystalline ice between two neighboring strong H-bonding sites. From the two-dimensional Einstein equation, $D = \langle|\mathbf{r}(t) - \mathbf{r}(0)|^2\rangle/4\tau$, the mean time for jumps between surface sites from the NMR results yields a diffusion coefficient of 0.9×10^{-10} m^2/s, which is also in remarkable agreement with the average

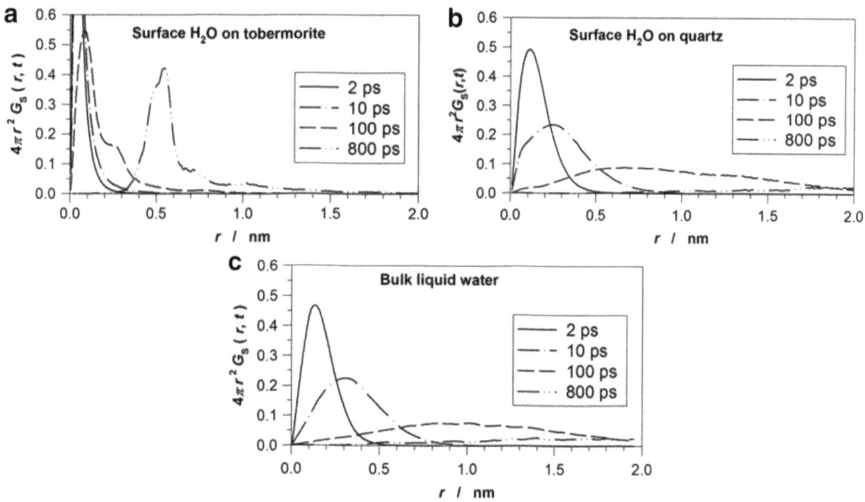

Fig. 9.4 Van Hove self-correlation functions for diffusing H_2O molecules at the surface of tobermorite (**a**) and quartz (**b**). The VHSCF behavior of H_2O molecules in bulk liquid water (**c**) is also shown for comparison

diffusion coefficient for all surface-associated H_2O molecules obtained from MD simulations (1.0×10^{-10} m^2/s). Here, **r** is the mean jump displacement (assumed to be 5.5 Å, from the results of the VHSCF calculations, Fig. 9.4a) and τ is the mean jump time, $\tau \sim \tau_m \sim 0.8$ ns. This level of agreement definitely provides strong support for the interpretation of the experimental results and strong encouragement for further application of atomistic computational modeling techniques to study the molecular scale properties of nano-confined water.

The hopping surface diffusion mechanism is not evident for the water molecules at an atomically smooth surface, such as that of quartz (Fig. 9.4b): the corresponding VHSCFs show only slight differences with the diffusional dynamics in bulk liquid water (Fig. 9.4c).

9.3.3 Effect of Metal Cations on the Membrane Fouling

Prediction of ultrafiltration membrane performance is not generally possible without performing pilot-scale tests because membrane fouling is related to membrane material and solution chemistry. The inadequate understanding of membrane fouling has in many cases hindered wider adoption of membrane processes in large-scale drinking water treatment plants. Computational molecular modeling of the model membrane-solution interfaces allowed detailed probing of the fouling process due to the interaction of dissolved natural organic matter (NOM) with various metal cations [39]. It was observed that divalent ions (Ca^{2+} and Mg^{2+}) may cause membrane

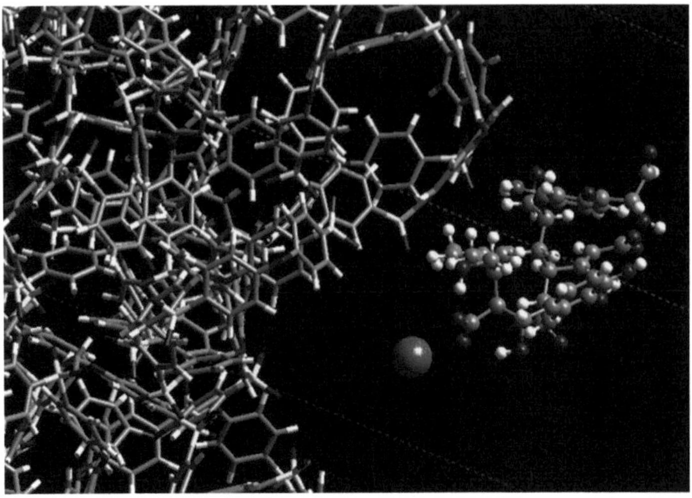

Fig. 9.5 A fragment of the MD simulation box containing the polyethersulfone membrane surface (*cylinder* representation), a calcium ion (*large dark ball*), and a molecule of natural organic matter (*ball* and *stick* representation) coordinating to Ca^{2+} by its deprotonated carboxylic group. All interfacial water molecules are hidden for clarity

fouling not by forming "ionic bridges" between the negatively charged functional groups on the membrane surface and the negatively charged functional groups of NOM, but rather by promoting the aggregation of NOM molecules in solution.

Thus, even though the presence of truly dissolved NOM is not important in the fouling process, it can contribute to fouling after being aggregated by divalent ions. The partially neutralized Ca^{2+}-NOM complex (Fig. 9.5) may be more easily adsorbed at the membrane surface than a negatively charged NOM molecule. However, it still remains unclear whether fouling is caused by the cation-mediated NOM aggregation in solution or by stronger NOM-surface interactions in the presence of divalent ions. Since only one NOM molecule was used in the present model, NOM aggregation could not be quantitatively studied in this work. More extensive MD simulations using multiple NOM molecules, employing much larger size of the simulated systems, and much longer simulation times (at least on the order of tens of nanoseconds) are necessary to quantitatively address these problems.

9.4 Conclusions

The methods of computational molecular modeling in application to the mineral-fluid systems have advanced dramatically in the past decade and have already emerged as very powerful quantitative tools in the studies of the structure and properties of interfacial and nano-confined aqueous solutions. There is a growing consensus, however, that the idealized mineral-fluid interfaces typically used in the present-day molecular simulations and the relatively small assemblages used in these models and in quantum calculations are at least an order of magnitude too small in characteristic length (three orders of magnitude in volume) to effectively address many critically important issues. A typical size of the systems currently simulated is on the order of a few nanometers, but can reach up to 20 nm [26] for classical force-field-based MD simulations which can probe the systems evolution over tens of nanoseconds. However, the application of quantum MD is still limited to the sizes <1.5–2 nm and time periods ~ 10 ps. The inability to effectively model realistic compositional and structural complexity is significantly impeding further progress. For instance, idealized, defect-free mineral surfaces are of little use in predicting local charge development, which is key to understanding surface reactivity in real systems [13, 16]. Particle edges, surface defects, the surface roughness on length scales of 10–100 nm play central roles in determining the energetics and kinetics of dissolution, precipitation, sorption and catalytic reactions at mineral interfaces, but these phenomena are currently modeled only very rarely and on a limited scale [19, 21, 34, 35]. There is no doubt, however, that further rapid development of supercomputing capabilities will allow such molecular simulations to achieve truly realistic system sizes and time scales, thus significantly increasing their practical value.

Acknowledgments This work was supported by the US Department of Energy, Office of Basic Energy Sciences, Division of Chemical Sciences, Geosciences, and Biosciences (grant number DE-FG02-08ER-15929) and by the industrial chair "Storage and Management of Nuclear Waste" at the Ecole des Mines de Nantes, funded by ANDRA, Areva, and EDF. The supercomputing resources of the NSF TeraGrid (grant number TG-EAR000002) and of the US DOE National Energy Research Scientific Computing Center (NERSC) were used for the simulations.

References

1. Brown GE (2001) Surface science – How minerals react with water. Science 294:67–69
2. Sverjensky DA (2006) Prediction of the speciation of alkaline earths adsorbed on mineral surfaces in salt solutions. Geochim Cosmochim Acta 70:2427–2453
3. Brown GE, Calas G (2012) Mineral-aqueous solution interfaces and their impact on the environment. Geochem Perspect 1(4–5):483–742
4. Fenter P, Sturchio NC (2004) Mineral-water interfacial structures revealed by synchrotron X-ray scattering. Prog Surf Sci 77:171–258
5. Wenk H-R (ed) (2006) Neutron scattering in earth sciences. Rev Miner Geochem 63:1–471
6. Shen YR, Ostroverkhov V (2006) Sum-frequency vibrational spectroscopy on water interfaces: polar orientation of water molecules at interfaces. Chem Rev 106:1140–1154
7. Cole DR, Mamontov E, Rother G (2009) Structure and dynamics of fluids in microporous and mesoporous earth and engineered materials. In: Liang L, Rinaldi R, Schober H (eds) Neutron applications in earth, energy and environmental sciences. Springer, New York
8. Arbogast T (ed) (2007) Computational needs for the subsurface sciences. Workshop report. U.S. DOE Office of Science, Washington D.C., April 2007, 291pp

9. Frenkel D, Smit B (2002) Understanding molecular simulation: from algorithms to applications, 2nd edn. Academic, San Diego
10. Cygan RT, Kubicki JD (eds) (2001) Molecular modeling theory and applications in the geosciences. Rev Miner Geochem 42:1–531
11. Kalinichev AG, Kirkpatrick RJ (2002) Molecular dynamics modeling of chloride binding to the surfaces of Ca hydroxide, hydrated Ca-aluminate and Ca-silicate phases. Chem Mater 14:3539–3549
12. Cygan RT, Liang J-J, Kalinichev AG (2004) Molecular models of hydroxide, oxyhydroxide, and clay phases and the development of a general force field. J Phys Chem B 108:1255–1266
13. Rustad JR, Felmy AR (2005) The influence of edge sites on the development of surface charge on goethite nanoparticles: a molecular dynamics study. Geochim Cosmochim Acta 69:1405–1411
14. Tossell JA (2005) Theoretical study on the dimerization of $Si(OH)_4$ in aqueous solution and its dependence on temperature and dielectric constant. Geochim Cosmochim Acta 69:283–291
15. Criscenti LJ, Kubicki JD, Brantley SL (2006) Silicate glass and mineral dissolution: calculated reaction paths and activation energies for hydrolysis of a Q^3 Si by H_3O^+ using ab initio methods. J Phys Chem A 110:198–206
16. Kerisit S, Rosso KM (2006) Computer simulation of electron transfer at hematite surfaces. Geochim Cosmochim Acta 70:1888–1903
17. Spagnoli D, Cooke DJ, Kerisit S, Parker SC (2006) Molecular dynamics simulations of the interaction between the surfaces of polar solids and aqueous solutions. J Mater Chem 16:1997–2006
18. Wang J, Kalinichev AG, Kirkpatrick RJ (2006) Effects of substrate structure and composition on the structure, dynamics and energetics of water on mineral surfaces: MD modeling study. Geochim Cosmochim Acta 70:562–582
19. Churakov SV (2007) Structure and dynamics of the water films confined between edges of pyrophyllite: a first principle study. Geochim Cosmochim Acta 71:1130–1144
20. Kalinichev AG, Wang J, Kirkpatrick RJ (2007) Molecular dynamics modeling of the structure, dynamics and energetics of mineral-water interfaces: application to cement materials. Cem Concr Res 37:337–347
21. Larentzos JP, Greathouse JA, Cygan RT (2007) An ab initio and classical molecular dynamics investigation of the structural and vibrational properties of talc and pyrophyllite. J Phys Chem C 111:12752–12759
22. de Leeuw NH, Cooper TG (2007) Surface simulation studies of the hydration of white rust $Fe(OH)_2$, goethite α-FeO(OH) and hematite α-Fe_2O_3. Geochim Cosmochim Acta 71:1655–1673
23. Predota M, Cummings PT, Wesolowski DJ (2007) Electric double layer at the rutile (110) surface. 3. Inhomogeneous viscosity and diffusivity measurement by computer simulations. J Phys Chem C 111:3071–3079
24. Allen JP, Gren W, Molinari M, Arrouvel C, Maglia F, Parker SC (2009) Atomistic modelling of adsorption and segregation at inorganic solid interfaces. Mol Simul 35:584–608
25. Cygan RT, Greathouse JA, Heinz H, Kalinichev AG (2009) Molecular models and simulations of layered materials. J Mater Chem 19:2470–2481
26. Suter JL, Anderson RL, Greenwell HC, Coveney PV (2009) Recent advances in large-scale atomistic and coarse-grained molecular dynamics simulation of clay minerals. J Mater Chem 19:2482–2493
27. Wang J, Kalinichev AG, Kirkpatrick RJ (2009) Asymmetric hydrogen bonding and orientational ordering of water at hydrophobic and hydrophilic surfaces: a comparison of water/vapor, water/talc, and water/mica interfaces. J Phys Chem C 113:11077–11085
28. Argyris D, Cole DR, Striolo A (2010) Ion-specific effects under confinement: the role of interfacial water. ACS Nano 4:2035–2042
29. Bourg IC, Sposito G (2010) Connecting the molecular scale to the continuum scale for diffusion processes in smectite-rich porous media. Environ Sci Technol 44:2085–2091

30. Malikova N, Dubois E, Marry V, Rotenberg B, Turq P (2010) Dynamics in clays – combining neutron scattering and microscopic simulation. Zeitschrift Phys Chem 224:153–181
31. Fenter P, Lee SS, Skelton AA, Cummings PT (2011) Direct and quantitative comparison of pixelated density profiles with high-resolution X-ray reflectivity data. J Synchrotron Radiat 18:257–265
32. Ferrage E, Sakharov BA, Michot LJ, Delville A, Bauer A, Lanson B, Grangeon S, Frapper G, Jimenez-Ruiz M, Cuello GJ (2011) Hydration properties and interlayer organization of water and ions in synthetic Na-smectite with tetrahedral layer charge. 2. Toward a precise coupling between molecular simulations and diffraction data. J Phys Chem C 115:1867–1881
33. Rotenberg B, Patel AJ, Chandler D (2011) Molecular explanation for why talc surfaces can be both hydrophilic and hydrophobic. J Am Chem Soc 133:20521–20527
34. Liu X, Lu X, Meijer EJ, Wang R, Zhou H (2012) Atomic-scale structures of interfaces between phyllosilicate edges and water. Geochim Cosmochim Acta 81:56–68
35. Tazi S, Rotenberg B, Salanne M, Sprik M, Sulpizi M (2012) Absolute acidity of clay edge sites from ab-initio simulations. Geochim Cosmochim Acta 94:1–11
36. Sahai N (2002) Is silica really an anomalous oxide? Surface acidity and aqueous hydrolysis revisited. Environ Sci Technol 36:445–452
37. Morrow CP, Yazaydin AO, Krishnan M, Bowers GM, Kalinichev AG, Kirkpatrick RJ (2013) Structure, energetics, and dynamics of smectite clay interlayer hydration: molecular dynamics and metadynamics investigation of Na-hectorite. J Phys Chem C 117:5172–5187
38. Duval Y, Mielczarski JA, Pokrovsky OS, Mielczarski E, Ehrhardt JJ (2002) Evidence of the existence of three types of species at the quartz-aqueous solution interface at pH 0-10: XPS surface group quantification and surface complexation modeling. J Phys Chem B 106:2937–2945
39. Ahn WY, Kalinichev AG, Clark MM (2008) Effects of background cations on the fouling of polyethersulfone membranes by natural organic matter: experimental and molecular modeling study. J Membr Sci 309:128–140
40. Wang JW, Kalinichev AG, Kirkpatrick RJ (2005) Structure and decompression melting of a novel, high-pressure nanoconfined 2-D ice. J Phys Chem B 109:14308–14313
41. Wang JW, Kalinichev AG, Kirkpatrick RJ (2004) Molecular modeling of the 10Å phase at subduction zone conditions. Earth Planet Sci Lett 222:517–527
42. Kalinichev AG (2001) Molecular simulations of liquid and supercritical water: thermodynamics, structure, and hydrogen bonding. Rev Mineral Geochem 42:83–129
43. Korb JP (2010) Multi-scales nuclear spin relaxation of liquids in porous media. Compt Rend Phys 11:192–203

Chapter 10
Two Generalizations of the Theory of Seismoelectric Effect: Parameterization Providing Suitability of Frenkel's Theory for Any Geometry of Soil's Pore Space. The Role of Thermoosmosis in Seismoelectric Effect

Vladimir N. Shilov

Abstract Seismic vibrations of water-saturated soils and rocks cause the fluid motion relative to the charged interface and thus to the flow potential. Thus a seismoelectrical wave is formed, in which the ratio of its sound and electrical components carries diverse information about water in soils and rocks. As a result electroseismic method is of importance for the problem of water resources and is widely used in hydrogeology. However, at present it is possible to explain only a small part of the information carried by the electroseismic signal. Empowering the method depends on the improvement of equipment and measurement as well as by means of the further development of the theory of seismoelectric effect. This paper makes two steps in the development of theory. First, it manages to express the seismoelectric field strength through the ζ-potential and through only the macroscopic parameters of Frenkel's classic theory, which makes it suitable for any geometry of soil's pore space. Second, for the first time the influence on the seismoelectric effect is considered of thermoosmotic flow arising under the action of temperature gradients which always accompany the sound wave.

10.1 Introduction

Seismoelectric (SE) effect accompanies the propagation of an acoustic wave through a soil and consists in the occurrence of an electrical potential difference between points in the soil at different distances from the source of the wave. Classic Frenkel's theory (Ref [1]) explains SE effect by the fact that vibration of the moist soil causes the movement of water together with the ions of diffuse part of the electric double layer with respect to the charged surface of soil's solid phase and

V.N. Shilov (✉)
Institute of Biocolloid Chemistry, Ukrainian Academy of Science, Vernadskogo 42, 03142 Kiev, Ukraine
e-mail: shilov11@yahoo.com

L. Mercury et al. (eds.), *Transport and Reactivity of Solutions in Confined Hydrosystems*, NATO Science for Peace and Security Series C: Environmental Security,
DOI 10.1007/978-94-007-7534-3_10,

the displacement of these ions creates an electric field and correspondence potential difference. Frenkel's theory comprises two parts. The first one, the significance of which goes far beyond description of SE effect, describes the interrelated vibrations of two phases of capillary-porous system modeling the moist soil. This model includes a solid skeleton – an elastic structure which is formed by solid particles rigidly attached to each other, and liquid phase (electrolyte solution) which fills the space within the skeleton. Frenkel developed first part of his theory at exclusively macroscopic level (with such parameters as porosity, specific surface of solid skeleton, its Young's modulus, compressibility modulus of skeleton and liquid, its density and viscosity; see below), without any kind of simplified models of the geometry of the pore space. As a consequence of uncertainty and irreproducibility of pore space geometry this macroscopic approach seems to be the most adequate level for description of real soil. Just on this macroscopic level, Frenkel obtained the expression for the relative velocity $\overline{v}$ of the liquid phase and solid skeleton. However, in the second part of his theory, using this $\overline{v}$ for the calculation of SE field strength, he goes beyond his macroscopic model and replaces it with a model "of a Poiseuille flow of the liquid through a capillary tube with a radius r". The first two sections of our paper are devoted to the justification of a simple procedure that allows us to express the SE field strength in terms of macroscopic parameters of Frenkel's model, without using an uncertain parameter r.

The last section of the paper is devoted to the consideration of the influence on the SE effect of temperature gradients that arise due to the known adiabaticity of the processes of compression-extension in the sound wave. As a result of the action of temperature gradient on the liquid in surface layer, where the specific entropy (see Refs. [2] and [3]) is different from that in the bulk liquid, thermoosmotic flow arises. The displacement of the ions entrained by this flow in the diffuse part of the electric double layer determines the contribution of thermoosmosis to seismoelectric effect. The magnitude and sign of this contribution is determined by the magnitude and sign of entropy excess in the surface layer. The latter, in turn, is highly dependent on the degree of hydrophilicity or hydrophobicity of the surface of soil's solid skeleton, and the latter, of course, is different for wet or drought-prone soils. These circumstances allow us to hope, that the inclusion of the contribution of thermoosmosis to SE signal will help expand opportunities for the use of the SE method to obtain information about the surface layers of the fluid in the soils of the arid zones.

10.2 Classic Formula for Seismoelectric Field Strength

Frenkel defined the strength E_e of SE field by means of Smoluchowski formula (see also Ref. [4]) connecting it with an effective pressure gradient, $\frac{\partial p_{ef}}{\partial x}$:

$$E_e = -\frac{\varepsilon_0 \varepsilon \cdot \zeta}{\mu \cdot \sigma} \frac{\partial p_{ef}}{\partial x} \tag{10.1}$$

here ε_0: electrostatic constant, ε and μ: dielectric constant and viscosity of liquid, ζ: electrokinetic potential of the surface of solid skeleton, σ: conductivity of the soil.

The problem of determining the expression for $\frac{\partial p_{ef}}{\partial x}$ is hampered by the fact that Smoluchowski formula was derived for the stationary case, when the fluid filtration through the solid skeleton is given by the action of applied pressure gradient in the absence of inertial forces, while the filtration which is responsible for SE effect arises under the action of inertial forces in vibrating soil. To solve this problem, Frenkel determines the velocity of fluid filtration through the solid skeleton (equation without number after Eq. (44b) from [1]):

$$\overline{v} = v_2 - v_1 \tag{10.2}$$

where v_1 and v_2 are the local velocities of macroscopic elements of solid skeleton and liquid phase.

Then he equates it with the bulk velocity of the fluid in the direct cylindrical capillary radius r using Poiseuille formula (Eq. (44a) from [1]):

$$\overline{v} = -\frac{r^2}{8\mu}\frac{\partial p_{ef}}{\partial x}, \tag{10.3}$$

by defining in such a way $\frac{\partial p_{ef}}{\partial x}$, he receives from Eq. (10.1) the expression for seismoelectric field strength (Eq. (44b) in [1]):

$$E_e = \frac{8\varepsilon_0\varepsilon \cdot \zeta}{\mu \cdot \sigma \cdot r^2}\overline{v} \tag{10.4}$$

However, this approach to the definition of p_{ef} has an essential shortcoming: it uses a simplified model of the geometry of the pore space where an uncertain parameter appears, capillary radius r, and so we lost the benefits of the exclusively macroscopic level description of the first part of Frenkel's theory.

10.3 Parametrization Providing Suitability of Frenkel's Theory for Any Geometry of Soil's Pore Space

To determine an alternative value of p_{ef}, we use the equation of motion of the liquid phase of the soil (Eq. (22a) from [1]), together with definition given by Eq. (10.2):

$$\rho_2\frac{\partial \boldsymbol{v}_2}{\partial t} = K_2\nabla\varphi - \frac{\mu}{\kappa \cdot f}\overline{v} \tag{10.5}$$

where $\varphi = -\frac{\Delta\rho_2}{\rho_2}$, ρ_2 and K_2: density and compressibility modulus of liquid phase, $\Delta\rho_2$: sound-induced variation of ρ_2, $k = \kappa \cdot f$: filtration coefficient of the soil, f: its porosity

Taking into account the equation of state of the liquid phase (Eq. (15a) of [1]) connecting φ with hydrostatic pressure p_2:

$$p_2 = -K_2 \varphi, \tag{10.6}$$

we rewrite the equation of motion (10.5) in the following form:

$$\rho_2 \frac{\partial \boldsymbol{v}_2}{\partial t} = -\nabla p_2 - \frac{\mu}{k} \overline{v} \tag{10.7}$$

This equation reflects the fact that to unit volume of liquid phase in a macroscopically small element of soil volume two forces are applied: the pressure force $-\nabla p_2$ and the viscous force $-\frac{\mu}{k}\overline{v}$ from the solid skeleton.

In our definition of the expression for ∇p_{ef} we rely on the fact that the value of "microscopic" current which arises when charged liquid in the diffuse part of thin EDL moves near a small area of the solid surface, as well as the value of the "microscopic" viscous force with which this area acts on a liquid, are both determined by the value of the velocity gradient near the surface. But the sum of such filtration-induced "microscopic" currents forms a macroscopic electric current, as well as the sum of "microscopic" viscous forces forms the macroscopic viscous force per unit volume, $F_{wisc} = -\frac{\mu}{k}\overline{v}$. So, the force F_{wisc} is directly connected with a macroscopic current density, and hence with the seismoelectric field. This macroscopic viscous force acting on the liquid from the solid skeleton, is presented in Eq. (10.1) by a balancing pressure force, $F_{pef} = -\nabla p_{ef}$, equal to F_{wisc} in magnitude and opposite in sign, $F_{pef} = -F_{visc}$.

The above considerations lead to the following definition of p_{ef}:

$$\nabla p_{ef} = -\frac{\mu}{k}\overline{v}, \tag{10.8}$$

After spending the following identity transformation on the right side of Eq. (10.7)

$$\rho_2 \frac{\partial \boldsymbol{v}_2}{\partial t} = -\nabla \left(p_2 - p_{ef}\right) - \nabla p_{ef} - \frac{\mu}{k}\overline{v},$$

and using the definition (10.8) we obtain the equation:

$$\rho_2 \frac{\partial \boldsymbol{v}_2}{\partial t} = -\nabla p_2 + \nabla p_{ef} \tag{10.9}$$

which is equivalent to Eq. (10.6), but without the viscous force, which was replaced by the effective pressure gradient. Thus, it can be argued that the definition of ∇p_{ef} by Eq. (10.8) is consistent also in the nonstationary regime (which is described by both Eqs. (10.6) and (10.9)).

Substituting Eq. (10.8) in Eq. (10.1) we obtain for the SE field strength, instead of Eq. (10.4), the following:

$$E_e = \frac{\varepsilon_0 \varepsilon \cdot \zeta}{\sigma \cdot k} \overline{v}, \tag{10.10}$$

Substituting the expression for the filtration velocity (Eq. (45) from [1]) in Eq. (10.10) we obtain the final expression for the strength of seismoelectric field, in which the ζ-potential and only macroscopic parameters of Frenkel's model appear:

$$E_e = \frac{\varepsilon_0 \varepsilon \cdot \zeta \omega^2 \rho_2}{\sigma} \left(\frac{K_2}{\rho_2} \frac{\beta}{\beta'} \frac{1}{w_0{}^2} - 1 \right) u \tag{10.11}$$

where

$w_0 = \sqrt{\frac{1}{\gamma_1+\gamma_2}\left[E + \frac{\beta}{\beta'} K_2 \left(1 - \frac{K}{K_0}\right)\right]}$: sound velocity in the soil (see Eq. (33a) in Ref. [1]); u: sound-induced displacement of small macroscopic element of soil volume; ω: cyclic frequency of the sound; ρ_1, ρ_2: true densities of solid and liquid phases; $\gamma_1 \equiv \rho_1(1-f)$, $\gamma_2 \equiv \rho_2 f$: average (per total volume, occupied by the soil) densities of solid and liquid phases; E, K: Young's and compressibility modulus of the soil; K_0 and K_2: compressibility modulus of solid and liquid phases, respectively; $\alpha = s\left(\frac{1}{f} - 1\right)$, $1 \le s < \frac{1}{f}$; $\beta = \frac{1}{f(1+\alpha)}$, $\beta' = 1 + (\beta - 1)\frac{K_2}{K_0}$.

Formula (10.11) has been obtained without the use of any simplified models of geometry of the pore space, and is valid for any geometry of the solid skeleton, as Smoluchowski formula (10.1) in the case of small thickness of the electric double layer and as the results of [1] that were obtained on the macroscopic level. This formula was first proposed in Ref. [5] but there was not given its justification.

10.4 The Role of Thermoosmosis in SE Effect

The previous results refer to the classical mechanism of SE effect when the fluid moves by inertia in the pores of oscillating solid skeleton, simultaneously moving the ions of diffuse part of electric double layer. The resulting electrical current is manifested in SE effect. Here we pay attention to an alternative mechanism associated with thermoosmosis when the liquid moves within limits of the diffuse part of the electric double layer under the influence of sound-induced temperature gradients rather than under the direct influence of the mechanical movements of the phases. Gradients of temperature always arise as a response to the fast variations of liquid density in sonic wave, corresponding to adiabatic conditions, where there is interrelation (see [6]) between the gradients of temperature and pressure in the liquid:

$$\text{grad}\, T = \frac{\beta_2 T}{c_p \rho_2} \text{grad}\, p_2 \tag{10.12}$$

Where c_p and β_2: specific heat at constant pressure and temperature coefficient of expansion for liquid phase.

To find grad p_2, we use the equation of state (10.6), expressing in it sound-induced variation of specific volume of the liquid phase, φ, through that of the soil, θ, $\varphi = \theta\frac{\beta}{\beta}$ (see Eq. (28) from [1]) and using the relations: $\theta = -iqu$ (see Eq. (28) from [1]), $q = \frac{\omega}{w_0}$ (see Attachment) and w_0 from Eq. (10.12). As a result of this procedure, we get after simplification:

$$p_2 = i\omega K_2 f\,(1+\alpha)\left\{\left[1+\frac{K_2}{K_0}\left(\frac{1-f\,(1+\alpha)}{f\,(1+\alpha)}\right)\right]^{\frac{3}{2}}\sqrt{\frac{E+K_2\,(K_0-K)}{K_0\,(\gamma_1+\gamma_2)}}\right\}^{-1}u \tag{10.13}$$

So, since $i\omega\cdot iq = -\,\omega^2/w_0$ (see Attachment), we get from Eqs. (10.12, 10.13):

$$\mathrm{grad}\,T = \frac{\partial T}{\partial x} = \frac{\omega^2}{w_0}K_2\frac{\beta_2 T}{c_p\rho_2}f\,(1+\alpha)\left\{\left[1+\frac{K_2}{K_0}\left(\frac{1-f\,(1+\alpha)}{f\,(1+\alpha)}\right)\right]^{\frac{3}{2}}\sqrt{\frac{E+K_2\,(K_0-K)}{K_0\,(\gamma_1+\gamma_2)}}\right\}^{-1}u \tag{10.14}$$

If the specific entropy (entropy per unit volume) of the water in the surface layer, s, is different from specific entropy beyond its borders, s_V, by the value $q = s - s_V$, then appears the bulk force equal to $f_b = -\,q\cdot\mathrm{grad}\,T$ (see Ref. [3]) giving rise to the movement of liquid with respect to a particle surface. For quasi-planar layer, when the characteristic size of the solid particles and the distance between them are much greater than the thickness of the surface layer, l_σ, the fluid velocity is directed along the surface, the Stokes equation, $\frac{d^2 v_T}{dy^2} = -\,f_b$, will have the following form:

$$\frac{d^2 v_T}{dy^2} = \frac{q(y)}{\eta}\nabla T, \tag{10.15}$$

where y -is the distance from the surface.

The surface conditions reflect "sticking" of liquid to a particle surface ($y = 0$):

$$v_T\big|_{y=0} = 0 \tag{10.16}$$

and the absence of bulk force outside ($y = \infty$) the surface layer:

$$\left.\frac{dv_T}{dy}\right|_{y\to\infty} = 0. \tag{10.17}$$

Let us accept the exponential dependence of the excess of specific entropy in the surface layer from q_0 at the interface to 0 outside ($y = \infty$) the surface layer:

$$q(y) = q_0 \cdot e^{-\frac{y}{l_q}}, \tag{10.18}$$

where l_σ – the thickness of the surface layer.

Substituting $q(y)$ from Eq. (10.18) to Eq. (10.15) and solving it with surface conditions (10.16) and (10.17) we get:

$$v_T(y) = -\frac{\nabla T}{\eta}\int_0^y q \cdot y \cdot dy = -\frac{\nabla T}{\eta} q_0 {l_q}^2 \left(1 - e^{-\frac{y}{l_q}}\left(\frac{y}{l_q} + 1\right)\right) \tag{10.19}$$

The movement of liquid in the diffuse part of thin EDL gives rise to the surface current:

$$I_{sT} = \int_0^\infty \rho(y) \cdot v_T(y) \cdot dy \tag{10.20}$$

From this it follows for the corresponding macroscopic electric current density: $I_T = s_p I_{sT}$ and for the contribution of thermoosmosis to the strength of the SE field:

$$E_T = \frac{1}{\sigma} I_T = \frac{s_p}{\sigma} I_{sT} \tag{10.21}$$

Where s_p: the specific surface of the solid skeleton of the soil.

In order to find the distribution of charge density ρ in the diffuse part of electric double layer, we express it through the potential distribution by the Poisson equation:

$$\rho = -\varepsilon_0 \varepsilon \frac{d^2\Phi}{dy^2} \tag{10.22}$$

The potential distribution is found from the Debye equation:

$$\frac{d^2\Phi}{dy^2} = \frac{1}{{R_D}^2}\Phi, \tag{10.23}$$

with surface conditions,

$$\Phi|_{y=0} = \varsigma, \quad \Phi|_{y\to\infty} = 0 \tag{10.24}$$

the first of which sets the potential at the inner boundary of diffuse part of electric double layer equal to the electrokinetic potential ς, while the second, as usual, follows from zero potential outside the electric double layer.

The well-known solution of the problem, given by Eqs. (10.23) and (10.24), is:

$$\Phi = \varsigma \cdot e^{-\frac{y}{R_D}} \tag{10.25}$$

Where R_D is the Debye screening length.

Substitution of (10.25) to (10.22) leads for $\rho(y)$ to:

$$\rho = -\frac{\varepsilon_0 \varepsilon \cdot \varsigma}{{R_D}^2} \cdot \exp\left(-\frac{y}{R_D}\right) \tag{10.26}$$

Further we use the well-known formula of the Debye screening length, for the case of a 1–1 electrolyte solution:

$$R_D = \sqrt{\varepsilon_0 \varepsilon RT / 2F^2 c_0} \tag{10.27}$$

where T: the absolute temperature, F: Faraday number, c_0: the concentration of 1–1 electrolyte solution in a liquid phase far from the surface (outside the surface layer and EDL). If we substitute Eqs. (10.19) and (10.26) in Eq. (10.20), and then the result of the integration in Eq. (10.21), we obtain the final expression for the contribution of thermoosmosis to the strength of the SE field:

$$E_T = \varsigma \cdot {l_q}^2 q_0 RT \frac{\omega^2 R_D \cdot s_p}{{w_0}^2 \eta c_p \sigma} \frac{1}{\rho_2} \frac{K_0 K_2}{f\,(1+\alpha)\,(K_0 - K_2) + K_2} u \tag{10.28}$$

All the designations are given above under the Eqs. (10.1), (10.5), (10.11), (10.12), (10.18), (10.21) and (10.27).

Analyzing this equation, we first of all should pay attention to the fact, that the fraction in its right side is always positive, since $f(1+\alpha) < 1$ (see in Ref. [1], at the end of page 880). For example, it means, that for the hydrophilic solid skeleton, when excess of entropy q_0 near its surface is negative (see. Ref. [2], after Eq. (10.86)), then the contributions of thermoosmosis and electroosmosis (given respectively by the Eqs. (10.28) and (10.11)) to the strength of SE field will have opposite signs.

From the figure below one can see, that for large enough magnitudes of both the values of the thickness of the surface layer l_q and of the specific surface of the solid skeleton of the soil s_p, the absolute value of E_T may be close to that of E_e, or even surpass it. So the influence of thermoosmosis in the case of hydrophilic solid skeleton may significantly reduce the measured value of the strength of SE field and this reduction carries information about the parameters of the surface layer of soil water.

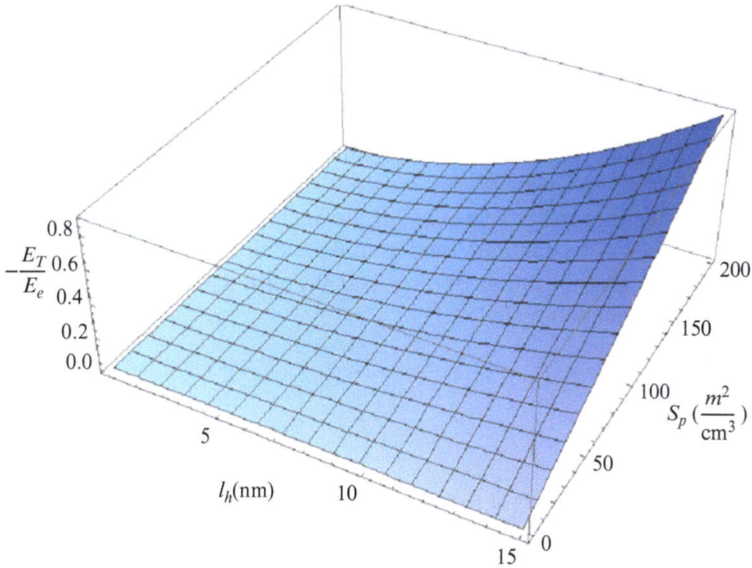

The following are values of parameters that were used in constructing the figure. All these data, except for those related to the parameters of the surface layer q_0 and l_σ, are common in heavy clay soils. For the value of q_0 and the range of the changes in l_σ, we used the results of experimental and theoretical studies presented in Chap. 10 of the monograph [2].

Liquid phase $\varepsilon = 80$, $\eta_m \approx 10^{-3}\frac{kg}{m \cdot \text{sec}}$, $K_2 = 2.17 \cdot 10^9 \frac{N}{m^2}$, $\beta_m \approx 2.09 * 10^{-4}\frac{1}{{}^{\text{o}}K}$, $c_p \approx 4200\frac{\text{J}}{\text{kg} \cdot {}^{\text{o}}K}$, $q_0 = -50\frac{J}{{}^{\text{o}}K \cdot m^3}$, $l_q = 5\ nm$, $c_0 = 10^{-1}N$. Solid phase: $\rho_m \approx 2.6\frac{kg}{m^3}$, $K_0 = 4 \cdot 10^6 \frac{N}{m^2}$. Soil: $K = 0.5K_0$, $E = 4 \cdot 10^6 \frac{N}{m^2}$, $f = 0.7$

10.5 Conclusion

The equation for the strength of the SE field in the classical mechanism of seismoelectric effect has been obtained without the use of any simplified models of geometry of the pore space, and valid for any geometry of the solid skeleton of the soil. Furthermore the existence of a new effect is here theoretically predicted: the contribution of thermoosmosis to the strength of the SE field. Accounting for this effect in hydrogeological applications of the SE method offers the prospect of obtaining information about the surface layers of the water contained in the soil and rocks.

Appendix

It should be noted that here we consider the case of a weakly damped plane longitudinal sound wave propagating along the x-axis. Respectively, the time and space dependence of every sound-induced value may be represented by a multiplayer $e^{i(\omega t - qx)}$, where ω and $q = 2\pi\lambda$ are the cyclic frequency and wave number; the sound speed is $w = \omega/q$. All the vector quantities are represented by their x-components. Operators ∇ and grad denote differentiation by x, which, in turn, reduces to multiplication by $-iq$ and the differentiation with respect to t reduces to multiplication by $i\omega$.

References

1. Frenkel J (1944) On the theory of seismic and seismoelectric phenomena in a moist soil. J Phys USSR 3(5):230–241, Re-published, J Eng Mech 131(9): 879–887 (2005)
2. Derjaguin BV, Churaev NV, Muller VM (1987) Surface forces. Plenum Press, New York
3. Shilov VN (1998) Equations for the non-reversible hydrodynamics of the boundary layer in an electrolyte solution. Colloids Surf A 142:375–379
4. Hunter RJ (1981) Zeta potential in colloid science. Academic, New York
5. Dukhin AS, Shilov VN (2010) Seismoelectric effect: non-isochoric streaming current. 1. Theory and its experimental verification. J Colloid Interface Sci 345:248–253
6. Landau LD, Lifshitz EM (1987) Fluid mechanics, vol 6, 2nd edn. Butterworth–Heinemann

Part III
In-Pores/Channels Cavitation

It is common knowledge that at atmospheric pressure water is liquid in the 0–100 °C range; below that range it turns into solid ice while above it, it turns into gaseous steam. However, it is "not-so-common-knowledge", that the previous statement is false; water at atmospheric pressure can stay liquid outside of the 0–100 °C range. Below zero, it can be undercooled liquid, above 100 °C it can be overheated liquid. In both conditions, the liquid state will be metastable, i.e. it might turn – even spontaneously – into the corresponding stable phase by freezing or by vapourization (cavitation) [1, 2]. This introductory paper and the following chapter deals with liquid water (and aqueous solutions) metastable for the liquid-vapour phase transition as well as with the corresponding relaxation process, known as cavitation.

The fluid phase diagram (ice phases are omitted) can be seen in Fig. 1. One can see three different curves. The medium one (dashed) is the vapour pressure curve; this is the traditional "border" between liquid and vapour phases. The other two curves are the so-called stability lines (or spinodals) [1, 3]. Reaching these lines by changing the pressure and/or temperature, the initial phase (liquid or vapour) cannot stay stable anymore and phase transition (boiling or condensation) have to happen spontaneously. The upper spinodal marks the limit where vapour has to condensate into liquid, while the lower one marks the limit where liquid has to boil into vapour. Between the vapour pressure curve and the two spinodals one finds the metastable region; we are interested in metastable liquid water, therefore only the region below the vapour pressure curve will be introduced in detail. The region marked as "forbidden states" (and sometimes erroneously marked as "instable liquid states") are too hot for solid (ice), too negative for liquid water and it is impossible to have gas (vapour phase) under negative pressure, therefore these states cannot exist in Water-World.

Let's start with water at room temperature and atmospheric pressure; under these conditions, the liquid water is stable. By increasing the temperature (and keeping the pressure at constant value), the system can reach the vapour pressure curve at 373.15 K (100 °C). Here, usually one can see very intense bubbling even inside of the water and the liquid will turn into steam (vapour). This is the usual scenario,

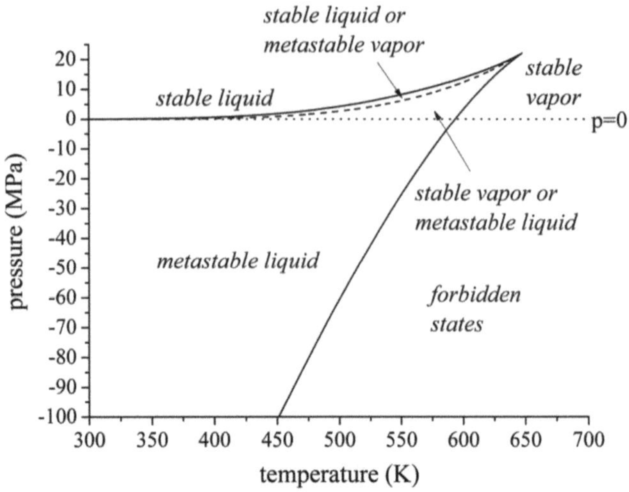

Fig. 1 Phase diagram of water without the solid (ice) phases, calculated by the IAPWS phase diagram. Possible phases of the various regions are marked

but when the water is pure and external disturbances (mechanical vibrations, voices or even cosmic radiation) is minimalized, one can see that the thermometer goes above the 100 °C, without observing boiling. Even in a high-school chemistry lab one can overheat the water in some extent, usually with 4–5 °C; under very special laboratory conditions one can even overheat it by 200 °C [4]. The overheated water is a metastable liquid; at the same pressure and temperature the vapourous state (steam) would be at lower free energy, therefore any small disturbance (like knocking the container) would initiate a very rapid, explosion-like boiling to relax to the stable vapour phase. The boiling might happen even without external trigger. Inside of the metastable liquid the density will fluctuate; at some places the liquid will be more dense than the average density, while at other places it will be less dense; the fluctuations are stronger as one goes deeper into the metastable region. Waiting long enough, at least one fluctuation will produce a region, where the density is low enough to be called "vapour" instead of "liquid" and the size of this region might be big enough to call it "bubble"; this is the point, where the spontaneous boiling starts and the metastable liquid splits into a stable vapour and also stable liquid phases. The new pressure will be equal to the corresponding (and always positive) vapour pressure. Because the bubble is a "cavity" inside of the liquid, the phenomena is also called "cavitation".

In bio- and partly in geo-sciences (including soil-science) usually the "lower" metastable liquid region ($p < 0$) is the important one; the "higher" one ($p > 0$) can be neglected, due to the required high temperature. In biological systems as well as in soils those temperatures usually do not appear; in geo-sciences they can be relevant, but in this chapter we are dealing with systems (fluid inclusions) with lower temperatures. The low-temperature metastable region has a very unique and strange feature, namely the negativity of the pressure. Negative pressure states are hardly known, physicist and chemists can finish their education without knowing about them; biologists at least mention them. This negligence has two reasons. First, the way that pressure is usually explained in university-level education, does not allow the existence for $p < 0$ states. Pressure is caused by the bounce of particles on the wall; it is proportional to the (positive) number of particles and the (also positive) speed of the particles. This is a good picture for gases, but not for liquids and solids. The other reason is the metastable nature of these states, which makes experiments – and even demonstration – hardly reproducible.

Huygens was the first who demonstrated the existence of negative pressure states [5]. He used a Toricelli-tube (a glass-tube, closed at one end and filled by mercury); when he turned the tube upside down – while the open end was immersed to a pool of mercury – usually he obtained the fall-back of mercury column to 76 cm, corresponding to the atmospheric pressure. But occasionally the liquid stuck; he obtained a 1.5 m long mercury column, which fell back to the "normal" 76 cm only after knocking the tube. This "usually" and "occasionally" was too much for a scientist, living in an age, where everything was believed to be deterministic. He asked the help of Boyle, Papin and others [6], but after a couple of years, he gave up these experiments without reaching any answer. More than a century later the new science of cohesion (between molecules) and adhesion (between the wall and the liquid) was able to give the proper explanation for the phenomena, but the new experiments started only in the mid-nineteenth century [3]. Since then, various methods have been developed to generate negative pressure, usually applying some external force to "pull" the liquid isotropically [7].

At the end of the nineteenth century, biologists realized that tall trees need some extra force to transport sap to the top. One plausible theory was the "cohesion theory", where the sap was sucked up by moderate negative pressure [8]. This theory has been supported by several direct sap-pressure measurements [9], as well as by the observation of the embolism, the appearance of a cavitational bubble [10] in the xylem capillaries, which is a very characteristic feature of metastable liquid states. The mechanism of the creation of negative pressure and especially the mechanism to re-fill the capillary after cavitation is still in need of some study; in some of the following papers the recent state of these studies are summarized.

While most of the methods generating negative pressure apply some external force for stretching the water, in narrow tubes and pore spaces this force is internal; it is provided by the strongly curved liquid-vapour interface. It can be shown that some of the properties of this capillary water differ significantly from the bulk water [11]. Pore spaces similar to the xylem in size but more random in organization can be found in soil; there are tiny water-bridges between soil particles. Due to the negativity of their pressure, they can work as an adhesive and hence they can provide stability for the soil and for other wet granular materials, like wet sand. Obviously, cavitation can be enhanced by adding some extra external force to stretch the capillary liquid; this force can be mechanical as well as electromagnetic [12]. On the other hand, for nano-capillaries the size is not sufficient to have a proto-bubble to start cavitation – at least not inside of the capillary - therefore in these systems the cavitation process itself might differ from the cavitation in bulk liquid [13].

Finally we should mention one more geo-system, where negative pressure appears. In the sixties, Roedder [14] found some microscopic fluid inclusions in rock samples. He observed that the fluid inside was still in liquid state. When he tried to freeze the liquids in the inclusions, he obtained an enormous freezing point shift which was explainable only by assuming that the liquid was under deep negative pressure, meaning several hundreds of atmospheres. Since then, the study of inclusions became an important tool to study deeply metastable aqueous solutions [15].

The following chapters deal with properties of aqueous systems under negative pressure, with the relaxation process (cavitation) from metastable to stable state and with the effect of cavitation on some complex bio- and geo-systems.

Attila R. Imre and John S. Sperry

References

1. Debenedetti PG (1996) Metastable liquids: concepts and principles. Princeton University Press, Princeton
2. Skripov VP, Faizullin MZ (2003) Crystal-liquid-gas phase transitions and thermodynamic similarity. Wiley-VCH, Weinheim
3. Trevena DH (1987) Cavitation and tension in liquids. Adam Hilger, Bristol
4. Imre AR, Maris HJ, Williams PR (eds) (2002) Liquids under negative pressure, NATO science series. Kluwer, Dordrecht
5. Huygens M (1672) An extract of a letter of M. Hugens to the author of the Journal des Scavans of July 25. Attempting to render the cause of that odd phaenomenon of the quickSilvers remaining suspended far above the usual height in the Torricellian experiments. Philos Trans 7:5027–5030
6. Kell GS (1983) Early observations of negative pressures in liquids. Am J Phys 51:1038–1041
7. Imre AR (2007) How to generate and measure negative pressure in liquids. In: Rzoska SJ, Mazur VA (eds) Soft matter under exogenic impacts: fundamentals and emerging technologies, NATO science series. Springer, Dordrecht, pp 379–388
8. Dixon HH, Joly J (1894) On the ascent of sap. Ann Bot 8:468–470
9. Alder NN, Pockman WT, Sperry JS, Nuismer S (1997) Use of centrifugal force in the study of xylem cavitation. J Exp Bot 48:665–674
10. Pittermann J, Sperry JS (2006) Analysis of freeze-thaw embolism in conifers. The interaction between cavitation pressure and tracheid size. Plant Physiol 140:374–382
11. Mercury L, Tardy Y (2001) Negative pressure of stretched liquid water. Geochemistry of soil capillaries. Geochimica et Cosmochimica Acta 65:3391–3408
12. Janssen KGH, Eijkel JCT, Tas NR, de Vreede LJ, Hankemeier T, van der Linden HJ (2011) Electrocavitation in nanochannels. In: Proceedings MicroTAS, W2E. Seattle WA, USA
13. Duan C, Karnik R, Lu M, Majumdar A (2012) Evaporation-induced cavitation in nanofluidic channels. Proc Natl Acad Sci 109:3688–3693
14. Roedder E (1967) Metastable superheated ice in liquid-water inclusions under high pressure. Science 155:1413–1417
15. Shmulovich KI, Mercury L, Thiéry R, Ramboz C, El Mekki M (2009) Experimental superheating of water and aqueous solutions. Geochimica et Cosmochimica Acta 73:2457–2470

Chapter 11
Evaporation-Induced Cavitation in Nanofluidic Channels: Dynamics and Origin

Chuanhua Duan

Abstract In this chapter, we report a new mode of heterogeneous cavitation, i.e. evaporation-induced cavitation in water-filled hydrophilic nanochannels under enormous negative pressures. The liquid menisci in nanochannels are observed to be pinned at the entrance while vapor bubbles form and expand inside during the evaporation. It is observed that the growth rate of the vapor bubbles is controlled by water evaporation at the channel entrance, which is actually significantly enhanced due to absence of vapor diffusion along the nanochannel. We also report previously unexplored bubble nucleation, growth, stability, translational symmetry and dynamics that seem to be unique at the nanoscale.

11.1 Introduction

Water is known to be able to maintain metastable liquid state when its pressure is below the corresponding saturation vapor pressure or even becomes significantly negative (liquid in tension) [1–4]. However, if the negative pressure is large enough and/or there are available nuclei, vapor bubble will form and expand inside the liquid, a well-known phenomenon referred to as cavitation. Although the real rupture of liquids, i.e. homogeneous cavitation, is hard to achieve in general, heterogeneous cavitation has been widely seen in various areas and scales For example, heterogeneous cavitation has been observed in bulk liquids near rotating ship propellers and other hydraulic machines due to centrifuge force-induced negative pressure and the presence of dissolved gas [2]. It has also been reported in plant xylems (water transport microchannels) due to large evaporation-induced negative pressure [1–4].

C. Duan (✉)
Department of Mechanical Engineering, Boston University, 110 Cummington Street, Boston, MA 02215, USA
e-mail: duan@bu.edu

L. Mercury et al. (eds.), *Transport and Reactivity of Solutions in Confined Hydrosystems*, NATO Science for Peace and Security Series C: Environmental Security, DOI 10.1007/978-94-007-7534-3_11, © Springer Science+Business Media Dordrecht 2014

Despite these abundant observations in macro and microscale, heterogeneous cavitation in nanoscale confined liquids, where surface tension offers large negative pressure, has not been reported. In contrast, it has been claimed that heteorogeneous cavitation is unlikely to occur in nanofluidic channels as the required critical size of the nucleation site has to be equal or larger than the smallest dimension of the channel [5, 6]. Here we show that this argument is not necessarily true and heterogeneous cavitation indeed can occur in certain nanochannels during an evaporation process. The resulting enhanced evaporation and dynamics of such cavitation are studied and quantitatively compared with several analytical models.

11.2 Experimental Section

This cavitation phenomena were observed in nanochannel devices fabricated using sacrificial layer releasing method [7, 8]. As show in Fig. 11.1, 1-D silica nanochannels are buried underneath a microchannel (1 mm long, 40 μm deep and 500 μm wide). All nanochannels are 120 μm long, 4 μm wide. The SEM cross-sectional view indicates that these nanochannels have a small expansion at both entrances: the nanochannel height at the channel entrance was about 20 nm larger than that in the centre and gradually decreased to the centre value within a few microns. Such local expansions were produced by SF6 plasma etching that was used to remove native oxide of the sacrificial polysilicon layer before it was released in XeF2 gas. Since SF6 not only etched silicon dioxide but also etched silicon, the channel height

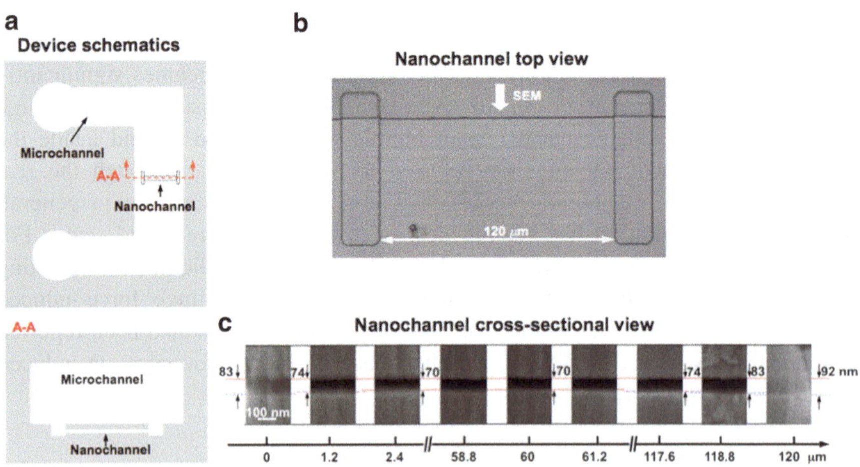

Fig. 11.1 Embedded nanochannel device. (**a**). Schematic of the nanochannel device. (**b**). Microscopic image of ten embedded nanochannels. (**c**). SEM cross-sectional view of a nanochannel

close to the entrance was thus larger than the rest of the channel. Although the nanochannel height can be tuned by KOH etching, such local expansions remained in all nanochannels with a central height from 20 to 120 nm.

In a typical experiment, deionized (DI) water was first introduced into nanochannels through the microchannel. The device was then left in the atmosphere for evaporation to occur. The drying process was recorded using a CCD camera mounted on a Nikon TE 2000-U inverted microscope in the bright-field mode.

11.3 Experimental Results and Discussion

11.3.1 Cavitation Phenomena

In these nanochannels with local expansions at both entrances, it was observed that the menisci were pinned at the channel entrance and a small bubble emerged at one end of the channel. This bubble then rapidly propagated toward the center of the channel until it stopped after a short time period in a certain position between the emerging end and the center. Once the bubble become stationary, it started gradually growing toward both ends of the nanochannel. During this process, if one end of the bubble reached the center of the nanochannel, the bubble tended to break into two, both of which then occupied in symmetric locations in the nanochannels. These bubbles then continuously expanded inside the nanochannel until they reached at the entrance of the channel.

Such bubble appearance/expansion during water evaporation is totally different from a typical evaporation process in micro/nanochannels, where menisci recede from the channel entrance till water is evaporated. Interestingly, the occurrence of this phenomenon also did not depend on the temperature, ambient pressure and gas concentration in water. Same phenomena were observed using degassed water at various temperatures ($25 \sim 70$ °C) and under varied pressure (20 Pa ~ 1 atm) (Fig. 11.2).

11.3.2 Cavitation Origin

There are many questions raised by the above observations. The most intriguing ones are where the initial bubbles come from and whether this process is a cavitation process.

We found that the appearance of the initial bubble and the subsequent bubble expansion are intimately connected with the local expansions at the nanochannel entrances. In nanochannels without local expansion, bubbles were never observed and menisci were seen receding from one side of the nanochannel till they reached the other side. In nanochannels with local expansion, high-speed imaging revealed

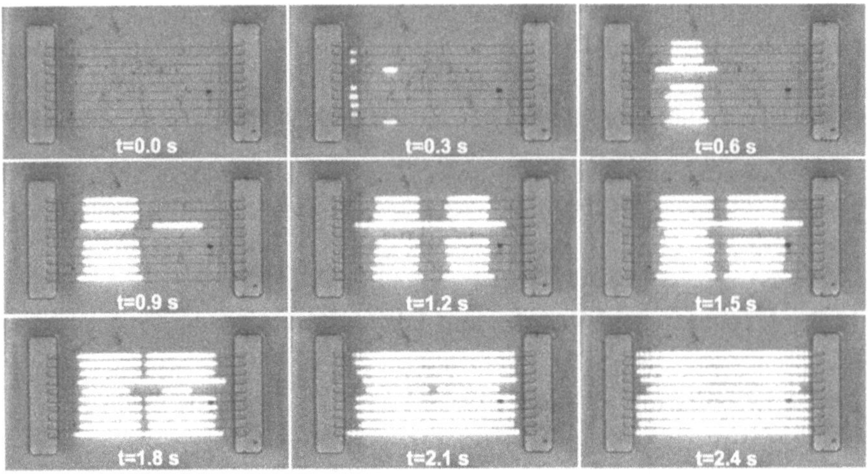

Fig. 11.2 Cavitations in 41-nm nanochannels during a evaporation process. Instead of menisci recession, vapor bubbles occurred at the *left* entrance and then moved toward the *center* of the channel before starting expansion. There are two symmetric stationary positions in each channel that bubbles prefer to stay

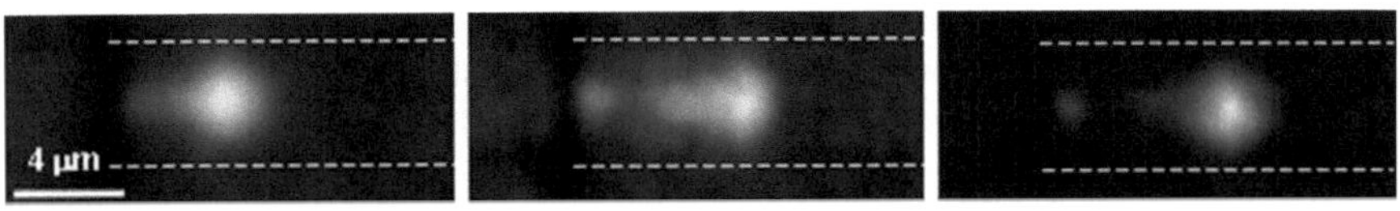

Fig. 11.3 Trapped gas bubble at the nanochannel entrance during evaporation

that the meniscus first receded a short distance (up to several microns) into the nanochannel. Afterwards, water thin film around the meniscus then remerged at the channel entrance (see Fig. 11.3), entrapping a bubble. It is still not clear at this point how the local expansion caused water to remerge (possibly due to Plateau-Rayleigh instability and capillary condensation [9, 10]), but the trapped bubble definitely has dimensions comparable with the channel feature size, and thus can be served as a stable nucleus.

The following bubble expansion is believed to be facilitated by the negative pressure of water during evaporation. The small radius of the menisci at the channel entrance results in a pressure drop $\sim {}^{2\sigma}\!/_{h}$ at the liquid/air interface [5, 11], where σ is the surface tension of water and h is the nanochannel height. The corresponding water pressure in 20–120 nm 1-D confined nanochannels varies from -7 to -1 MPa. Water is metastable at such negative pressures, which may trigger heterogeneous cavitation and result in bubble expansion with the entrapped bubbles acting as the nuclei [3]. In fact, entrapped bubbles have been considered as one of the four possible mechanisms of cavitation introduction [3, 12]. To further ensure this bubble expansion process is a cavitation process, a control experiment was performed

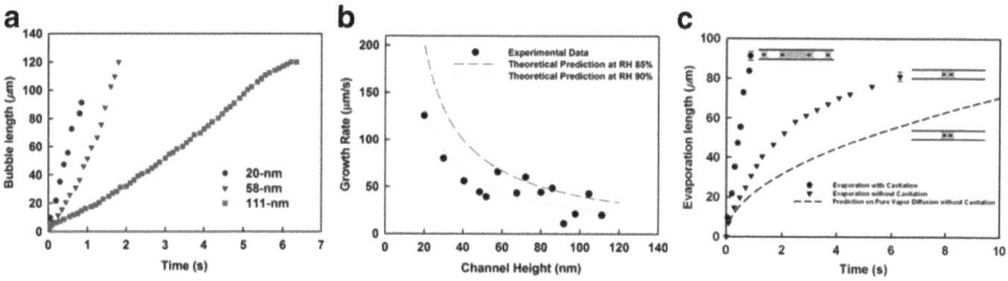

Fig. 11.4 Vapor bubble growth rate (water evaporation rate) in the cavitation process. (**a**) Observed linear bubble growth rate in 20, 58 and 111 nm nanochannels (**b**) Experimental results and theoretical prediction of bubble growth rate in nanochannels with heights ranging from 20 to 120 nm. (**c**) Enhanced evaporation rate due to the presence of cavitation

where water was re-introduced into the system before bubbles reached the end of the nanochannel. Since the re-introduction of water removed the evaporation-induced menisci at the nanochannel entrance, water should no longer be under negative pressure and vapor bubbles should collapse. This was indeed what we observed, i.e. all bubbles disappeared immediately after water introduction, suggesting that the final bubble content is water vapor and this bubble expansion process is indeed evaporation-induced cavitation.

11.3.3 Bubble Growth Rate and Enhanced Evaporation

Another interesting aspect regarding this cavitation phenomenon is the bubble growth rate. Figure 11.4a plots the total bubble length as a function of time for 20-nm, 58-nm and 111 nm nanochannels. Clearly, the total bubble length increases linearly with time in all three channels and the bubble growth rate decreases with the increasing channel height. This linear growth behavior and its height dependence were further confirmed in all other nanochannels that have been investigated in this work (see Fig. 11.4b). Such linear growth behavior suggests that the bubble growth is not governed by conventional momentum-driven theory [13]; otherwise a gradual increase of growth rate should be seen as there are less and less water needs to be pushed towards the entrance. More likely, it is determined by water evaporation at the channel entrance. In fact, although bubble expansion indicated that water was continuously vaporizing in the bubble/water interface, most of the disappeared water should still evaporate out from the entrance because the vapor bubble can only hold at most 0.002 % of the disappeared liquid water due to the density difference when menisci were pinned at the entrance. Consequently, the bubble growth rate is the same as the water evaporation rate at the nanochannel entrance.

The evaporation processes can be treated as a steady-state vapor diffusion problem using the concept of diffusion resistances [14]. Since the vapor does not need to diffuse along the nanochannel, which is the case when mensci recede, the dominant diffusion resistance is actually the one along the microchannel. Consequently, the evaporation rate can be expressed as a function of channel geometry

$$U = \frac{4Dw_m h_m \rho_v (\phi_n - \phi_\infty)}{N l_m w_n h_n \rho_w} \tag{11.1}$$

where D is the diffusion coefficient of water vapor in air; w_m, h_m, and l_m are the microchannel width, height and length, respectively; w_n and h_n are the nanochannel width and height; N is the total number of nanochannels; ρ_v is the saturated water vapor density and ρ_w is the liquid water density; ϕ_n is the relative humidity at the nanochannel entrance and ϕ_∞ is the relative humidity of ambient atmosphere. Equation 11.1 indicates that bubble growth rate (water evaporation rate) U is a constant for a given geometry and humidity condition. Furthermore, it clearly shows that U is inversely proportional to the height of the nanochannel, $U \propto 1/h_n$. Figure 11.4b also plots the corresponding theoretical prediction of the evaporation rate based on Eq. 11.1. The quantitative agreement between the experimental data and the theoretical analysis indicates that this bubble expansion process is indeed governed by evaporation at the nanochannel entrances.

It is worth noting that although evaporation limits the bubble growth rate, it is actually significantly enhanced by the presence of this cavitation phenomenon. Figure 11.4c shows the total evaporation length as a function of time in 20-nm nanochannel device with/without cavitation. Clearly, there are totally two different time dependence in these two cases. In the nanochannel where cavitation was absent, although evaporation occurred faster than the theoretical prediction based on pure vapor diffusion as a result of water thin film flow and/or corner flow, the total evaporation length still showed $L \propto \sqrt{t}$ [15]. Compared with this square-root time dependence, the linear time dependence in the nanochannel with cavitation resulted in much fast evaporation, which may provide a new mechanism to explain fast drying of nanoporous materials.

11.3.4 Bubble Dynamics

Besides bubble growth rate, it is also of interest to understand the bubble motion and stationary locations during the cavitation process. Apparently there are two symmetric locations on either side of the central line which the bubbles prefer to move to. Such stationary positions also showed a height-dependence (Fig. 11.5) but they did not depend on the evaporation rate and local humidity conditions. We observed exactly same stationary positions in 70-nm nanochannels even when the evaporation rate changed three times due to change of ambient humidity.

The occurrence of these stationary locations suggests that there are two competing motions. One of them definitely stems from the evaporation-induced

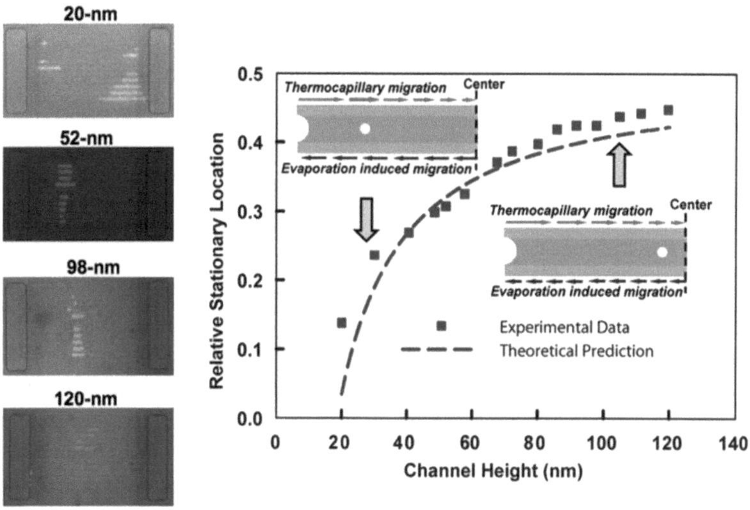

Fig. 11.5 Bubble stationary locations in 120-μm-long nanochannels

hydrodynamic flow, but this one can only drive the bubbles towards the entrance. Therefore, there must be another motion that drives the bubbles towards the center of the nanochannels. This motion also must have the same dependence on evaporation rate as evaporation-induced hydraulic flow. Otherwise, the overall evaporation effect cannot be cancelled out and the final stationary position will depend on the evaporation rate. These requirements regarding the motion direction and evaporation-rate dependence exclude all motion mechanisms based on surface or geometry-induced Marangoni flows, but leave temperature-induced thermocapillary flow as a possible explanation. We thus examined whether the balance between thermocapillary flow and evaporation-induced hydraulic flow could predict the stationary location. A 1-D steady state heat transfer equation was used to estimate the temperature distribution along the nanochannel. It turns out there is roughly a 0.03 k temperature difference between the channel entrance and the center. The corresponding temperature gradient $\frac{dT}{dx} = \frac{(l_n - 2x)}{l_n} \frac{U \rho_w h_n h_{latent}}{k_o d_o}$ reaches its maximum value (1,000 K/m) at the channel entrance and linearly decreases to zero at the center of the channel. Such a temperature difference around the bubble can generate non-uniform surface tension, yielding a net water flow toward the cold side (channel entrance) and thus pushing the bubble toward the hot side (channel center). For the circular-disk-shaped bubble, the corresponding thermocapillary flux is $J_{Thermocapilary} = C \frac{r}{\mu} \frac{d\sigma}{dT} \frac{dT}{dx} w_n h_n$, where C is a geometry-related constant, μ is the viscosity of water, r is the radius of the bubble and T is the temperature. Note this flux changes with the bubble location because of the varying temperature gradient. Ideally, such water flow would push the bubble all the way to the center of the channel if there is no water loss at the channel entrance. However, water continuously evaporates there with a flux $J_{Evaporation} = Uw_n h_n$. If the thermocapillary flux exactly compensates evaporation-induced flux at a certain location, the bubble will stop moving. The stationary location position thus can be predicted by

$$X = \frac{1}{2}\left(1 - \frac{\mu k_o d_o}{C r \rho_w h_{latent} h_n \frac{d\sigma}{dT}}\right) \tag{11.2}$$

where k_o is the thermal conductivity of silica, d_o is the thickness of the capping silica layer and h_{latent} is the latent heat of water evaporation. $X = {}^{x}\!/_{l_n}$, is the relative stationary position with $X = 0.5$ the center of the channel.

Equation 11.2 clearly shows that the predicted stationary position is only related to the device geometry and material properties. In other words, for a given nanochannel, such stationary positions are pre-determined, consistent with the experimental results. Assuming a confined bubble with a diameter of $2\,\mu$m, theoretical predictions based on Eq. 11.2 even show excellent agreement with the experimental results (Fig. 11.5). Therefore, it can be concluded that the bubble motion and stationary position indeed result from a balance between a constant evaporation-induced hydrodynamic flow and a thermocapillary flow. As both flows actually originate from evaporation, this self-controlled motion is rather amazing.

11.4 Conclusion

In summary, evaporation-induced cavitation has been directly observed in transparent nanofluidic channels in the presence of large negative pressure. We found that local expansions at the nanochannel entrances result in entrapped bubbles that act as nuclei for such cavitation. The growth rate of vapor bubbles during this cavitation process is determined by water evaporation at the nanochannel entrance, but water evaporation in nanochannel actually benefits from the presence of the cavitation and exhibits a much faster evaporation rate than evaporation without cavitation. There are certain stationary positions for vapor bubble inside the nanochannel, which can be explained by two competing flows driven by evaporation. Our visual study combined with theoretical analysis can provide insight into cavitation in plants and fast drying of micro/nanoporous materials.

Note: This work is a revision of a previous paper titled "Evaporation-induced cavitation in nanofluidic channels", which has been published in Proceedings of the National Academy of Sciences 109, 3688–3693, 2012.

References

1. Dixon HH (1914) Transpiration and the ascent of sap in plants. Macmillan and Co. Limited, London
2. Trevena DH (1987) Cavitation and tension in liquids. A. Hilger, Bristol/Philadelphia
3. Tyree MT (1997) The cohesion-tension theory of sap ascent: current controversies. J Exp Bot 48:1753–1765
4. Tyree MT (2003) The ascent of water. Nature 423:923–923
5. Tas NR, Mela P, Kramer T, Berenschot JW, van den Berg A (2003) Capillarity induced negative pressure of water plugs in nanochannels. Nano Lett 3:1537–1540
6. Zhang RJ, Ikoma Y, Motooka T (2010) Negative capillary-pressure-induced cavitation probability in nanochannels. Nanotechnology 21:105706
7. Karnik R, Castelino K, Fan R, Yang P, Majumdar A (2005) Effects of biological reactions and modifications on conductance of nanofluidic channels. Nano Lett 5:1638–1642
8. Karnik R, Castelino K, Duan C, Majumdar A (2006) Diffusion-limited patterning of molecules in nanofluidic channels. Nano Lett 6:1735–1740
9. Iwamatsu M, Horii K (1996) Capillary condensation and adhesion of two wetter surfaces. J Colloid Interface Sci 182:400–406
10. Tuller M, Or D, Dudley LM (1999) Adsorption and capillary condensation in porous media: liquid retention and interfacial configurations in angular pores. Water Resour Res 35:1949–1964
11. Wheeler TD, Stroock AD (2008) The transpiration of water at negative pressures in a synthetic tree. Nature 455:208–212
12. Wheeler TD, Stroock AD (2009) Stability limit of liquid water in metastable equilibrium with subsaturated vapors. Langmuir 25:7609–7622
13. Kenning DBR, Wen DS, Das KS, Wilson SK (2006) Confined growth of a vapour bubble in a capillary tube at initially uniform superheat: experiments and modelling. Int J Heat Mass Transf 49:4653–4671
14. Deegan RD, Bakajin O, Dupont TF, Huber G, Nagel SR, Witten TA (1997) Capillary flow as the cause of ring stains from dried liquid drops. Nature 389:827–829
15. Eijkel JCT, Dan B, Reemeijer HW, Hermes DC, Bomer JG, van den Berg A (2005) Strongly accelerated and humidity-independent drying of nanochannels induced by sharp corners. Phys Rev Lett 95:256107

Chapter 12
Electrocavitation in Nanochannels

Daniel S. van Schoot, Kjeld G.H. Janssen, Niels R. Tas, Thomas Hankemeier, and Jan C.T. Eijkel

Abstract A novel method has been developed to cavitate aqueous solutions, which is called electrocavitation. An axial voltage is applied in a nanochannel containing an aqueous solution with a stepwise conductivity gradient. A combination of electrical and viscous forces then generates a tension in the solution which, at sufficiently low pressures, causes it to cavitate. Measurement of the current during the experiment as well as optical observation provides knowledge on the time and axial position of cavitation, after which the pressure at the cavitation position can be calculated from a theoretical model in which also the ζ-potential is inserted, which is separately determined from electroosmotic flow experiments. It is found that generally the cavitation position coincides with the position of the conductivity step. In several experiments the cavitation pressure in successive experiments on the same channel became increasingly lower, suggesting a gradual removal of cavitation nuclei from the system. We calculated that pressures as low as -1630 bar ± 10 % have been reached, close to theoretically predicted pressures for homogeneous cavitation. The platform performs reliably and can be easily controlled.

D.S. van Schoot • J.C.T. Eijkel
BIOS-Lab on a Chip Group, Mesa+ Institute for Nanotechnology, University of Twente, Enschede, The Netherlands
e-mail: j.c.t.eijkel@utwente.nl

K.G.H. Janssen • T. Hankemeier
Department of Analytical Biosciences, Leiden/Amsterdam Centre for Drug Research (LACDR), Leiden University, Leiden, The Netherlands

N.R. Tas (✉)
Department of Transducers Science and Technology, MESA+ Institute for Nanotechnology, University of Twente, Enschede, The Netherlands
e-mail: n.r.tas@utwente.nl

L. Mercury et al. (eds.), *Transport and Reactivity of Solutions in Confined Hydrosystems*, NATO Science for Peace and Security Series C: Environmental Security,
DOI 10.1007/978-94-007-7534-3_12, © Springer Science+Business Media Dordrecht 2014

12.1 Introduction

The cavitation of water is a subject of considerable interest for water transport in confined spaces such as in soil and trees and thus is studied both theoretically and experimentally [1]. Water drainage from soil for example can be facilitated by cavitation events [2]. Thermodynamics predicts pure water to become unstable and cavitate at the spinodal pressure, which at 300 K is estimated to lie between −1,500 and −2,000 bar [1, 3]. Experimentally, the spinodal pressure cannot be reached since cavitation occurs at lower pressures. In general during experiments both heterogeneous and homogeneous cavitation can occur. Homogeneous cavitation occurs in the water itself by loss of cohesion between its molecules, and thus can provide information on the spinodal pressure. The homogeneous cavitation pressure expected in a certain experiment can be calculated by assuming that cavitation nuclei are continuously generated by thermal movement and that after nucleation surface tension strives to close the nucleus while the pressure difference between nucleus and water strives to expand it. Cavitation then occurs when the latter prevails. In such calculations the homogeneous cavitation pressure of pure water depends on temperature, observation time and volume and at 300 K lies between −1,400 and −1,900 bar [4]. Heterogeneous cavitation on the other hand occurs at the interface of water and surfaces such as container walls, or at the interface of water and impurities such as particles or dissolved molecules. When it occurs it makes it impossible to obtain the homogeneous cavitation pressure. The properties of the surface or the impurity are here of central importance. Most often pre-existing bubbles are thought to act as nucleation sites, which can for example be stabilized in surface cracks or by layers of adsorbed substances [1].

In cavitation studies metastable water is created either by superheating or by stretching [1]. Mainly three experimental platforms exist for the study of cavitation by stretching: centrifugation of a tube filled with liquid perpendicular to its axis [5], an acoustic wave generated by a piezo-electric transducer [3] and heating of water inclusions in quartz inclusions followed by cooling until cavitation occurs [6]. The former two methods quite reproducibly yield cavitation pressures around −260 bar at 0 °C, while the latter method shows much less reproducibility but has yielded cavitation pressures down to −1,400 bar at 40 °C, close to the theoretical estimates [5]. To explain the discrepancy between the different methods, ubiquitous stabilizing impurities have been assumed for the quartz inclusion experiments (such as hydroxyl ions) or ubiquitous destabilizing impurities for the other methods (such as hydronium ions) [7]. Shmulovich et al. explicitly studied the (de)stabilizing influence of different electrolytes on the cavitation pressure of water in quartz inclusions [8].

Recently we introduced a new platform to generate negative pressures which we called electrocavitation [9]. The method relies on the generation of strong electrical and restoring shear forces in an aqueous electrolyte solution filling a nanometer-scale channel. Since our experimental procedure relies on the presence of dissolved ions, no data on pure water can be obtained. The procedure however is eminently useful to investigate cavitation of aqueous salt solutions and especially the influence of different salts. Since the procedure furthermore is rapid, a sequence of experiments can be performed on the same channel yielding additional valuable information.

12.2 Experimental Section

Channels of nanometer scale depth ('nanochannels') were manufactured in the cleanroom of the University of Twente in borofloat glass wafers using lithography and wet HF etching. The height of the channels was determined using a mechanical surface profiler (Dektak 8, Veeco Instruments Inc. Plainview, USA) and found to be uniform over the entire wafer at approximately 53 nm. The channel wafer was thermally bonded to a second wafer and annealed. Subsequently through holes were supersonically drilled. Each wafer contained 19 single channels, six 20 μm wide, seven 10 μm wide and six 5 μm wide. The channel length was 35 mm and all channels had rulers and a unique identifier. In this study only the 10 μm wide channels were utilised.

A custom-built chip interface provided tubing inlets from the top and below, aligned with the nanochannel access holes (Fig. 12.1). The table with the interface

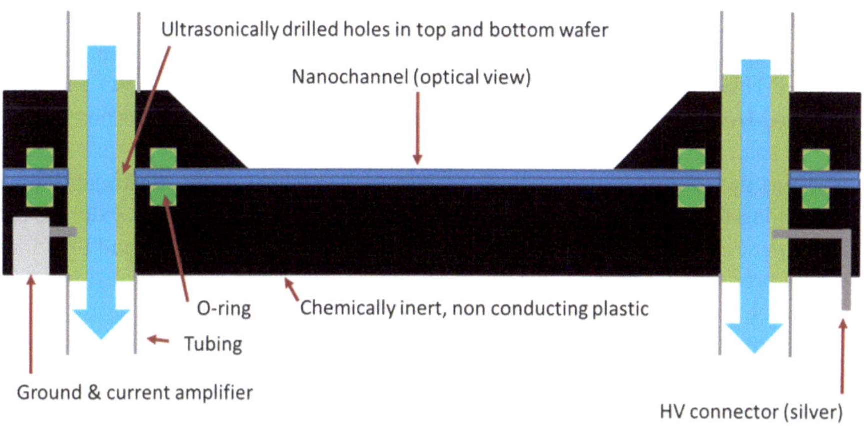

Fig. 12.1 Cross section of the interface holding the wafer. The wafer can be aligned such that one nanochannel can be selected, as is the case in the picture. The drilled holes were then aligned with the tubing through which liquid can be pumped in the direction indicated by the *blue arrows*. The liquid can make its way into the nanochannel by either capillary filling or by applying a voltage between the High Voltage (*HV*) connector (*on the right*) and ground (*on the left*), creating electroosmotic flow (*EOF*). These two connectors were connected to the power source by a coax-cable. The nanochannels can be optically viewed from above as indicated by the *arrow*. Elastic O-rings were manually placed into the setup to prevent any leakage. Silver/silver chloride electrodes were employed for voltage application and current measurement

Table 12.1 The solutions used in the cavitation experiments

Salt used	Concentration (mM)	Ionic strength (mM)	Approximate conductivity (mS/cm)
NaHEPES	5	5	0.39
NaCl	9	9	1.35
	10	10	1.35

can be moved to optically determine the exact position of cavitation. The liquids were pumped through the wafer through-holes by a syringe pump (Harvard Apparatus, USA). Voltages between −3 and +3 kV were applied using a HSV488 6000D power supply (LabSmith inc., Livermore, USA) controlled by a LabView script. This same script was used to start and stop measurements. Data was gathered at the ground electrode, which is connected to a current-to-voltage (IV) converter and then to a DAC-PCI card at the PC. This IV converter converts the current into a voltage and amplifies it by a factor 10^8 for the PCI-card. A specific voltage source provides the IV converter circuit with the necessary voltages (+2.5 and −2.5 V). The PCI-card has an input range of about −2.2 to +2.2 V, which translates to a current between −22 and +22 nA. Digital (pre)processing and analysis of the data was performed using Matlab (MathWorks Inc., Natick, USA). The syringes, tubing and waste containers were extensively shielded and connected to ground. An Olympus BX51WI microscope was used for optical inspection with bright field and fluorescence imaging, equipped with a long pass filter cube (488 nm excitation, 518 nm emission), using a 10X magnification objective with a numerical aperture of 0.25 (Olympus Corporation, Tokyo, Japan). Images and movies were recorded using a Hamamatsu Orca-ER CCD camera and included HOKAWO software (HAMAMATSU Photonics K.K., Japan) for reference and checking purposes.

The solutions used are shown in Table 12.1. We chose both salts to have sodium (Na^+) as the cation, so that a single counterion is present in the electrical double layer along the channel axis and the axial uniformity of the ζ-potential during the experiment is increased. The ζ-potential will generally be <-25 mV and therefore will be proportional to the square root of the ionic strength which is also given in Table 12.1.

The proper functioning of the current measurements was checked with high resistance, high voltage resistors (Hymeg Corporation, Addison, USA). To determine the ζ-potential, the electroosmotic flow (EOF) velocity was determined for every channel prior to cavitation experiments, using the same salt solutions. By applying a relatively low voltage (under 500 V), a low conductivity solution was replaced by a high conductivity solution. From the time needed for the current to increase to a new plateau value the EOF velocity can then be derived, which yields the ζ-potential [10].

A schematic overview of a typical measurement cycle is shown in Fig. 12.2. After the initial filling of the channel from the high voltage side by capillary action (2A and B), both through holes were flushed with their respective solutions. After this initial filling or a cavitation experiment the channel was filled entirely by EOF with

Fig. 12.2 Experimental procedure for one cycle of the electrocavitation experiments (see text)

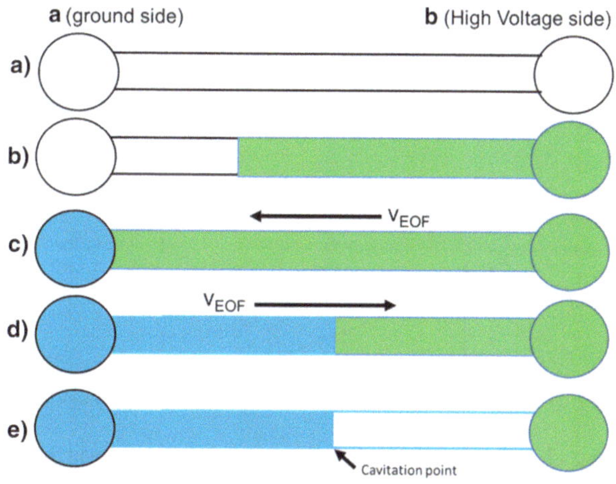

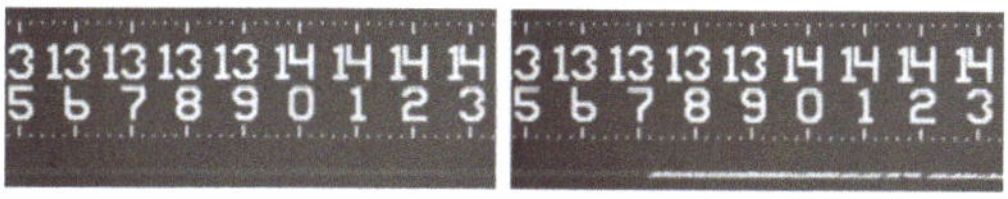

Fig. 12.3 A cavitation event. In this case the low conductivity fluid is drained to the right. The difference between the two pictures is one frame of 0.2 s. The cavitation front stays in place at 13.72 mm. This position can be compared to the position of the interface between the two solutions of different conductivity which can be theoretically predicted from the ζ potential and applied voltage

low conductivity solution by applying 500 V over the channel for 60 or 90 s (2C). The voltage was then lowered to 0 V for 20 s and liquid was pumped through the through holes. The voltage was subsequently reversed to −1,000 V, −2,000 V or −3,000 V for 60 s. The high conductivity solution will now move into the channel by EOF (2D), causing the pressure to decrease and leading to cavitation (2E). Cavitation was detected both optically and electrically. The cavitation time was taken directly from the current measurements since a substantial current decrease was observed. The position was also determined optically as usually one of the solution fronts remained in place. The voltage was then turned off again for 20 s and the channel started to refill by capillary action while fresh fluid was pumped through the through holes. The cycle from C to E was then repeated by setting the voltage to 500 V again for 60 or 90 s. Each experiment consisted of four cycles.

12.3 Experimental Results and Discussion

In each channel, after determining the ζ-potential, electrocavitation experiments were performed. Two subsequent recorded frames of a cavitation event are shown in Fig. 12.3.

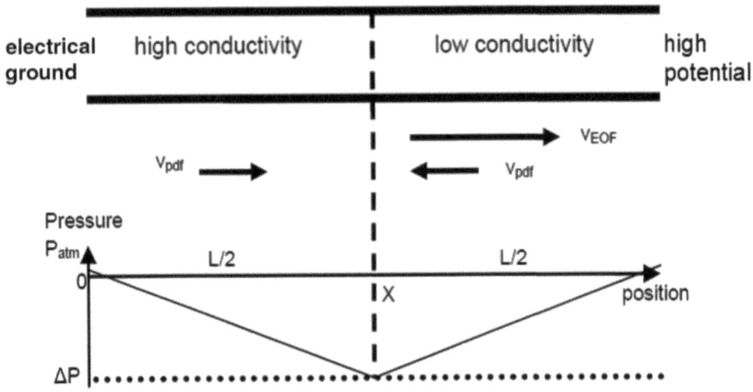

Fig. 12.4 Schematic drawing of a nanochannel containing two adjacent electrolyte solutions, each occupying half of the channel. The difference in conductivity between both solutions is assumed to be very large, limiting the axial electrical field to the low-conductivity half where it generates EOF. To assure a constant axial volume flow (as imposed by mass conservation), a pressure gradient is generated as indicated, generating flow against the EOF at the right-hand side and with the EOF at the left-hand side (*arrows*), and creating a negative pressure in the channel

The cavitation generation mechanism is thought to be the following. When a nanochannel is filled with adjacent solutions of high and low conductivity (Fig. 12.4), an applied axial potential difference will drop predominantly over the low-conductivity solution, inducing a locally stronger EOF. To satisfy a continuous mass flow a pressure gradient arises, generating liquid flow against the EOF in the low-conductivity part and with the EOF in the high conductivity part. Due to the high hydrodynamic resistance of a nanochannel this pressure gradient is very large, resulting in a very low pressure at the interface between the two solutions.

An order of magnitude estimate of the generated negative pressure is obtained by assuming that each electrolyte fills half of the nanochannel and that the potential drop over the high-conductivity part can be neglected. It then follows from the equations of pressure-driven flow and EOF [11] that the generated pressure drop, ΔP, from the reservoirs to the midpoint in the channel is

$$\Delta P = -\frac{6\epsilon\zeta V}{h^2}$$

where ε (F/m) is the dielectric constant of the solution, ζ (V) the zeta potential, V (V) the applied potential difference and h (m) the channel height. For $\varepsilon = 7.5 \cdot 10^{-10}$ F/m, $\zeta = -50$ mV, h = 50 nm and V = −3,000 V a negative pressure of −2,750 bar results. This very low value indicates the potential of this method for liquid cavitation studies. Using a similar but more elaborate theory, we theoretically predicted the time dependent pressure profile in the channel in the remainder of this chapter.

A representative example of an electrocavitation current measurement at 1 kV is shown in Fig. 12.5, and shows one of the four cycles of which an experiment

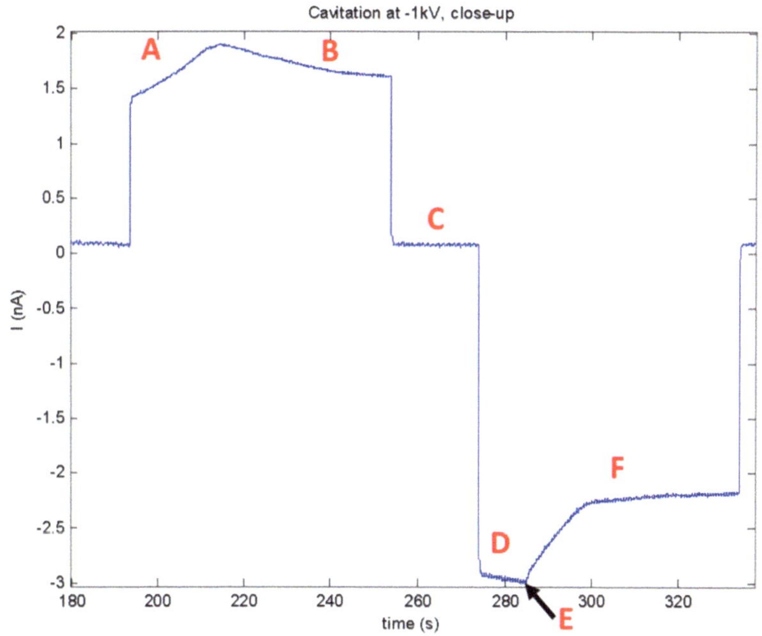

Fig. 12.5 One cycle of an electrocavitation measurement at −1 kV. See the text for the annotated characters A–F

consists. The applied voltage is positive at first, so the NaHEPES solution will start refilling the emptied (from a previous cavitation event) channel both by capillary forces and EOF. This causes the current to rise as the conductivity greatly increases from very low (i.e. empty channel) to that of the NaHEPES solution (A and Fig. 12.2A and B). At some point, depending on where the cavitation in the previous experiment took place, the channel will be fully filled with solution again. Now the EOF starts to displace the NaCl solution with NaHEPES solution. This lowers the current since NaHEPES has a lower conductivity than NaCl. (B) As the whole channel gets filled with NaHEPES, the current stabilises. The voltage is then turned off to pump and refresh the two solutions through the holes (C). Next, −1 kV is applied over the channel. The channel starts filling with NaCl solution again, making the current rise (D and Fig. 12.2D). At some point the pressure can become so low that the solution cavitates (E and Fig. 12.2E). The channel drains and a residual electrical current remains, most likely due to surface conduction (F). The voltage is then turned off and the channel fills by capillary action while new solution is pumped through the holes. The cycle then starts again.

The cavitation position is noted as the location where the liquid meniscus is located after cavitation. Numerous optical observations confirmed that this is the same place as where cavitation takes place, within a distance of maximally 50 μm, which defines the error on the cavitation position. The cavitation time from the

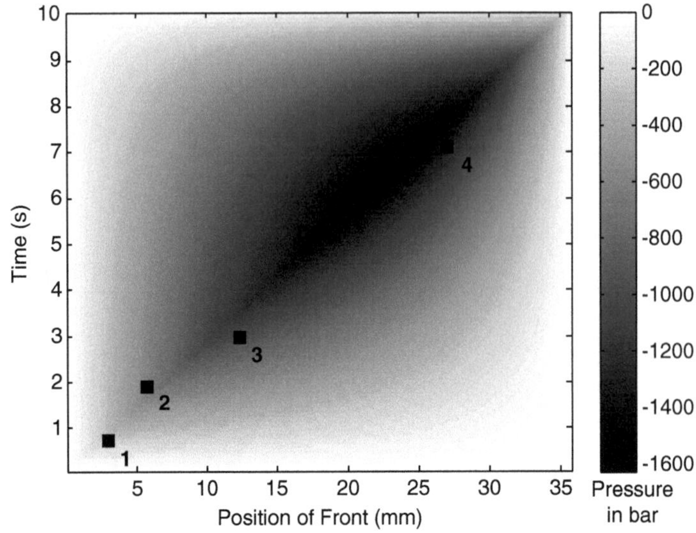

Fig. 12.6 The theoretical axial pressure in the nanochannel (*horizontal axis*) against time (*vertical axis*), generated using an experimentally determined ζ-potential of −60 mV, an applied voltage of −3 kV and solution conductivities of $K_{high} = 0.135$ S/m and $K_{low} = 0.0387$ S/m and measured channel height of 53 nm. The number next to each point represents the cycle number. The pressures at the cavitation locations were (from *left* to *right*): −280, −720, −1,050 and −1,320 bar. Due to an uncertainty in the ζ-potential of 10 % they can vary by 10 %

electrical current measurements is slightly influenced by the 50Hz-filter used and reading errors and is taken as 0.1 s.

Several subsequent four-cycle experiments were performed in three different channels at three applied voltages: −1, −2, and −3 kV, and the cavitation positions and times were recorded. In some channels cavitation always occurred at approximately the same position. In other channels the cavitation position in subsequent cycles or experiments moved to positions further down the channel, whereby a linear relationship was found between cavitation position and time. The first observation suggests there may be local fixed anomalies in the channels such as discontinuities in the ζ-potential caused by surface defects or particles. The second type of observations suggests anomalies that can be removed or displaced on cavitation, such as nanobubbles [12] or mobile particles.

Using the ζ-potential retrieved from the separate EOF measurements a prediction of the movement of the front between both solutions can be made, as well as of the spatiotemporal pressure distribution in the channel. Figure 12.6 plots these data, with superimposed the results of the cavitation experiments. In the four cycles recorded it seems that the cavitation location approximately coincides with the theoretical position of the front. This implies that the cavitation takes place at the point of lowest pressure which is the interface between the two solutions moving with the same velocity as the EOF. The lowest pressure where cavitation occurs appears to be at data point number four, which corresponds to a pressure of −1,320 bar ± 10 %, based on a measured ζ-potential of −60 mV ± 10 %. The errors on the time and position also influence the exact value by a few tens of bar, though this error is much smaller. It is interesting that the last cavitation event (point number 4) seems to take place after a lower pressure was already reached elsewhere in the channel. This can be explained by assuming that cavitation still is heterogeneous, and that the first anomaly in this cycle is encountered at this position. In subsequent experiments at −3 kV on the same channel no more cavitation events were observed. This implies that the homogeneous cavitation pressure of this aqueous solution is below the lowest pressure reached in this experiment, which was calculated as −1,630 bar ± 10 %. Interestingly, this value is very close to the theoretical value using the thin wall approximation [3] of −1,600 bar for homogeneous cavitation in an observation volume of $\sim 10^{-13}$ m^3 and an observation time of ∼1 s [3].

12.4 Conclusion

A new method for controlled liquid cavitation is described, called electrocavitation. It represents an experimentally flexible and reliable method for liquid cavitation studies and pressures below −1,000 bar can be easily reached. Experimental results suggest interesting data can be obtained both on cavitation at liquid impurities as on homogeneous cavitation. Indications are found that the homogeneous cavitation pressure of the aqueous solution used is below −1,630 bar ± 10 %.

Acknowledgements We gratefully acknowledge the help of Raphaël Zwier (Fine Mechanical department, Leiden University), Raymond Koehler and Peter van Veldhuizen (LION, Leiden University) for their work on the chip interface, and of L.J. de Vreede (BIOS-lab on a chip group, University of Twente) and Martijn Witlox (Fine Mechanical Department, Leiden University) for chip fabrication.

References

1. Caupin F, Herbert E (2006) Cavitation in water: a review. Compte Rendu Phys 7:1000–1017
2. Or D, Tuller M (2002) Cavitation during desaturation of porous media under tension. Water Resour Res 38:1061
3. Rajamani S, Truskett TM, Garde S (2005) Hydrophobic hydration from small to large length-scales: understanding and manipulating the crossover. Proc Natl Acad Sci USA 102:9475–9480
4. Herbert E, Balibar S, Caupin F (2006) Cavitation pressure in water. Phys Rev E 74:041603
5. Briggs LJ (1950) Limiting negative pressure of water. J Appl Phys 21:721–722
6. Zheng Q, Durben D, Wolf G, Angell C (1991) Liquids at large negative pressures: water at the homogeneous nucleation limit. Science 254:829–832
7. Davitt K, Rolley E, Caupin F, Arvengas A, Balibar S (2010) Water at the cavitation limit: density of the metastable liquid and size of the critical bubble. J Chem Phys 133:174507

8. Shmulovich KI, Mercury L, Thiéry R, Ramboz C, Mekki ME (2009) Experimental superheating of water and aqueous solutions. Geochim Cosmochim Acta 73:2457–2470
9. Janssen KGH, Eijkel JCT, Tas NR, de Vreede LJ, Hankemeier T, van der Linden HJ (2011) Electrocavitation in nanochannels. Proceedings MicroTAS 2011. Seattle WA, USA
10. Huang X, Gordon MJ, Zare RN (1988) Current-monitoring method for measuring the electroosmotic flow rate in capillary zone electrophoresis. Anal Chem 60:1837–1838
11. Bruus H (2007) Theoretical microfluidics. Oxford University Press, Oxford
12. Borkent B, Dammer SM, Schonherr H, Vancso GJ, Lohse D (2007) Superstability of surface nanobubbles. Phys Rev Lett 98:204502

Chapter 13
Stability and Negative Pressure in Bulk and Confined Liquids

Attila R. Imre

Abstract Negative pressure in liquids – especially in confined systems, like capillaries – often acts as a cohesive force between the solid walls, surrounding the liquid. These forces are responsible for various processes and phenomena, like sap transport in trees or mud stability/mud slides of granular systems (like soil). Due to the metastability of the liquids under negative pressure, different properties (including the limit of stability) cannot be measured directly because the metastable state might equilibrate back to stable ones (liquid + vapour) by cavitation, before the end of the measurement. Therefore it would be crucial to have an equation of state to describe the behavior of liquids (especially for water) in this region. We are going to present some result – comparing experimental data, molecular dynamic simulations and some analytical calculations -, showing which equations could be used in the metastable region and which should be the special pre-cautions taken during their use.

When we are talking about liquid water, probably most of us visualize "bulk" water; the ocean, a lake, a river, maybe a cup of water. The bulk water is the most visible, but not necessarily the most frequent form of this liquid. One might find huge amounts of water – still in liquid phase – from the droplets forming a cloud through the confined, capillary-like spaces in the soil, down to the tiny liquid inclusions in the rocks. Even in living systems, water – although not pure one – can be in strongly confined conditions; one should remember to the sap in the micrometer-size xylem [1–4]. In arid regions – like deserts – one can find all liquid water confined into the capillaries formed between the grains of the soil/sand. Although the pure bulk water

A.R. Imre (✉)
HAS Centre for Energy Research, H-1525 Budapest, POB 49, Hungary

Institute of Physical Chemistry, University of Cologne, Luxemburger Str. 116, D-50939 Köln, Germany
e-mail: imre.attila@energia.mta.hu

L. Mercury et al. (eds.), *Transport and Reactivity of Solutions in Confined Hydrosystems*, NATO Science for Peace and Security Series C: Environmental Security,
DOI 10.1007/978-94-007-7534-3_13,

and the pure confined water are chemically the same substance, they can have very different properties. There are two main reasons for these differences:

- Within the bulk water, almost all of the water molecules are in the neighborhood of the other water molecules, while in confined systems a considerable amount of water molecules interact with the confining substance (like the wall in capillary liquids or the air or other material in droplets).
- Confined water can be easily in metastable liquid state (superheated or under negative pressure). Metastable states can be generated in bulk liquids too, but they are more frequent – and sometimes even inherent – in confined ones. This is one of the reason, while capillary water in arid regions can remain in liquid state, instead of vapour.

In this chapter, we are dealing with the second problem, namely with the metastability of liquid water. Liquid water should boil (i.e. turn to vapours) at 100 °C, but might stay in liquid state too. The liquid state (overheated in this case) will be metastable, which means that under those conditions, vapour would be energetically more preferable state for the system. Metastability can be reached either by heating (overheated states) or by decreasing the pressure. The pressure decrease for a liquid does not have to stop at zero; it can goes to the negative region (stretched liquids). In confined systems, pressure drop can be generated mainly by the presence of strongly curved liquid-vapour interface (capillaries) or by the cooling of an isochoric – constant volume – system (inclusion).

Properties of the metastable liquids differ from their stable counterparts, although the transition is not sharp. It means that stable water at $p = 1$ bar and $T = 99.9$ °C physically – density, compressibility, etc. – very similar to the metastable one at $p = 1$ bar and $T = 100.01$ °C. On the other hand, going deeper to the metastable region, the difference can be more characteristic; according to the most popular water reference equation of state (IAPWS) [5], going from 99.9 to 299.9 °C ($p = 1$ bar) the density changes from 0.9584 to 0.6900 g/cm^3, while the isothermal compressibility increases a whole order of magnitude, from $4.90*10^{-4}$ 1/MPa to $4.51*10^{-3}$ 1/MPa. Additionally, upon reaching 594.6 K (321.5 °C), we can meet a so-called spinodal, which is the limit for stability; liquid has to turn into vapour here.

There are two basic problems with metastable liquids. First, one should know the properties. It seems easy, just a few lines before we presented density and compressibility data for slightly and deeply metastable liquid water. But that easiness was just an illusion. The IAPWS equation of state (EoS) is a mathematically complex equation, mainly to fit measured data in stable liquid and vapour phases. As all fitting, it is good only in the fitted interval and lose the accuracy by leaving it [5]. One might say, that then we should measure those properties directly in the metastable region, but due to the nature of metastability, any invasive method – like inserting a pressure gauge – can cause immediate break of the metastability [1–4], therefore direct measurements are almost impossible. Therefore it is quite common to use "common" equation of states in the metastable region, although theoretically it is not a correct way. We would like to show here, what might be the problem with the "best" equation of state, the IAPWS and we will try to find some solution too.

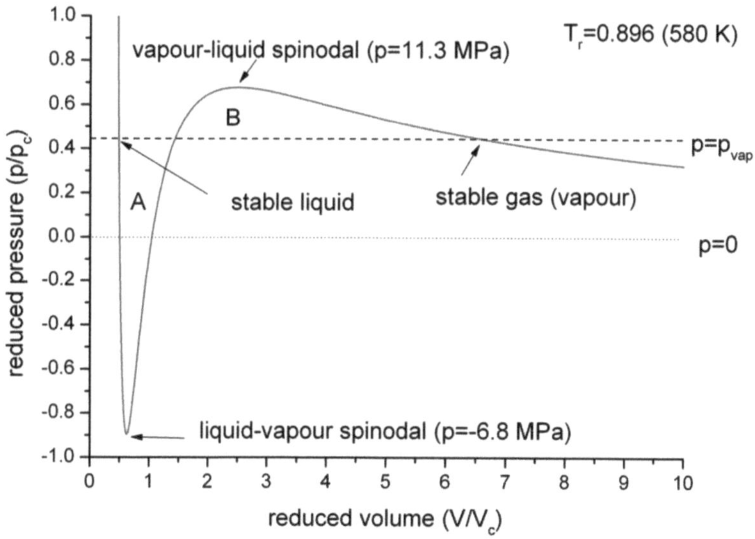

Fig. 13.1 A subcritical water isotherm calculated from the Peng-Robinson equation of state, showing two extrema (*spinodals*). Reduced quantities (*subscript r*) are calculated by dividing the actual value with the critical value

First of all, one should be familiar with the concept of "stability" in fluid – particularly in liquid – systems. An isotherm of a "traditional" equation of state can be seen on Fig. 13.1. Traditional means that the equation is cubic in volume, i.e. it might cross the zero three times and exhibit a maximum and minimum [6]. The most known cubic equation of state is the van der Waals EoS, but for water it is very inaccurate. Therefore for demonstration, we are going to use another cubic EoS, the Peng-Robinson (PR) one [6, 7]. On Fig. 13.1, one can see a so-called "van der Waals loop", describing the volume (density) dependence of the pressure. The slope of the curve is proportional with the isothermal compressibility. A system cannot be stable with negative compressibility, because for a pressure increase (pushing) it would react with volume increase (explosion). Therefore the part between the minimum and maximum describes a physically non-existent "state". The two extrema itself are the two stability limits. The maximum is the stability limit where vapour MUST turn into liquid, while the minimum is the stability limit, where liquid MUST turn into vapour; this is the interesting one for us. One should keep in mind, that the scales on the figure are reduced values; to get back the real spinodal pressure, they should be multiplied by the critical pressure of the water, which is 22.06 MPa. The stable, equilibrium conditions (calculated by the Maxwell-construction, when the A and B areas are equal) are marked by arrows. The region between the equilibrium and the spinodal represent the metastable states.

On Fig. 13.2, one can see the isotherms for the same temperature, but now not only by the PR EoS, but also by the IAPWS EoS. One can see, that the two curves are not only quantitatively, but also qualitatively are different. The two lines run

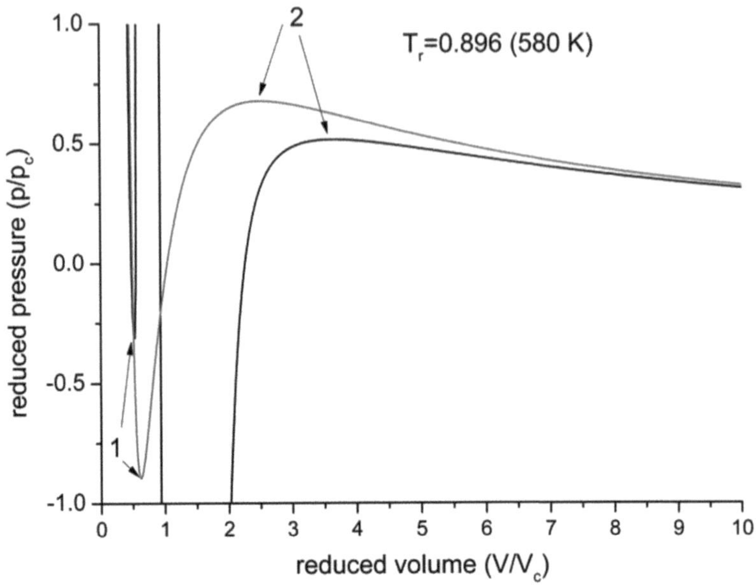

Fig. 13.2 Subcritical water isotherms calculated from the Peng-Robinson (*grey*) and IAPWS (*black*) equation of states in the same pressure and volume ranges shown in Fig. 13.1. The two extrema (marked *1* and *2*) can be seen on both curves, but on the IAPWS line one can see something else too

together only at the very beginning, as well as at the end of the scale, showing that the stable liquid and vapour phases can be described by both equation with more or less the same accuracy. The two extrema – and hence the spinodal values – by the IAPWS EoS are only half of the ones given by the PR EoS, but this is only the smaller problem. The big problem is the presence of two other extrema, which can be seen better on Fig. 13.3, covering a much wider pressure range. These two extrema can be characterized by two new spinodals, showing extreme values in the gigapascal range. Here we should mention, that at lower temperatures, these values can shift to the terapascal range [8]. These values corresponds to forces exceeding the strength of molecular interactions, therefore they are obviously artificial. The classical way to deal with these second peaks are the complete omission of them [7]. Although this is a good way to eliminate most of the problem, but one can easily see that "inserting" these huge peak-and-valley combination between the other, much smaller peak and valley (see Fig. 13.3), the curve around the small ones will be greatly distorted. Probably this is the reason, while the spinodal pressures by the IAPWS EoS are only half of the spinodal pressure by the PR EoS (see Fig. 13.4).

One might still say, that concerning the different spinodals, the IAPWS is the "good" one, because in the stable region it is more accurate, but being theoretically incorrect (having four spinodal, instead of two), this reasoning is not very convincing. Fortunately there are some more convincing reasoning. At the low-temperature

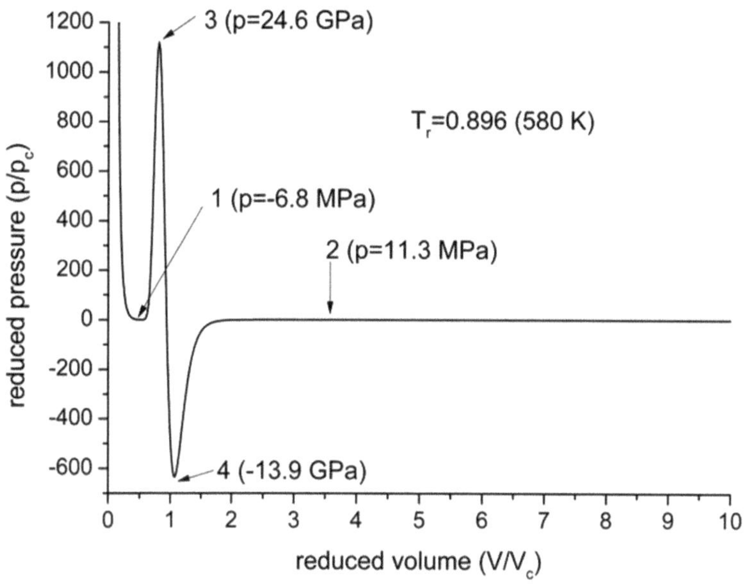

Fig. 13.3 The full subcritical water isotherm calculated from the IAPWS equation of state. One can see that this curve has four extrema; along with the two real spinodals (*1* and *2*) one can see two artificial ones (*3* and *4*), reaching pressure values in the gigapascal ranges

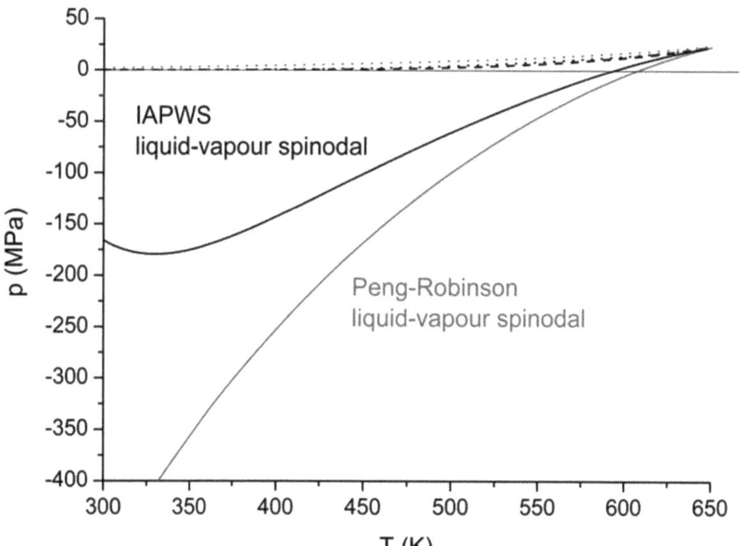

Fig. 13.4 Phase diagram of water without the solid (*ice*) phases, calculated by the IAPWS (*black*) and Peng-Robinson (*grey*) equation of states. While the vapour pressures (*dashed lines*) are more or less the same, one can see very big difference in the predicted liquid-vapour spinodals (*solid lines*). Vapour-liquid spinodals (*dotted lines*) are plotted for the completeness

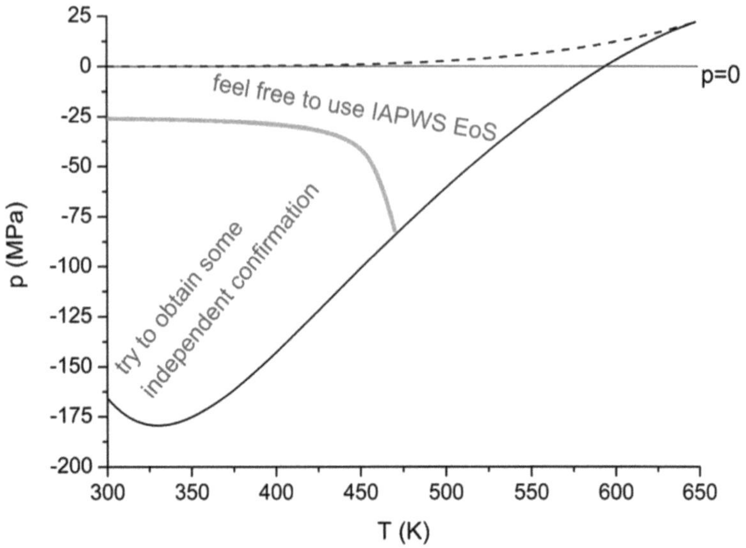

Fig. 13.5 Some advices to use the IAPWS EoS in the metastable region

region, Caupin and his co-workers – using state-of-art methods – measured some liquid properties down to −26 MPa and compared them to the ones predicted by the IAPWS EoS, finding very good agreement [9]. Using a bit less convincing, but also promising method, Imre et al [8]. used a novel molecular dynamic method [10] to obtain spinodal values from interfacial properties [11] and found quite good agreement on higher temperature, down to −100 MPa. For more confirmation, several mathematically independent EoS was compared concerning the predicted spinodal pressures, and while generally the values were very scattered, there was a group of EoS (the IAPWS one among them) giving very similar values at higher temperatures [8]. This might be explained by the fact, that at higher temperatures, the second, gigantic loop is smaller than at lower temperatures, therefore it might influence the real spinodals in lesser extent.

On Fig. 13.5 the IAWS liquid-vapour spinodal is presented (the vapour pressure is also plotted). The region, where the IAPWS EoS might be used without problem is marked. For most of the capillary systems in bio- and geo-sciences, this will cover the relevant metastability region. In nano-capillaries one might generate deeper negative pressure, but in that case the water is already so much influenced by the presence of confining wall that it cannot be considered as bulk one. In inclusions, one might reach deeper negative pressures, where the usability of the IAPWS EoS is more questionable; in that case it would be recommended to obtain some independent confirmation about the estimated spinodal values [12, 13].

In this short chapter we demonstrated that equation of states used routinely to estimate properties of metastable water might not be a sufficient tool to do that in case of deep metastability. We also tried to find the region, where the validity is still fair. Fortunately this covers the range, relevant in most biological and geological problems, including problems related to water confined into pores.

Acknowledgements The Author wish to express his appreciation to Prof. U. K. Deiters and Dr. T. Kraska for their advices and for the German Humboldt Foundation for supporting his stay in Cologne.

References

1. Debenedetti PG (1996) Metastable liquids: concepts and principles. Princeton University Press, Princeton
2. Imre AR, Maris HJ, Williams PR (eds) (2002) Liquids under negative pressure (NATO science series). Kluwer, Dordrecht
3. Trevena DH (1987) Cavitation and tension in liquids. Adam Hilger, Bristol, Philadelphia
4. Skripov VP, Faizullin MZ (2006) Crystal-liquid–gas phase transitions and thermodynamic similarity. Wiley-VCH, Weinheim
5. Wagner W, Pruβ A (2002) The IAPWS formulation 1995 for the thermodynamic properties of ordinary water substance for general and scientific use. J Phys Chem Ref Data 31:387–536
6. Deiters UK, Kraska T (2012) High-pressure fluid-phase equilibria – phenomenology and computation. Elsevier, Oxford
7. Deiters UK (2006) ThermoC Project website, http://thermoc.uni-koeln.de/
8. Imre AR, Baranyai A, Deiters UK, Kiss PT, Kraska T, Quiñones Cisneros SE (2012) Estimation of liquid-vapour spinodal of bulk water from interfacial properties, Fluid Phase Equilibria. Int J Thermophys, in press. doi:10.1007/s10765-013-1518-8
9. Davitt K, Arvengas A, Caupin F (2010) Water at the cavitation limit: density of the metastable liquid and size of the critical bubble. Europhys Lett 90:16002
10. Kiss PT, Baranyai A (2012) On the pressure calculation for polarizable models in computer simulation. J Chem Phys 136:104109
11. Imre AR, Mayer G, Házi G, Rozas R, Kraska T (2008) Estimation of the liquid–vapor spinodal from interfacial properties obtained from molecular dynamics and lattice Boltzmann simulations. J Chem Phys 128:114708
12. Mercury L, Jamme F, Dumas P (2012) Infra-red imaging of bulk water and water-solid interfaces under stable and metastable conditions. Phys Chem Chem Phys 14:2864
13. El Mekki-Azouzi M, Ramboz C, Lenain J-F, Caupin F (2013) A coherent picture of water at extreme negative pressure. Nat Phys 9:38–41. doi:10.1038/NPHYS2475

Chapter 14
Experimental Superheating and Cavitation of Water and Solutions at Spinodal-Like Negative Pressures

Lionel Mercury and Kirill I. Shmulovich

Abstract The superheated liquids are metastable with respect to their vapour, what means they can exist under arid conditions whatever the temperature: capillary liquid residing in arid soils (desert shrubs, Mars sub-surface, ...), solutions in the deep Earth crust, or water involved in rapid disequilibrium events (terrestrial or submarine geysers). The superheating state changes the solvent properties of liquids, and so modifies phase transitions (solid–liquid, liquid–vapor) P-T-X conditions. The synthetic fluid inclusion (SFI) enables to fabricate micro-volumes of hand-made liquid dispersed inside quartz, which readily superheat. Volumes of SFI are intermediate between macro-systems, in which superheating is restricted to around −30–35 MPa with very short lifetime, and nanosystems, wherein confinement effects predominate and in which the host size is similar to the one of the critical nucleus of vapour phase (huge nucleation barrier). This volume-to-metastability relationship is still to be defined quantitatively, and we are targeting to combine thermometric classical measurements with spectrometric characterizations, enabling to establish the threshold between micro- and nano-systems precisely. Meanwhile, the experiments performed so far illustrate the diversity of contexts and situations that could be modelled by superheating issues.

L. Mercury (✉)
Institute of Earth Sciences, University of Orléans, Orléans, France
e-mail: lionel.mercury@univ-orleans.fr

K.I. Shmulovich
Institute of Experimental Mineralogy, Russian Academy of Science, 142432 Chernogolovka, Russia

L. Mercury et al. (eds.), *Transport and Reactivity of Solutions in Confined Hydrosystems*, NATO Science for Peace and Security Series C: Environmental Security,
DOI 10.1007/978-94-007-7534-3_14,

14.1 Introduction

Liquid water presents a remarkable ability to stay liquid against evaporative demand by becoming superheated with respect to water vapor. Water can be readily superheated by increasing the temperature higher than 100 °C at constant 1 atm pressure, or by decreasing its internal pressure, it becomes superheated even at low temperature (Fig. 14.1a). Meanwhile, the driving force is toward evaporation and so the liquid is prone to nucleate bubbles at any time, except if some reasons, notably the trapping in a host solid, make the nucleation probability to decrease and the lifetime to lengthen (Fig. 14.1b). It is why superheating is a feature closely interwoven with capillarity and confinement, in which cases the closeness of the solid walls make the infilling liquid to behave differently than the bulk.

As a consequence, the conditions of occurrence and the features of superheated water are relevant to understand the dynamics of liquid water in arid (deserts) and hyper-arid (Mars surface) settings, and its possible volume and influence as a function of the environment. Experimentally, the most efficient way to produce superheated liquid is to carry out an isochoric cooling on a liquid trapped in a strictly closed container with rigid walls. With water, as the isochores are very steep (around 20 bars per degree) and the saturation curve is very flat, a slight cooling below liquid–vapor equilibrium puts the occluded liquid under significant superheating degree, even down to negative pressures (Fig. 14.1a). Negative pressure can be sustained by condensed matter (liquid and solid materials) which is able to undergo internal stretching or tensile stress without losing the cohesion of its structure (e.g., [4]).

Capillary water, common in non-saturated soils, has got superheated properties without its limited lifetime because the capillary channel/pore has the size of the

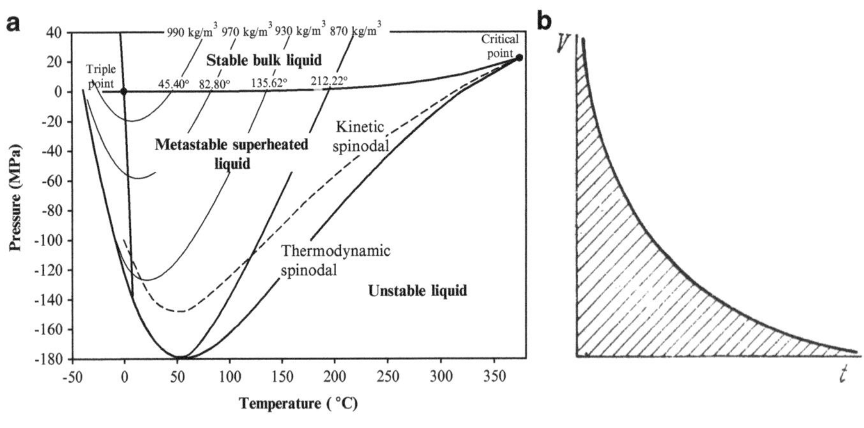

Fig. 14.1 (**a**) Phase diagram of pure water in P-T space, calculated with the IAPWS-95 equation of state [1]. Kinetic spinodal is drawn after [2]. (**b**) Qualitative relationship between space volume allowed to liquid and lifetime of its superheating (potential) state (After [3])

nucleating bubble, meaning that the boiling restoring equilibrium cannot start. In this sense, capillary water is a superheated water "de-metastabilized" or "re-stabilized" by geometrical constraints defined by the so-called Young-Laplace equation. At the liquid-air interfaces, capillary water is at equilibrium with dryer atmosphere than the bulk (Kelvin law), which is the driving force to "capillarization". Actually, lower vapor pressure than the saturation value drives a liquid–vapor system to either evaporation or capillarity.

Water located in restricted volume ("confined water") has also peculiar properties due to the additional thermodynamic term afforded by the surface fields from the close walls (disjoining pressure effect). In many respect, confined water has got similar properties than the capillary or superheated water. Interestingly, confinement experimentally exhibits metastable behaviors (e.g., [5, 6]).

The present contribution reports experimental data obtained with synthetic fluid inclusions on the extreme tensile strengths (negative pressure) that pure water and aqueous solutions can attain and on the lifetime over which they can sustain them. It has also the ambition to enlist the possible consequences of superheating (and so capillarity, and/or confinement) in terms of macroscopic behaviors. Putting the liquid under extreme conditions of tension gives pathway to measure directly how the superheating properties, may drive specific behavior in the corresponding natural settings.

14.2 Experimental Section

Synthetic Fluid Inclusions (SFI) are made in internally-heated pressure vessels (IHPV, or "gas bombs"). The hydrothermal syntheses were performed at 750 MPa in the 530–700 °C range, during 8–13 days. For each run, the IHPV is loaded with three or four Pt-capsules filled with quartz (2×2×12 mm, with the longest dimension parallel to the quartz c-axis), the selected liquid (pure water or aqueous solutions at the chosen concentration) and amorphous silica. Sealed capsules were placed on a Ta or Mo holder wherein the temperature is controlled by two Pt-Pt 13 % Rh thermocouples (certified to ± 1), and pressure measured by manganin gauges after calibration at the mercury melting point. Once the required synthesis time elapsed, the power supply is turned off and temperature decreases to 100 °C in 1 min, while pressure is decreased by intensifier to 300 MPa. In a second step, the IHPV is allowed to cool down on its own to room temperature and is finally opened to recover the capsules and quartz samples the healing of which trapped the liquid.

The microthermometric procedure consists of progressive heating on a "Linkam" stage of one sample, in general containing initially the biphasic liquid + vapour assemblage (L + V).

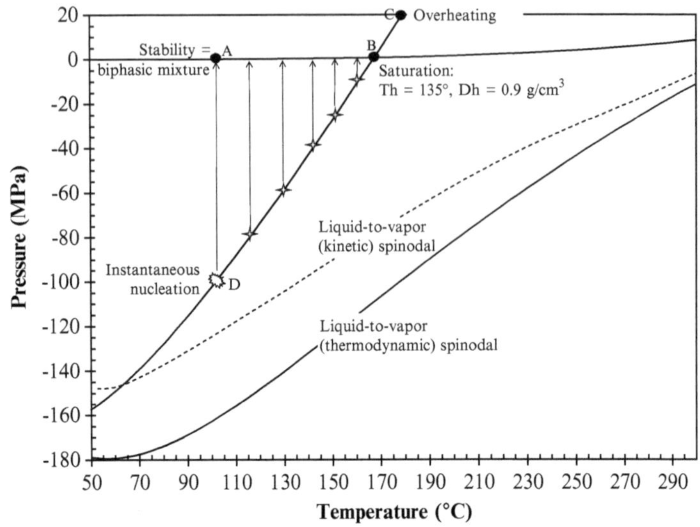

Fig. 14.2 Classic microthermometric path from the ambient conditions (see text). Nucleation can appear instantaneously (Tn commonly measured by inclusionists, sparky point) or after a delay (starry points) if the temperature is stopped before reaching the instantaneous Tn

The first step corresponds to the displacement along the saturation line, with a progressive change of temperature, liquid density and bubble pressure (A to B path, Fig. 14.2). At a particular density, the trapped liquid invades the whole inclusion space: there is no more vapour and the inclusion is said to be homogenized (B point, Fig. 14.2). The temperature at which this filling appears is then the homogenization temperature, noted Th. Just before Th, vapour bubble is so small that it can move along the minor thermal gradients within the inclusion, which results in the appearance of a mobile shadow. Further heating drives the P-T conditions in the inclusion along the isochore inside the stable domain of liquid (B to C path, Fig. 14.2). Special care must be taken because overheating above Th may lead to the decrepitation of the inclusion, especially at overheating greater than 30 °C (but 10 °C can be sufficient; see also [7]): this depends also on (at least) the shape of the inclusion, its thickness and the distance to neighboring inclusions. One decrepitation event can be detected when a further Th measurement on the same inclusion increases by more than 1 °C with respect to antecedent measurements. The second step is the progressive cooling of the sample, which follows the isochoric path as long as the inclusion remains homogeneously filled with liquid (C to D path without nucleation at B, Fig. 14.2). The temperature of the bubble appearance is called the nucleation temperature (Tn) and is always located within the tensile domain: nucleation in fluid inclusions often disobeys the saturation conditions. The bubbling at Tn produces either one-two big bubbles or a large amount of intermediate-sized bubbles which explode violently.

14.3 Superheating Limits and Lifetime

14.3.1 Measurements and Interpretation

The basic measurements are then of two types: first the (Th – Tn) gap at which instantaneous nucleation occurs, and the time needed for a nucleation to occur for any temperature set between Th and Tn (Fig. 14.3). The present dataset is reported in Fig. 14.3, and lists hundreds of inclusions measured in more than 40 samples. (Th – Tn) gap are strongly liquid- and temperature-dependent, varying the most for CsCl electrolytic solutions at low T (scattering over 80 °C) and denser pure water (scattering over 70 °C). The smallest gaps are encountered with dilute NaOH solutions whatever the temperature, and for each type of solutions, the gap is narrowing with increasing temperature. This last feature fits well the expected behavior considering the narrowing of the saturation to spinodal gap with temperature (Fig. 14.1). A close discussion of the different trends is outside the scope of this paper, some aspects are developed elsewhere [8].

The conversion of micro-thermometric measurements into internal pressure of superheated liquids requires using an equation of state (EoS). Basically, measuring Th enables to define the density of the trapped liquid at homogenization (Dh). Assuming a perfectly isochoric cooling, the couple (Tn, Dh) can be readily turn into (Tn, Pn), with Pn stands for the internal pressure inside liquid just before nucleation (Fig. 14.4).

Two questions are actually embedded in such procedure: (i) applicability of EoS to superheating state, and (ii) having the EoS characteristic of the measured solutions. About the first issue, pure water EoS can be readily used for dilute solutions, in which most of the molar content is made of water molecules (Fig. 14.4). However, studying the saline solutions require to extrapolate the available EoS, not always internally-consistent, toward the metastable state (e.g., [9]). As a consequence, inter-

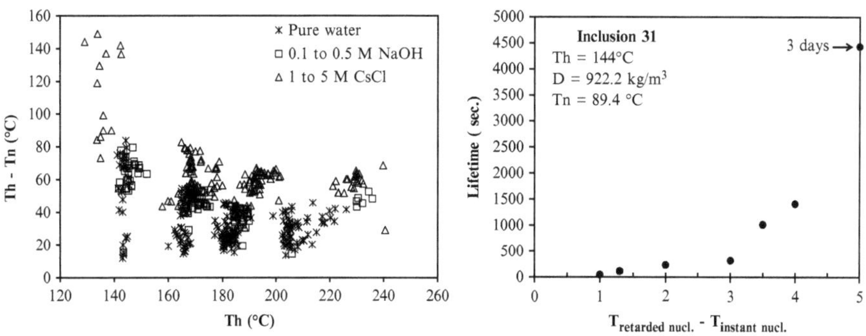

Fig. 14.3 Measurements classically performed on liquids trapped in SFI by microthermometry: temperature gap to observe instantaneous nucleation, and lifetime of superheated state

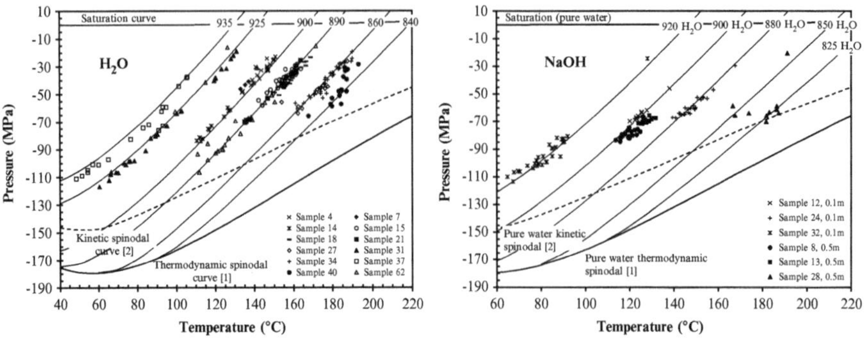

Fig. 14.4 P-T phase diagram, including the (Th, Tn) measurements of different datasets turned into (P,T) data points (see text) using the IAPWS-95 equation of state for water [1]

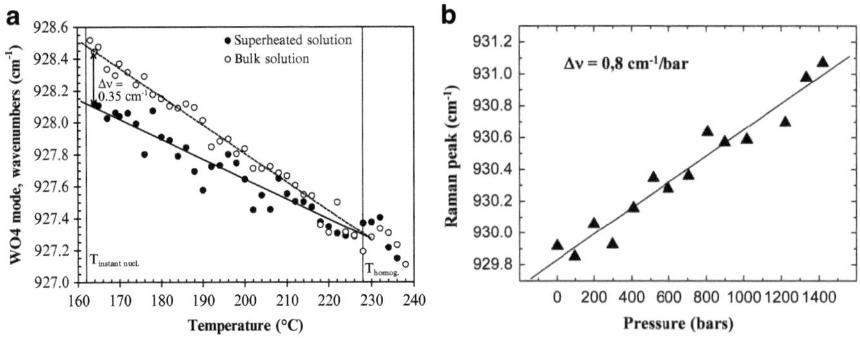

Fig. 14.5 (**a**) Spectroscopic pressure measured by the Raman shifts of the tungstate dissolved in tensile solutions trapped in SFI. (**b**) Calibration at positive pressure

pretation are certainly correct in terms of global trends, but cannot be considered as quantitatively precise for the time being. About the second issue, the general feeling is toward admitting the continuity of behavior from stability to metastability, meaning that superheated water is no more than a "particular" state of otherwise-bulk liquid (e.g. [10–13]), but, here again, some doubts still need to be lifted.

A current experiment tries to quantify the internal stress in fluid inclusions, by recording direct signals of the pressure change directly inside the inclusion, avoiding any use of EoS. To do this, an aqueous solution of Na_2WO_4 has been trapped in inclusion, because the Raman peak of this anion is pressure-dependent. Running classic micro-thermometry path into a Raman spectrometer resulted in a continuous Raman shift that demonstrates an increasing tension inside the trapped solution. This experiment solved the challenge on the measurement itself (the shift is lower than the absolute precision of the spectrometer) and provided direct evidence on the internal tension of the occluded liquid: extrapolating the positive pressure behavior (Fig. 14.5b), the tension comes to −432 bars. However, this experiment has not

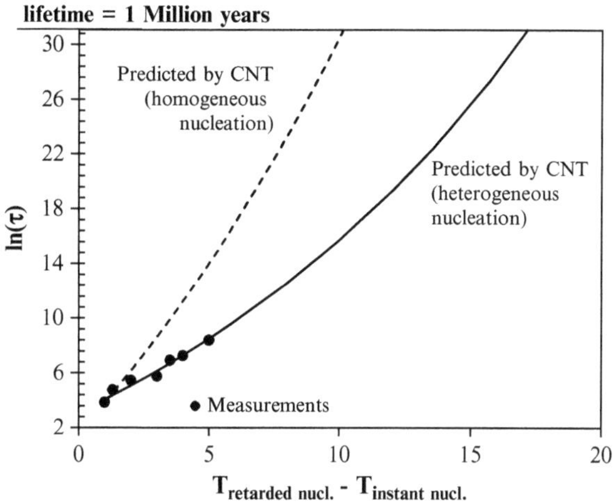

Fig. 14.6 Lifetime of superheated pure water SFI in quartz (50×20 μm; [14]), plotted against the distance to temperature of instantaneous nucleation (Fig. 14.2). The heterogeneous CNT equation is from [15], compared to the usual homogeneous CNT one (e.g., [16])

fully answered the question, since these Raman measurements cannot be compared to EoS calculations, not existing for Na_2WO_4. Complementary experiments are under way.

14.3.2 *Lifetime and Cavitation*

In terms of lifetime, the measurements have been fitted by the Classical Nucleation Theory (CNT) which considers the work of creating LV surface compared to the work brought about by the higher stability of the vapor embryo. Interestingly, the CNT equation for homogeneous nucleation (the source of work in the system is the thermal fluctuations) does not fit the data, but that for the heterogeneous nucleation on a hydrophobic (and pure quartz is) surface does (Fig. 14.6).

These are preliminary results since the data are only acquired over 3 days for restricted distance to instantaneous nucleation ($\Delta T_{max} = 6$ °C). Interestingly, the extrapolation of the fitted trend indicates that a very long metastable lifetime is possible even for significant superheating state ($\Delta T \approx 18°$, possibly corresponding to a tension down to −500 bars). The present target is to confirm the validity of the heterogeneous CNT equation toward extrapolation, by performing longer experiments increasing the range of distance to equilibrium.

One interesting feature should be additionally addressed: the role of the host volume to stabilize the superheating as evoked in Fig. 14.1. This question is beyond a simple academic interest and corresponds to understand the size threshold below which the confinement effect would suppress the nucleation event, allowing to safely produce highly superheated liquid.

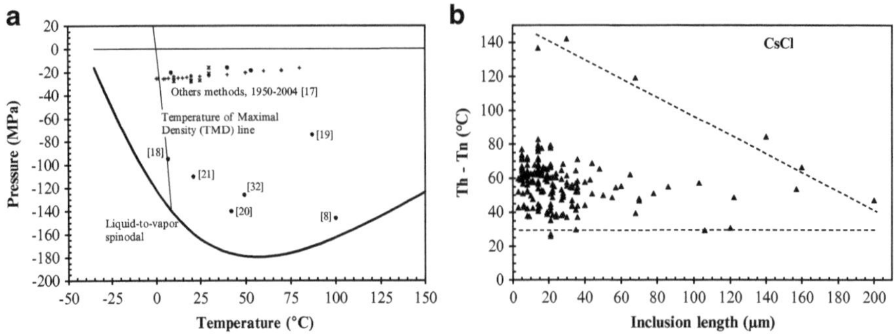

Fig. 14.7 (**a**) Extreme tensile strength obtained with different static and dynamic methods [17–21]. *Black dots* are Berthelot tubes and SFI. (**b**) Relationship in CsCl-filled samples (1 and 5 m CsCl) between (Th – Tn) and the volume of the inclusions (1D simplification) (*dotted lines* are only guides for eyes)

A first clue comes from an open question in the physics of metastability [17]: among the different techniques available to produce superheating, only the SFI enables to reach high level of metastability (Fig. 14.7a). A key difference between SFI and all other technique is the way the liquid volume occupy the experimental cell. In case of dynamic/capillary techniques there is only one experimental water volume, so that the first nucleation event makes all the volume to boil. In case of inclusions samples, liquid is dispersed as numerous micrometric cavities in the same solid fragment, and the vapour nucleates at P,T conditions peculiar to each inclusion. That means the geometrical confinement, even over hundreds of micrometers (which make a pretty large volume), should help in pushing the superheating state at larger value in inclusions than in the other experimental cells without walls.

A second clue is afforded by the common relationship between the (Th – Tn) metastability gap and the host volume (e.g., [8], see their figure 8a,b). However, as apparent in Fig. 14.7b, this relationship is common but has exceptions or limits (see also figure 8b in [8]), demonstrating that other factors are embedded in this size factor, for instance the internal shape of the inclusions that reveals its degree of equilibrium ("negative quartz shape") [8]. Even Th itself can vary (over 1–3 °C) with the size of inclusion due to surface tension effects [22]. In another direction, the size of the inclusion and the internal faceting of the space void, partly control the properties of the SFI, notably the modes of their non-elastic deformation (stretching, leakage, decrepitation) [7].

Here again, complementary experiments are needed to decipher how the size of inclusions combine with other critical factors to control the behavior of inclusions, notably the extent of metastability and the deformability.

14.4 Expected Behaviour of Superheated (Capillary and Confined) Water

14.4.1 Geochemistry of Phase Transitions

Capillary liquids have the same properties of superheated water, and most probably confined water as well. Superheated water is known to be a very good solvent for organics compounds as pesticides, making it the "ultimate green solvent for separation science" (e.g., [23, 24]). This experimental observation can be easily illustrated extrapolating the solubility line of different metals sulfides in aqueous solution beyond the saturation LV line, down to the spinodal (Fig. 14.8a). Additionally, the diagram shows that the cavitation which restores the equilibrium changes the saturation conditions and contributes to precipitate the excess dissolved mass. Interestingly, a recent study showed the ice nucleation just following the bubble cavitation in superheated water [25].

In the second part of the figure, it appears that the gases are more soluble in superheated/capillary water than in bulk water. Actually, this in-gassing effect could be of significant importance to understand the global carbon cycle: the immobile water trapped under tension in the capillary clusters of all the arid and semi-arid soils of Earth are pumping in more atmospheric CO_2 than expected without considering the capillary role on dissolution. These findings are mitigated by the small volume of capillary water at dry conditions that limits the total mass that can be actually stored in soils.

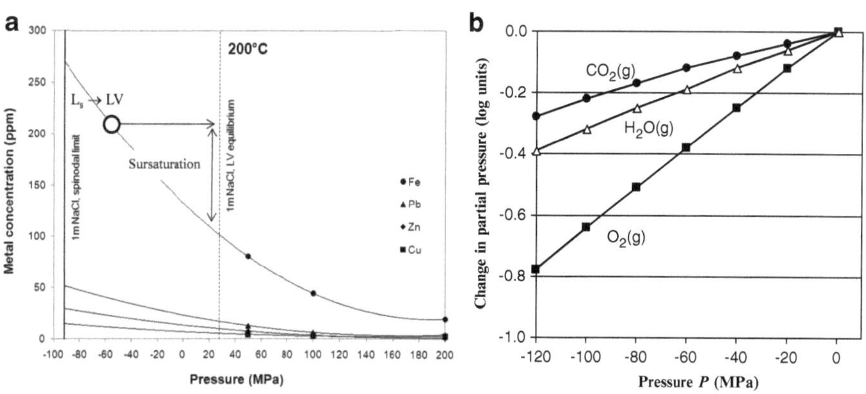

Fig. 14.8 (**a**) Solubility line, empirically extrapolated at negative pressures, illustrating the solvent effect of superheated (capillary/confined) liquid (After [8]). (**b**) Gases solubility as a function of the internal tension in superheated/capillary water (Modified from [26])

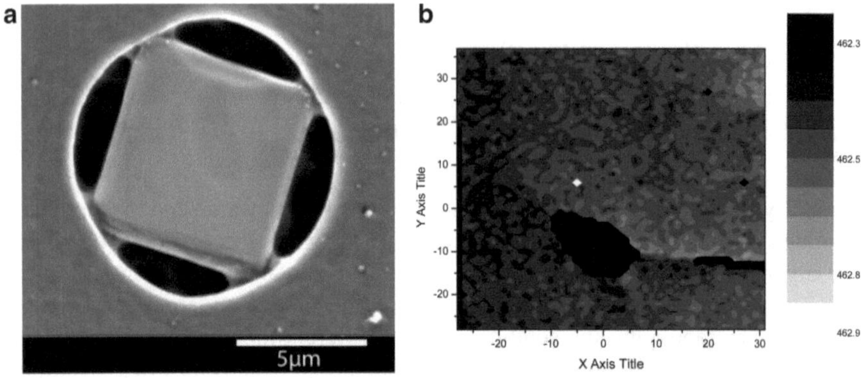

Fig. 14.9 (**a**) Cubic NaCl precipitation in round pore (radius = 5 μm), secondary precipitation of concave-curved NaCl at the prior salt-pore contacts, and resulting deformation. (**b**) Raman shift inside the quartz matrix around an inclusion filled with 5 m CsCl, and put at (Th − Tn) = 90 °C (unpublished results; Collab. CEMTHI, P. Simon)

14.4.2 Geochemistry and Geomechanics

The liquid under tension when trapped in solids exerts traction on the host solid, and so contributes to the geomechanical balance at the pore scale [27]. Drying experiments performed in micrometric round pore showed that a first precipitation of cubic salt allowed the arising of capillary bridge at the salt-pore contacts that generated a deformation of the host (Fig. 14.9a).

Another experiment is presently running, searching to record the constraints induced in the surrounding solid by the tension inside a highly superheated electrolytic aqueous solution (Fig. 14.9b). Interestingly, part of the surrounding quartz is under slight tension (darkest part, to the left of the inclusion), while another part is under compression (dark/light grey, to the right of the inclusion). These two compartments are easily distinguished by two lines making a 120° angle between, classic in crystalline quartz. The current interpretation is that the in-inclusion tension pulls up the right compartment that moves as a wedge and so is put under overpressure, while the other rigid compartment does not move and undergo the tensile elastic deformation. This first experiment demonstrates the complexity of the stress field all around an inclusion, active as long as the liquid tension is effective, and evidences the compensation effects due to the heterogeneity on the matrix itself.

14.4.3 Explosivity

A superheated liquid is metastable with respect to its vapour and then able to nucleate at any moment (Fig. 14.10a). During the eighteenth century, Rev. Dayton

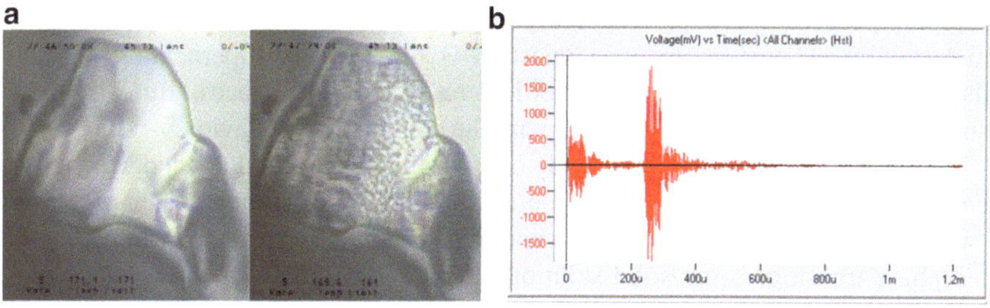

Fig. 14.10 (**a**) Intense and brutal boiling in 5 m NaCl: Th = 221 °C; Th − Tn = 51 °C (also at http://www.iem.ac.ru/staff/kiril). (**b**) Soundwave accompanying the trapped liquid cavitation recorded by an acoustic emission sensor (Euro Physical Acoustics)

already reported on the easiness to superheat a liquid and the associated explosive danger [28]. Nowadays, this is a well-known class of industrial spills accidents, connected to the Boiling Liquid Expanding Vapour Explosion (BLEVE) type (for instance, [29]). As thermodynamically demonstrated [9], the superheating state of the mother liquid does not increase so much the energy release at the LV phase transition, but dramatically increases the rate of this transition. The power of one explosion is the product of the energy release by the time of release, it comes to the superheating boiling can be qualified as explosive because it is a rapid-phase transition whose rapidity is depending on the superheating "level" [29].

The cavitation event produces a sound wave that can be recorded to characterize the intensity of the prior metastability and estimate the mass that cavitates (Fig. 14.10b). Additionally, several sensors can be used to detect the exact location of cavitation which is not so valuable when working with transparent material like quartz SFI, but could become interesting if using this system to record cavitation event in heterogeneous porous systems.

Interestingly, the boiling of large bubbles in porous soils was evoked as a potentially important factor to the transport in porous media (e.g., [30, 31]). Actually, this boiling occurs in water immobilized against gravity due to the capillary forces. The boiling breaking out this capillary suction makes the water body able to flow again according to the gravitational field. This de-metastabilization event can also be expected to move an initially continuously water body into a sum of dispersed flowing/not-flowing water clusters, shaping a discontinuous water content in soils.

14.5 Conclusion

Superheated water is a special state which appears for different reasons (decreasing vapor pressure, increasing temperature) and can be useful for different purposes. First, the properties of superheated water are of interest for those situations during which this special metastable state arises. Second, these properties also fit the capillary properties, because capillarity is no more than a superheated state without metastability. However, an open question is the role the superheating plays in the confinement of liquid in geometrical restricted pore spaces. But again, the superheated properties could fit well with the confinement behavior.

On another point of view, this study outlines the shortage of knowledge about the consequences in natural settings to contain liquid water with superheated properties (whatever its actual form: confined, capillary, or strictly superheated). In particular, the phase transitions (gas/solid-solution interactions, liquid–vapor explosive boiling) are drastically changed that could impact significantly our understanding of the mass transfer in the corresponding pore networks. All the results showed here need to be extended and generalized, and relativized to the surface area of the liquid–solid interface, and also to the volume of liquid involved and available for the reaction. A completely overlooked aspect is the retention/flow duality and more generally the consequences of the superheating on the transport functions of the infilling pores liquid.

Acknowledgements This work has received support from the French Agency for Research (Agence Nationale de la Recherche, ANR) through the grants CONGE BLAN-61001 and Labex Voltaire ANR-10-LABX-100-01.

References

1. Wagner W, Pruss A (2002) The IAPWS formulation 1995 for the thermodynamic properties of ordinary water substance for general and scientific use. J Phys Chem Ref Data 31(2):387–535
2. Kiselev SB, Ely JF (2001) Curvature effect on the physical boundary of metastable states in liquids. Physica A 299:357–370
3. Skripov VP, Sinitsyn EN, Pavlov PA, Ermakov GV, Muratov GN, Bulanov NV, Baidakov VG (1988) Thermophysical properties of liquids in the metastable (superheated) state. Gordon and Breach Science Publishers, New York
4. Imre A, Martinas K, Rebelo LPN (1998) Thermodynamics of negative pressures in liquids. J Non-Equilib Thermodyn 23(4):351–375
5. Restagno F, Bocquet L, Biben T (2000) Metastability and nucleation in capillary condensation. Phys Rev Lett 84(11):2433–2436
6. Morishige K, Yasunaga H (2006) Tensile effect on a confined phase. J Phys Chem B 110:3864–3866
7. Bodnar RJ, Binns PR, Hall DL (1989) Synthetic fluid inclusions. VI. Quantitative evaluation of the decrepitation behavior of fluid inclusions in quartz at one atmosphere confining pressure. J Metamorph Geol 7:229–242
8. Shmulovich KI, Mercury L, Thiéry R, Ramboz C, El Mekki M (2009) Experimental superheating of water and aqueous solutions. Geochim Cosmochim Acta 73(9):2457–2470
9. Thiéry R, Mercury L (2009) Explosive properties of water in volcanic and hydrothermal systems. J Geophys Res 114, B05205. doi:10.1029/2008JB005742
10. Skripov VP (1992) Metastable states. J Non-Equilib Thermodyn 17:193–236
11. Mercury L, Azaroual M, Zeyen H, Tardy Y (2003) Thermodynamic properties of solutions in metastable systems under negative or positive pressures. Geochim Cosmochim Acta 67:1769–1785

12. Mercury L, Jamme F, Dumas P (2012) Infrared imaging of bulk water and water-solid interfaces under stable and metastable conditions. Phys Chem Chem Phys 14:2864–2874
13. Davitt K, Rolley E, Caupin F, Arvengas A, Balibar S (2010) Equation of state of water under negative pressure. J Chem Phys 133:174507
14. El Mekki M, Ramboz C, Perdereau L, Shmulovich KI, Mercury L (2009) In: Rzoska SJ, Drozd Rzoska A, et Mazur V (eds) Metastable systems under pressure, NATO science for peace and security series A, chemistry and biology. Springer, Dordrecht, pp 279–292
15. Blander M, Katz JL (1975) Bubble nucleation in liquids. AIChE J 21(5):833–848
16. Debenedetti PG (1996) Metastable liquids. Concepts and principles. Princeton University Press, Princeton
17. Caupin F, Herbert E (2006) Cavitation in water: a review. CR Phys 6:1000–1017
18. Roedder E (1967) Metastable superheated ice in liquid-water inclusions under high negative pressure. Science 155:1413–1417
19. Green JL, Durben DJ, Wolf GH, Angell CA (1990) Water and solutions at negative pressure: Raman spectroscopic study to −80 megapascals. Science 249:649–652
20. Zheng Q, Durben DJ, Wolf GH, Angell CA (1991) Liquids at large negative pressures: water at the homogeneous nucleation limit. Science 254:829–832
21. Alvarenga AD, Grimsditch M, Bodnar RJ (1993) Elastic properties of water under negative pressures. J Chem Phys 98(11):8392–8396
22. Fall A, Rimsditch JD, Bodnar RJ (2009) The effect of fluid inclusion size on determination of homogenization temperature and density of liquid-rich aqueous inclusions. Am Mineral 94:1569–1579
23. Smith RM (2006) Superheated water: the ultimate green solvent for separation science. Anal Bioanal Chem 385:419–421
24. And CO, Su-in P (2006) Modified superheated water extraction of pesticides from spiked sediment and soil. Anal Bioanal Chem 385:83–89
25. Barrow MS, Williams PR, Chan H-H, Dore JC, Bellissent-Funel M-C (2012) Phys Chem Chem Phys 14:13255–13261
26. Mercury L, Pinti DL, Zeyen H (2004) The effect of the negative pressure of capillary water on atmospheric noble gas solubility in groundwater and palaeotemperature reconstruction. Earth Planet Sci Lett 223:147–161
27. Bouzid M, Mercury L, Lassin A, Matray J-M, Azaroual M (2011) In-pore tensile stress by drying-induced capillary bridges inside porous materials. J Coll Interface Sci 355:494–502
28. Dayton J (1739) An experience to prove, that water, when agitated by fire, is infinitely more elastic than air in the same circumstances. Philos Trans 41(1):162–166
29. Reid RC (1976) Superheated liquids. Am Sci 64:146–156
30. McManus KJ, Davis RO (1997) Dilation-induced pore fluid cavitation in sands. Géotechnique 47(1):173–177
31. Or D, Tuller M (2002) Cavitation during desaturation of porous media under tension. Water Resour Res 38(5):1061. doi:10.1029/2001WR000282

Chapter 15
Plant Water Transport and Cavitation

Teemu Hölttä and John Sperry

Abstract Water transport in the xylem of plants from the roots to leaves occurs under negative pressure. This makes the xylem sap vulnerable to cavitation. Cavitation is a common phenomenon in plants, and it induces major consequences on plant function by limiting the ability of the plant to extract water from drying soils and transport it to the leaves. Decreased water conductive capacity due to cavitation leads to decreased carbon assimilation rates by photosynthesis, and in extreme conditions, to plant mortality. Many plant species have the capability to refill xylem conduits, which have become air-filled due to cavitation.

15.1 Introduction

Water scarcity is probably the most important factor limiting the growth, yield and survival of plants throughout the world [1]. Plants use water in metabolic processes, but much larger quantities of water are lost in transpiration to the atmosphere. Transpirative water loss is an inevitable consequence of carbon dioxide assimilation in photosynthesis, as the exchange of both water and carbon dioxide between the leaves and atmosphere occurs by diffusion through the same stomatal openings on leaf surfaces. Large trees can use hundreds of liters of water per day (e.g. [2]). Water vapor transpired from the world's forests has a significant role in the global water and carbon cycles, as well as on the energy balance of the lower atmosphere [3].

T. Hölttä (✉)
Department of Forest Sciences, University of Helsinki, P.O. Box 24, FI-00014
Helsinki, Finland
e-mail: teemu.holtta@helsinki.fi

J. Sperry
Department of Biology, University of Utah, 257 South 1400 East, Salt Lake City,
UT 84112, USA

L. Mercury et al. (eds.), *Transport and Reactivity of Solutions in Confined Hydrosystems*, NATO Science for Peace and Security Series C: Environmental Security,
DOI 10.1007/978-94-007-7534-3_15,

The driving force for lifting water from the soil is created by capillary wicking forces in cell walls that are exposed to the ubiquitous air spaces inside the plant body [4]. The non-lignified cell walls are hydrophilic and retain water like a nano-porous sponge despite evaporation into the air spaces. Surface tension and adhesion of water to the cell wall material create air-water menisci that pull water up to replace evaporative losses. The pull is transmitted to the pipe-like-conduits in xylem which provide relatively wide (micro-scale), rigid, and low-resistance flow paths extending to the distant roots. At the roots, the pulling force extends out to the water-filled pore spaces of the soil. The plant functions as a highly sophisticated wick that is ramified skyward to replace foliar water loss during CO_2 uptake, and ramified underground to draw from soil water reserves. As long as the capillary forces in the cell walls exceed those of the soil pore space and the force of gravity, and as long as the xylem conduits remain water-filled, water will move towards the leaves down a gradient in sub-atmospheric water pressure.

A major threat to plant water transport is the problem of keeping the xylem conduits water-filled. Xylem transport is a unique phenomenon in nature in the sense that the transported water is under negative pressure, i.e. in a meta-stable state where the liquid pressure has decreased below the saturation vapor pressure. This leads to a constant threat of phase change of water to gas by cavitation. The xylem conduits develop from living cells, and begin to function after protoplast disintegration and cell death. Hence they are water filled from their inception. Subsequent cavitation in the conduit water is a common occurrence and results in a gas-filled, or embolized conduit that is unable to conduct water (Fig. 15.1). When embolized, the xylem conduits are too wide to generate much suction by capillary action, hence as we discuss later they can only refill with water if their water supply is near or above atmospheric pressure.

Plants differ considerably in their ability to avoid cavitation. Plants from consistently moist habitats do not have to pull very hard to extract the readily available soil water, and some will start cavitating at pressures above (less negative than) -1 MPa. Conversely, plants growing in seasonally drier habitats must be able to pull harder to extract more tightly-bound soil water, and some can continue to transport water to pressures below (more negative than) -10 MPa [5]. Avoiding cavitation must come at a competitive cost; otherwise it would occur at pressures negative enough to make it a rare event. Research in the area includes the mechanism of cavitation and the basis of trade-offs with plant performance, and the process by which many plants are able to reverse cavitation and refill empty xylem conduits.

15.2 Negative Water Pressure and the Cavitation Mechanism

Pressures in the xylem typically remain higher than -2 MPa, although pressures to -10 MPa and below are not uncommon in desert plants [6]. Water pressure in the soil is already typically below atmospheric, as water is held in capillary pores between soil particles by matric forces. However, the majority of the pressure

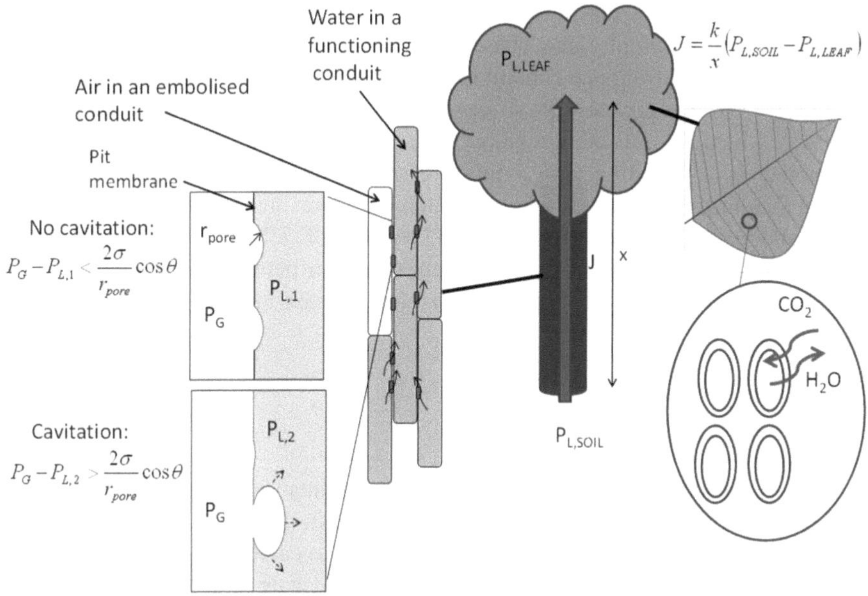

Fig. 15.1 Water flows through the xylem of a plant according to the water pressure gradient. The exchange of water vapor and CO_2 occur through the stomata in the leaves. The mechanism for cavitation in the xylem is air-seeding, where air spreads from an adjacent conduit through the pores in the pit membrane when water pressure is low enough. The pits are depicted as found in angiosperm species. Pit structure in gymnosperms is slightly different. Symbols: J is water flow rate through the xylem (~ rate of water uptake from soil ~ rate of whole tree level transpiration), k is hydraulic conductivity, x is transport distance (~ tree height), $P_{L,LEAF}$ is leaf water pressure, $P_{L,SOIL}$ is soil water pressure, $P_{L,1}$ is water pressure in the stem xylem at location 1, $P_{L,2}$ is water pressure in the stem xylem at location 2, P_G is gas pressure (~ atmospheric pressure), r_{pore} is radius of pores in the pit membrane, σ is surface tension of water, θ is the contact angle between water and the cell wall pore surface ($\theta = 0$ refers to a perfectly wettable, i.e. hydrophilic, surface), CO_2 is the exchange of carbon dioxide through stomata, H_2O is the exchange of water vapor through stomata

decrease of water on its way from the soil to the leaves is typically caused by viscous pressure losses in the laminar flow in the xylem. Only in very tall trees, pressure losses due to gravity (0.01 MPa m^{-1}) can be comparable in magnitude to viscous pressure losses.

In general, cavitation occurs by homogeneous nucleation, i.e. by thermodynamic fluctuations in the bulk of the liquid without being catalyzed by impurities or by external disturbances, at pressures of approximately −140 MPa [7]. Cavitation in the xylem of plants cannot occur by homogeneous nucleation as such low water pressures are never reached in xylem conduits. "Classical" heterogeneous nucleation is also likely to play a rather minor role in xylem cavitation, as its occurrence at the pressures found in the xylem would require the existence of very hydrophobic surfaces at conduit walls. According to both theoretical considerations and experimental manipulations, cavitation is likely to occur by air-seeding [8].

In air-seeding the gas phase is not created by thermodynamic fluctuations as in "classical" nucleation of a new phase. Instead, air, which is already present within the xylem, spreads from one conduit to another (Fig. 15.1).

Plants cannot completely avoid cavitation, but the xylem tissue is constructed in a way which restricts the gas phase from spreading from one embolised conduit to another by air-seeding until xylem pressure decreases low enough; Individual xylem conduits are separated from one another by thin areas in the intervening cell wall called pits. The thin "pit membrane" consists of cell wall material with small pores. These pores constrict the movement of the air between conduits by surface tension as long as the water pressure does not decrease too low. The condition for air-seeding is

$$P_g - P_L > \frac{2\sigma}{r_{pore}} \cos\theta \qquad (15.1)$$

The experimental evidence supporting air-seeding as the primary cavitation mechanism come from manipulation experiments testing Eq. 15.1. It has been found that Eq. 15.1 describes the vulnerability of xylem to cavitation even in conditions where (A) the surface tension is lowered by surfactants (e.g. [8, 9] and (B) the pressure difference on the left side of the equation is induced by increasing the air pressure, instead of decreasing the liquid pressure, e.g. [8, 10].

Notably, for air-seeding between conduits to be initiated, there must be an air-filled conduit to begin with. A certain fraction of gas filled conduits appear to be inevitable consequences of plant development. Protoxylem in growing tissues is stretched to the rupture point, creating embolized conduits that serve as air-entry points into the vascular system. Abscission of parts, whether leaves, roots, fruits will similarly create air-filled conduits, as will damage from breakage or herbivory. Any of these inevitable events introduce air into the vascular system, with only thin pit membranes to prevent its further propagation throughout the conduit network.

An additional cause of cavitation occurs during freeze-thaw cycles. Gases that are dissolved in the xylem sap are insoluble in ice, and gas bubbles are frequently created as the sap freezes. If these bubbles are large enough and the negative pressures low enough (i.e., Eq. 15.1, but with no contact angle and r_{bubble}, i.e. bubble radius, instead of r_{pore}), the bubble will nucleate cavitation. Much of the often extensive cavitation that occurs during the winter season can be attributed to this mechanism [11].

15.3 Consequenses of Cavitation for Plant Function

Cavitation within the xylem has large consequences for plant function. Excessive cavitation renders the plant incapable of extracting water from the soil and transporting it to the leaves. Cavitation during drought is a major cause of declined plant productivity, and a likely mechanisms of tree drought associated mortality worldwide [12]. Plants can avoid cavitation by restricting water loss from the leaves, which increases xylem water pressure. However, cavitation avoidance comes at the cost of reduced CO_2 assimilation by photosynthesis, since CO_2 and water are exchanged between plants and the ambient atmosphere though the same small holes in the leaves of plants, the stomata (Fig. 15.1). Water loss and carbon gain are constantly being balanced by plants by regulation of the stomatal opening. Stomatal closure is typically coordinated so that water pressure remains above levels where massive amounts of cavitation would be induced. Longer term adjustments at the whole plant level include reductions in leaf to sapwood area to decrease water loss in relation to transport capacity (hydraulic conductivity, see Fig. 15.1) of the xylem.

The capability of the xylem to transport large amounts of water while maintaining it in the liquid phase is thought to be one of the main constraints on tree function, growth and the maximum height which trees can attain [13, 14]. Xylem structure can be considered a compromise between the efficiency at transporting water, and the safety from cavitation. Transport efficiency can be quantified in terms of hydraulic conductivity, and in laminar flow, hydraulic conductivity is proportional to the radius of the conduits raised to the fourth power. Thus, wide and long xylem conduits increase water conducting capacity. However, large conduits also apparently decrease safety from cavitation. In the case of cavitation from freeze-thaw cycles, this is because wider conduits produce larger bubbles when frozen, and larger bubbles trigger cavitation at less negative pressures. Cavitation by air-seeding, however, is not mechanistically constrained by conduit size but by the size of pores of the pit membranes (see Eq. 15.1). Nevertheless, data and theory suggest that the size of the widest pit membrane pores will scale roughly with conduit size (e.g. [15, 16]). These conflicting requirements for plant hydraulic architecture in terms of xylem conduit size lead to differing hydraulic designs for different environmental conditions in terms of water availability. The hydraulic conductivity of the xylem and its capability to avoid cavitation thus ultimately limit the rate of plant metabolic processes, such as CO_2 assimilation by photosynthesis, similarly as the flow rate of blood through animal blood circulatory system sets limits for their metabolism rates [17].

Another aspect of cavitation in the xylem is to serve as a water release mechanism. When a xylem conduit embolises with gas, nearly all of the liquid water inside that xylem conduit is released, which causes a temporary release of water tension in the rest of the conduits. As xylem tissue is quite rigid, only a small amount of extra water released from an embolising conduit will bring about a relatively large increase in water pressure of the surrounding xylem tissue [18].

15.4 Plant Water Uptake from Air

In some cases the soil is not the only source of water for plants. Plants are also capable of taking up water from the air, and this may be particularly important in areas of regular fogs such as the Californian coast, where the largest trees in the world, the Giant Redwood, live [19]. These trees may draw up to 20 % of their water from the air. Water may condense on plant surfaces, or the direction of water vapor movement could be reversed to go from ambient air to air spaces within plant leaves. This may happen since the negative pressures in the leaves of plants can reduce the chemical potential of water below that of saturated vapor pressure in the ambient air even at a relative humidity slightly below 100 % [4], i.e. before the condensation on plant outside surfaces is expected to occur.

15.5 Measurements of Plant Water Pressure and Cavitation

Negative pressure in the xylem is difficult to measure directly since direct contact of a pressure probe with the xylem water under negative pressure is likely to induce cavitation. Perhaps the most common technique for measuring plant water pressure is the so called pressure chamber technique [20]. In this technique, the counter (positive) pressure required to push water out of plant leaves or branches is measured, and this positive pressure is equated to be reciprocal of the actual xylem (negative) water pressure. Thermocouple psychrometry is also commonly used to measure the water potential of air that is in vapor equilibrium with the xylem water potential; in the usual absence of significant solute concentration, sap water potential is identical with the negative sap pressure [21]. Another less used method of measuring xylem water pressure is to observe its effect on the minute changes in xylem or plant stem diameter. Lowered water pressure within the xylem causes water withdrawal from the living tissues within the plant, and also the xylem itself contracts slightly when pressure in it is lowered [22]. The relation between xylem water pressure and radial dimensional changes are very close to linear at the water pressure ranges typically encountered in the xylem of trees, making this method suitable for monitoring xylem pressure dynamics in field conditions. Cavitation in the xylem is typically quantified by its effect on plant hydraulic conductivity, i.e., the ratio of flow rate to pressure gradient [23]. Cavitation can also be detected by monitoring the ultra-acoustic emissions with sensors attached to the plant, as the release of water and xylem conduit wall tension in connection with cavitation events are large enough to be detectable [24].

15.6 Reversal of Cavitation

Hydraulic conductivity of the xylem that is lost by cavitation is replaced by growth of new xylem conduits and by the refilling of the embolised xylem conduits. Because the conduits are typically tens of microns in diameter, they can only generate very weak capillary uptake when gas-filled, and so xylem pressures must rise near to or above atmospheric pressure for refilling to occur. These higher xylem pressures collapse the embolism by driving the gas into solution (or otherwise expelling it from the conduit). Such pressures can be generated in the absence of transpiration by osmotically driven root pressures. Spring root pressures in woody plants (prior to bud-burst) is important for refilling often extensive embolism that accumulates over the winter [25]. Weak root pressures in root systems damaged by soil freezing events have been implicated in birch dieback events [26]. Nightly root pressures in herbaceous plants, visible as droplets of xylem sap exuded from leaf margins (guttation), are associated with the reversal of day time cavitation events [27].

Surprisingly, refilling of embolised xylem conduits has also been detected even when transpiration is occurring and xylem pressures are too negative. This observation implies there is an active pumping mechanism for increasing the pressure in refilling conduits above that of the surrounding vascular system. [28]. The details of the pumping mechanism are not known, but its action has been observed in intact plants by high-resolution computed tomography [29]. Water appears to be secreted by living cells into the embolized vessels, gradually refilling them. The process appears to be linked to phloem transport, starch hydrolysis, aquaporin regulation, and other metabolic processes [30].

15.7 Experimental Results

Figure 15.2 shows a result from a manipulation experiment where a surfactant was added to the xylem of a transpiring Scots pine seedling. The results are from the study of Hölttä et al. [18]. As the surface tension of the xylem sap

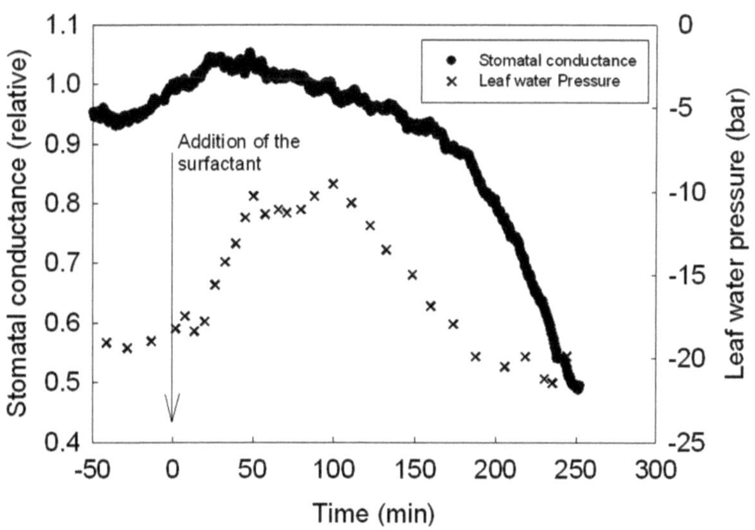

Fig. 15.2 Results from a manipulation experiment where the surface tension of the xylem was lowered by a surfactant in a Scots Pine (Pinus sylvestris) seedling. The surfactant caused a temporary increase in leaf water pressure and stomatal conductance, followed by a strong decrease in both

was lowered, massive cavitation was induced in the xylem. Cavitation caused a temporary increase in leaf water pressure as the embolising conduits released water to the surrounding tissue, including the leaves. After approximately 2 h, most of the water from the embolised conduits had been transpired, and the increase in leaf water potential ceased. Leaf water pressure started to decrease again, despite strong stomatal closure, since xylem hydraulic conductivity had been decreased due to the massive amount of cavitation (see equations in Fig. 15.1).

References

1. Chaves MM, Maroco JP, Pereira JS (2003) Understanding plant response to drought: from genes to the whole plant. Funct Plant Biol 30:239–264
2. Meinzer FC, Bond BJ, Warren JM, Woodruff DR (2005) Does water transport scale universally with tree size? Funct Ecol 19:558–565
3. Bonan GB (2008) Forests and climate change: forcings, feedbacks, and the climate benefits of forests. Science 320:1444–1449
4. Nobel PS (2005) Physiochemical and environmental plant physiology, 3rd edn. WH Freeman and Company, New York
5. Maherali H, Pockman WT, Jackson RB (2004) Adaptive variation in the vulnerability of woody plants to xylem cavitation. Ecology 85:2184–2199
6. Tyree MT (1997) The cohesion-tension theory of sap ascent: current controversies. J Exp Bot 48:1753–1765
7. Debenedetti P (1996) Metastable liquids. Princeton University Press, Princeton
8. Sperry JS, Tyree MT (1988) Mechanism of water stress-induced xylem embolism. Plant Physiol 88:581–587
9. Cochard H, Hölttä T, Herbette S, Sylvain D, Mencuccini M (2009) New insights into the mechanism of water-stress induced cavitation in conifers. Plant Physiol 151:949–954
10. Cochard H, Cruiziat P, Tyree MT (1992) Use of positive pressures to establish vulnerability curves: further support for the air-seeding hypothesis and implications for pressure-volume analysis. Plant Physiol 100:205–209
11. Tyree MT, Sperry JS (1989) Vulnerability of xylem to cavitation and embolism. Annu Rev Plant Physiol Plant Mol Biol 40:19–38
12. McDowell N (2011) Mechanisms linking drought, hydraulics, carbon metabolism, and vegetation mortality. Plant Physiol 155:1051–1059
13. Bond B, Kavanagh K (1997) Stomatal behavior of four woody species in relation to leaf-specific hydraulic conductance and threshold water potential. Tree Physiol 19:503–510
14. Koch GW, Sillett SC, Jennings GM, Davis SD (2004) The limits to tree height. Nature 428:851–854
15. Sperry JS, Hacke UG (2004) Analysis of circular bordered pit function. I. Angiosperm vessels with homogenous pit membranes. Am J Bot 91:369–385
16. Hölttä T, Mencuccini M, Nikinmaa E (2011) A carbon cost–gain model explains the observed patterns of xylem safety and efficiency. Plant Cell Environ 34:1819–1834
17. West GB, Brown JH, Enquist BJ (1999) A general model for the structure and allometry of plant vascular systems. Nature 400:664–667
18. Hölttä T, Juurola E, Lindfors L, Porcar-Castell A (2012) Cavitation induced by a surfactant leads to a transient release of water stress and subsequent 'run away' embolism in Scots pine (Pinus sylvestris) seedlings. J Exp Bot 63:1057–1067
19. Burgess SSO, Dawson TE (2004) The contribution of fog to the water relations of Seqoia sempervirens (D. Don): foliar uptake and prevention of dehydration. Plant Cell Environ 27:1023–1034

20. Scholander PR, Hammel HT, Bradstreet ED, Hemmingsen EA (1965) Sap pressure on vascular plants. Science 148:339–346
21. Dixon MA, Tyree MT (1984) A new temperature-corrected stem hygrometer and its calibration against the pressure bomb. Plant Cell Environ 7:693–697
22. Sevanto S, Hölttä T, Holbrook NM (2011) Effects of the hydraulic coupling between xylem and phloem on diurnal phloem diameter variation. Plant Cell Environ 34:690–703
23. Sperry JS, Donnelly JR, Tyree MT (1988) A method for measuring hydraulic conductivity and embolism in xylem. Plant Cell Environ 11:35–40
24. Jackson GE, Grace J (1996) Field measurements of xylem cavitation: are acoustic emissions useful? J Exp Bot 47:1643–1650
25. Sperry JS (1993) Winter embolism and spring recovery in *Betula cordifolia, Fagus grandifolia, Abies balsamifera*, and *Picea rubens*. In: Water transport in plants under climatic stress. Cambridge University Press, Cambridge, UK, pp 86–98
26. Cox RM, Malcom JM (1997) Effects of duration of a simulated winter thaw on dieback and xylem conductivity of *Betula papyrifera*. Tree Physiol 17:389–396
27. Stiller V, Lafitte HR, Sperry JS (2005) Embolized conduits of rice (*Oryza sativa* L.) refill despite negative xylem pressure. Am J Bot 92:1970–1974
28. Zwieniecki M, Holbrook NM (2009) Confronting Maxwell's demon: biophysics of xylem embolism repair. Trends Plant Sci 14:530–534
29. Broderson CR, McElrone AJ, Choat B, Matthews MA, Shackel KA (2010) The dynamics of embolism repair in xylem: in vivo visualizations using high-resolution computed tomography. Plant Physiol 154:1088–1095
30. Nardini A, LoGollo MA, Salleo S (2011) Refilling embolized xylem conduits: is it a matter of phloem unloading? Plant Sci 180:604–611

Part IV
Crystallization Under Confinement

Only about 3 % of the water on Earth is fresh water and more than two thirds of it is present as frozen water in polar ice caps and in glaciers. The remaining freshwater is mainly found as sub-surface water and is a major, yet not infinite, water resource on Earth. Sub-surface water is found in the pore space of rocks and soil and the movement of water and dissolved species through these porous media has a large impact on the environment. Understanding the flow and transport processes is essential in managing, utilizing and protecting the natural sub-surface water reservoir. The porous media containing sub-surface water are variably saturated covering the whole range from full saturation to partially saturated soils. In unsaturated porous media water is present both as capillary water in small pores, as liquid films on mineral surfaces and as capillary bridges in corners of angular pores and in crevices. Surface forces strongly affect the properties of confined water. Transport properties of confined water in variably saturated porous media at different length scales are discussed in the various contributions to part I of this book. Here, we consider additional influences on the transport properties of water in porous media caused by dissolved species.

In arid regions where evaporation dominates the water balance, increasing salinity and the enrichment of pollutants may cause significant long-term adverse effects limiting the future availability of sub-surface water resources. Dissolved species also have a strong influence on various properties of water that control its transport through porous media. First of all, the water activity decreases with increasing concentration of dissolved species. In a concentrated solution of sodium chloride (NaCl), a ubiquitous salt quite commonly found in natural waters, the water activity can drop to values as low as 0.75. Such highly saline solutions have a strong influence on the vapor–liquid equilibrium both in bulk and in confined water and, consequently, on the rate of evaporation (or condensation) and the rate of water vapor diffusion in a partially saturated porous media. Other thermophysical properties of water that are affected by dissolved salts include viscosity, surface tension, density and heat capacity. For example, in a saturated NaCl solution, the viscosity, compared to pure water, is nearly doubled, the surface tension is increased by about 10 %, the heat capacity is reduced by about a factor of two and the solution density is increased by about 20 %. It is obvious that dissolved salts strongly affect various thermophysical properties of water controlling its transport through porous media.

These influences are most pronounced at high salt concentrations that can evolve in solutions containing very soluble salts at low relative humidity, i.e., under conditions of rapid evaporation. However, even very soluble salts such as sodium chloride may reach saturation in evaporating solutions. Depending on the exact location of the evaporation front, crystalline deposits may then form on the surface of the porous media (efflorescences) or within the pore space (subflorescences). Crystallization within the existing pore space creates new solid–vapor and solid–liquid interfaces and also secondary porosity. Hence, new water films and capillary bridges may evolve. Crystalline deposits may also significantly reduce the total porosity leading to a reduction of permeability due to pore-clogging. If crystals growing in a pore may become confined between the pore walls, an additional interface between two solid surfaces, the mineral surface of the pore wall and the surface of the growing crystal, is formed. Due to the repulsion of the two solid surfaces they are usually separated by a thin liquid film. Such solution films at the solid–liquid–solid interface play a fundamental role in understanding such processes in porous media as crystallization pressure and pressure solution.

Pollutants and other dissolved species in pore solutions in soils and rocks may be also involved in various chemical reactions. Homogeneous reactions in the aqueous phase include acid–base reactions, complex formation or redox reactions. Important heterogeneous processes include sorption equilibria on mineral surfaces, the precipitation of new solid phases as mentioned before, or, the dissolution of a mineral constituent of the porous material in the pore solution. In order to fully understand the transport of both water and dissolved species including the fate of various pollutants (such as nutrients, organic toxicants, trace metals or radionuclides) in unsaturated soils requires that all individual processes and their interactions are carefully studied. Consequently, the modeling of the dynamics of reactive transport is particularly challenging.

The contributions to part IV of this book cover various aspects of some of the processes mentioned before. It is not the intention of this part to cover all aspects in detail but rather to provide some insight to specific problems related to the general topic of the workshop, i.e., confinement and capillarity.

The first two contributions start from a theoretical treatment of the thermodynamics of stressed solids at the solid–liquid–solid interface at grain contacts. The first contribution (Steiger) focuses on the generation of stress by growing crystals (crystallization pressure) and discusses the influence of the curvature of the solid–liquid interface of very small crystals in nanometer-sized pores on this crystallization pressure. In the second contribution Dysthe focuses on the dissolution at grain interfaces (pressure solution) and demonstrates that the thermodynamics of crystallization pressure and pressure solution are the same, only the direction of the process is reversed. Dysthe also provides an analysis of the kinetics of pressure solution. Considering both dissolution rate and transport (diffusion) rate limited kinetics he concludes that depending on the interface structure the rate variation may cover orders of magnitude.

Geochemical modeling in mixed pore solutions is dealt with in the contributions of Lassin et al. and Steiger. It is demonstrated that a mixed electrolyte thermodynamic model can be used to predict the thermodynamic properties (e.g. water activity, ion activity coefficients, density) in mixed multi-component electrolyte solutions and to accurately calculate complex phase diagrams (Steiger). Lassin et al. focus on the application of such geochemical models to predict the properties in capillary aqueous solutions, i.e. solutions under negative pressure. They provide two examples of successful model applications to understand experimental results obtained with capillary systems, i.e. the calculation of NaCl solubility in capillary bridges and the calculation of equilibrium CO_2 partial pressures over confined pore water.

The following chapters are concerned with salt and moisture transport and salt crystallization during evaporation of saline solutions from porous media. Prat reports on experiments to study the evaporation of a sodium chloride solution from a model porous medium (using glass beads). A major focus in these studies is to better understand the factors controlling the location of salt precipitation (efflorescence vs. subflorescence) and the nature of the efflorescences that are formed. At least two different types of efflorescences (patchy and crusty) are distinguished. While patchy efflorescences do not seem to have a strong influence on evaporation rates, crusty efflorescences significantly reduce evaporation rates. The author also identifies several open questions to be addressed in future research. Weisbrod et al. largely focused on the formation of subflorescences during evaporation of a sodium chloride solution from a porous media. X-ray microtomography (CT) and 3D-imaging of the porous media (dune sand) is used to correlate sodium chloride precipitation during evaporation to changes in porosity and permeability. The latter were calculated using lattice Boltzmann modeling. A significant reduction in permeability of about 60 % due to the precipitation of NaCl in the pore space and the corresponding decrease in total porosity is observed.

The following chapter deals with reactive transport modeling. Scher and Berkowitz report on an implementation of particle tracking into a continuous time random walk model. This model is applied to simulate different reaction transport scenarios, including the dynamics of a homogeneous reaction (formation of a $CuEDTA^{2-}$ complex) and a complex system of heterogeneous reactions (dedolomitization, dolomite dissolution and calcite precipitation). It is demonstrated that fluctuations of several key variables play a dominant role. Subtle changes in the patterns of water flow may have significant effects on water quality.

While the previous chapters were largely concerned with evaporation of water vapor from saline solutions in porous media, the final chapter of this section deals with water vapor flow in the opposite direction. Gordeeva and Aristov report on the extraction of water vapor from the atmosphere in arid regions using sorbent materials. These materials consist of a porous host material and an inorganic salt. The sorption capacity of the materials is significantly higher than in typical porous adsorption materials (e.g. zeolites) due to additional water uptake by salt hydration and deliquescence. The authors discuss the use and optimization of such materials to utilize the huge atmospheric freshwater reservoir.

Michael Steiger and Marc Prat

Chapter 16
Crystal Growth and Phase Equilibria in Porous Materials

Michael Steiger

Abstract In numerous research areas there is considerable interest in the phase changes occurring in pore solutions. First, the pressure generated by crystal growth of salts in confined spaces of porous materials is generally recognized as a major damage mechanism. Second, dissolved salts strongly affect the moisture retention and transport properties of porous media. This report briefly reviews recent advances in the treatment of phase equilibria of salts in porous substrates. The first part deals with the theory of crystallization pressure. In the second part model approaches for the calculation of thermodynamic properties in mixed electrolyte solutions and the calculation of phase equilibria are discussed.

16.1 Introduction

Mixtures of different soluble salts are commonly present in natural rocks and soils but also in building materials, e.g. natural stone, brick or concrete. Most of these salts are highly soluble and hygroscopic. Depending on the environmental conditions they may be present in a porous material in dissolved form as pore solution and are then subject to transport. Under dry conditions, as water evaporates from the pore solution, salts may crystallize out and form efflorescences (on the surface) or subflorescence (within the pore space). Other relevant phase changes apart from evaporation–condensation at the air–solution interface and crystallization–dissolution at the crystal–solution interface are changes of the state of hydration of salts and the crystallization of ice, i.e. the freeze–thaw equilibrium.

M. Steiger (✉)
Department of Chemistry, University of Hamburg, Hamburg, Germany
e-mail: steiger@chemi.uni-hamburg.de

L. Mercury et al. (eds.), *Transport and Reactivity of Solutions in Confined Hydrosystems*, 187
NATO Science for Peace and Security Series C: Environmental Security,
DOI 10.1007/978-94-007-7534-3_16, © Springer Science+Business Media Dordrecht 2014

In numerous research areas there is considerable interest in the phase changes occurring in pore solutions. First, the pressure generated by crystal growth of salts in confined spaces of porous building materials such as stone, brick and concrete is generally recognized as a major cause of damage in ancient monuments and modern buildings [1–3]. Crystal growth is also considered as an important weathering mechanism of natural rocks in a variety of environments not only on earth, e.g. in deserts, dry Antarctic valleys and tropical coastlines, but also on Mars. Crystal growth in porous materials is the result of phase changes that are induced by variation of ambient temperature and relative humidity, RH. Unfavorable conditions of temperature and RH may result in repeated cycles of crystallization–deliquescence, hydration–dehydration and freezing–thawing, respectively. Under such conditions, building materials and natural rocks are subject to rapid decay.

Apart from these damage processes, dissolved salts also strongly influence the water uptake and moisture transport properties of building materials and natural rock or soil. For instance, due to their hygroscopicity salts strongly affect the sorption isotherm of a material and dissolved salts alter the capillary transport properties, e.g. solution density and viscosity, of the pore liquid. Finally, salts present in crystalline form change the pore structure, i.e. the pore size distribution and the permeability of materials. Therefore, in moisture transport models an accurate representation of the behavior of salts and salt mixtures is required. This includes (1) an appropriate model for the representation of the thermodynamic properties of aqueous pore solutions and (2) the ability to treat the relevant solid–liquid phase equilibria.

This report briefly reviews recent advances in the treatment of phase equilibria of salts in porous substrates. The first part deals with the theory of crystallization pressure. In the second part model approaches for the calculation of thermodynamic properties in mixed electrolyte solutions and the calculation of phase equilibria are briefly discussed.

16.2 Crystallization Pressure

It is generally accepted that the crystallization of salts is a major damage mechanism in stone. Though experimental evidence that growing crystals can exert pressure in porous materials was provided more than a century ago, until recently, there was no agreement among researchers regarding the nature of the process responsible for the generation of stress [4]. However, in recent years, there has been substantial progress in understanding the thermodynamics of confined crystal growth and the generation of crystallization pressure [5–8]. The following paragraphs present a brief summary of the current state of knowledge.

Pioneering experiments at the beginning of the last century provided experimental evidence that growing crystals can exert pressure against a constraint (e.g. [9, 10]). Important conclusions were drawn from these experiments. First, in order that a crystal can continue to grow on its loaded surface a solution film must exist separating the loaded face from its constraint. Otherwise deposition of matter and growth in this contact region is impossible. The solution film acts as a diffusion path allowing the exchange of ions between the solution and the crystal. It originates from repulsive forces between the crystal and its constraint [5]. The second important observation was that growth upon the loaded face of a crystal can only exert pressure if this face is in contact (via the solution film) with a supersaturated solution. A thermodynamic treatment of the situation of a growing crystal that is subject to non-hydrostatic, anisotropic stress yields an equation for the pressure that is generated by a growing crystal confined in void spaces of a rock or any other porous material [8]:

$$\Delta p = RT/V_m \ln (a/a_0) \tag{16.1}$$

Here, the crystallization pressure is defined as the difference between the pressure p_c upon the loaded face of a confined growing crystal and the liquid phase pressure p_l, i.e., $\Delta p = p_c - p_l$, while R is the gas constant, T the absolute temperature and V_m the molar volume of the crystalline solid. The supersaturation is expressed as the ratio of the activities of the dissolved species in the supersaturated (a) and the saturated solutions (a_0). For a dissociating solid of general composition

$$M_{\nu M}X_{\nu X} \cdot nH_2O \rightleftharpoons \nu_M M^{z+} + \nu_X X^{z-} + nH_2O \tag{16.2}$$

the activities a and a_0 are expressed as ion activity products

$$a = a_M^{\nu M} a_X^{\nu X} a_w^n \tag{16.3}$$

In Eq. (16.2) a_M and a_x are the activities of the cation M and the anion X, a_w is the activity of water and ν_M and ν_X are stoichiometric coefficients. The activity a_0 of a saturated solution is equal to the thermodynamic solubility product of the solid.

The crystallization pressure generated by confined growing crystals induces tensile stress within the solid matrix that might exceed the strength of the stone. As a first indicator crystallization pressures may be compared to the tensile strengths of natural stones that rarely exceed values of about 3–5 MPa. Crystallization pressures for several salts commonly found in natural rocks and in building materials were calculated using Eq. (16.1) [8]. Assuming realistic supersaturations, the resulting crystallization pressures far exceed the tensile strength of most building stones and may be sufficient to cause damage in most materials. A more sophisticated treatment is based on the theory of poromechanics [11]. Applying this theory, it was shown recently that there is good agreement between calculated crystallization pressures using Eq. (16.1) and the crystallization pressure derived from deformation measurements using theory of poroelasticity [12, 13].

Very high supersaturation can be easily realized in rewetting experiments such as in the widely used durability tests of building materials or natural rocks. Typically, in such tests a porous material is impregnated with a sodium sulfate solution and dried at enhanced temperature (e.g. 60–105 °C) such that anhydrous Na_2SO_4 is formed.

Subsequently, after cooling to room temperature, the specimen is impregnated again with a Na_2SO_4 solution leading to the hydration of Na_2SO_4. Repeating this procedure several times sodium sulfate proved to be extremely destructive. Most of the damage occurs during the impregnation phase and the destructive effect is attributed to the growth of $Na_2SO_4 \cdot 10H_2O$ (mirabilite) crystals from the highly supersaturated solutions originating from the dissolution of anhydrous Na_2SO_4 [6, 14–16].

Crystals growing in confined space are typically not under homogenous but rather under anisotropic stress. The derivation of Eq. (16.1) is based on the solubility increase of a crystal face under pressure. The crystallization pressure is then the load at which solution and loaded crystal face are just at equilibrium. Due to the solubility increase with pressure it is clear that a pore solution, which is in equilibrium with the loaded crystal face, must be supersaturated with respect to the unloaded faces. Therefore, a confined crystal under anisotropic stress represents a non-equilibrium situation and the crystal will always grow at its unloaded faces consuming the actual supersaturation. In effect, the pressure generated at the loaded faces cannot be maintained [8, 17].

As long as unloaded crystal faces are present, the evolution of crystallization pressure in a porous material is a dynamic process that is controlled by kinetic influences such as evaporation or cooling rates, the diffusion of ions in the free solution and in the liquid film, the availability of unloaded crystal surface, and the growth rates on the unloaded faces. Under such conditions it is very unlikely that crystallization pressure builds up and remains constant over long periods of time. More likely, high pressures occur as transients if high supersaturation in the pore solution evolves temporarily as a result of sharp temperature drops or rapid evaporation. Amplitude and duration of stress maxima depend on the parameters controlling the degree of supersaturation. Such a dynamic behavior is confirmed in experimental work on salt crystallization in porous stone during both cooling [17] and re-wetting [13, 15] experiments. The stress relaxation is controlled by the diffusion of ions from the loaded to the free crystal surface and the subsequent growth at this crystal face. Espinosa Marzal and Scherer [17] attributed the slow relaxation found in their experiments to the large diffusion distance in the solution film at the loaded crystal surface.

It should be noted that under certain circumstances a permanent crystallization pressure can also occur. Such an equilibrium situation can evolve if the unloaded face of a growing crystal, due to its curvature, exhibits the same solubility increase as the loaded surface of the crystal [18]. In this particular case, the loaded and the unloaded faces are both in equilibrium with the same activity of the pore solution. This is schematically illustrated in Fig. 16.1 showing a crystal growing in a large spherical pore with small cylindrical pore entrances. As the solubility of a crystal increases with decreasing size, the crystal exhibits different solubilities at different crystal surfaces. The large crystal face (corresponding to the radius r_P) has a lower solubility than the small satellite crystal growing into the pore entrance (radius r_E). Hence, growth of the crystal into the small pore entrances requires a higher concentration of the pore solution due to the greater curvature of the small satellite

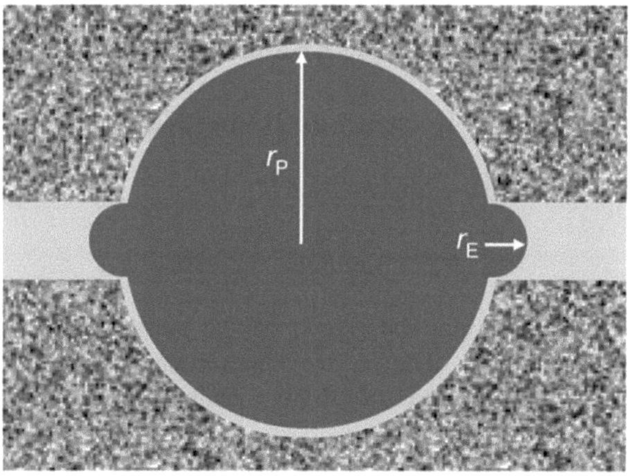

Fig. 16.1 Large crystal growing in a spherical pore with cylindrical pore entrances (Redrawn from ref. [18])

crystals. At equilibrium, the concentration in the surrounding solution is just equal to the solubility of the hemispherical crystal of size r_E. However, this solution is supersaturated with respect to the unloaded large crystal. Therefore, stress is generated due to growth of the confined crystal in the large pore until equilibrium is reestablished.

At equilibrium, the solution is saturated with respect to both crystal surfaces. However, due to their different size the particle in the pore entrance is under ambient pressure while the large particle must be under enhanced pressure. The pressure generated in the large pore can be calculated from the solubility increase [18]. For the particular geometry shown in Fig. 16.1 one obtains:

$$\Delta p = 2\gamma_{cl}\,(1/r_E - 1/r_P) \tag{16.4}$$

where γ_{cl} is the crystal–liquid interfacial free energy. The treatment of other special geometries and a general equation for Δp is provided in reference [18].

Surface energy effects are not limited to the crystal–liquid interface and small pores (i.e. <0.1 μm). Typically, the pores in natural rocks and soil are irregularly shaped and are not well represented by idealized geometries like spheres or cylinders. In contrast, angular pores with corners have to be considered. In unsaturated materials liquid remains in the corners of partially filled pores forming capillary bridges. The resulting curved liquid–air interfaces strongly affect both water transport [19, 20] and crystallization of salts [21, 22].

Several salts can exist in different hydrated forms. Hydrated salts that are commonly found in building materials or in natural rocks and soil include the series $Na_2SO_4 \cdot nH_2O$, $MgSO_4 \cdot nH_2O$ and $Na_2CO_3 \cdot nH_2O$ and these salts proved to be particularly damaging in salt crystallization tests. As already discussed, the damage mechanism in this type of experiments is essentially the formation of a solution supersaturated with a higher hydrated salt by dissolution of a lower hydrated or anhydrous salt. The general form of such a hydration reaction is given by

$$\mathrm{M}_{\nu_\mathrm{M}}\mathrm{X}_{\nu_\mathrm{X}} \cdot n_1\mathrm{H_2O} + (n_2 - n_1)\,\mathrm{H_2O} \rightleftharpoons \mathrm{M}_{\nu_\mathrm{M}}\mathrm{X}_{\nu_\mathrm{X}} \cdot n_2\mathrm{H_2O} \tag{16.5}$$

The equilibrium constant of this reaction is

$$K = p_{w,0}/p_{\mathrm{w,eq}} = 1/\mathrm{RH_{eq}} \tag{16.6}$$

where $p_{\mathrm{w,eq}}$ is the equilibrium water vapor partial pressure, $p_{\mathrm{w,0}}$ is the saturation water vapor pressure, and, $\mathrm{RH_{eq}}$ is the equilibrium relative humidity for the hydration–dehydration equilibrium of the two hydrates with n_1 and n_2 molecules of water.

In the conventional salt crystallization test the hydration reaction is triggered by impregnation of a porous material containing the anhydrous (or lower hydrated) phase with pure water or a salt solution. Dissolution of this phase yields a supersaturated solution and crystallization of the higher hydrated phase. According to Eqs. (16.5) and (16.6), the hydration reaction can be also triggered by simply increasing the relative humidity above the equilibrium value ($\mathrm{RH_{eq}}$). However, many hydration reactions are kinetically hindered in the absence of a liquid phase. This was confirmed recently in an in situ investigation of the hydration of $MgSO_4$âĹŹH_2O in a porous substrate [23]. It was found that hydration and formation of $MgSO_4$âĹŹ6H_2O was only rapid and complete at relative humidities above the deliquescence humidity of the monohydrate. Under such conditions a saturated solution of the monohydrate forms, which is far supersaturated with the hexahydrate. It was further shown by strain measurements that the hydration of $MgSO_4$âĹŹH_2O in confined space generates substantial stress resulting in considerable deformations of the porous substrate [23]. This study confirmed that hydration reactions are particularly damaging if they proceed via a two step dissolution–crystallization mechanism. Wetting and dissolution may be either triggered by direct contact with liquid water (by imbibition) or indirectly by water vapor transport and formation of a solution above the deliquescence humidity.

16.3 Modeling Thermodynamic Properties and Phase Equilibria in Mixed Pore Solutions

Moisture transport models for both building materials and soil require an accurate representation of the thermodynamic properties and phase equilibria in multicomponent salt solutions. Relevant thermodynamic properties include the activities of water and dissolved ions, which are both required for the calculation of phase equilibria and the prediction of salt and ice crystallization. Other relevant properties include thermal properties, e.g. heat capacities, and volumetric properties (densities and compressibilities) of mixed solutions.

In geochemical modeling the molality-based Pitzer formalism [24] was successfully applied in chemical equilibrium models [25] for the representation of thermodynamic properties of natural waters. Most of the available databases of model parameters are used to calculate sulfate and chloride solubilities in concentrated brines. However, only few models include nitrates, which are quite often found as contaminant salts in building materials. It was shown previously that existing parameters may cause erroneous results if applied for the prediction of solubilities in mixed solutions containing chloride and nitrate at high ionic strength [26]. The ion interaction approach has been also used for the calculation of volumetric properties [27] and heat capacities [28] in electrolyte solutions. Here, we report on a molality based Pitzer model for the calculation of activities and densities in mixed electrolyte solutions.

For the solubility reaction of a salt of general composition (Eq. (16.2)) the thermodynamic solubility product is given by

$$\ln K_{\mathrm{MX}} = \nu_{\mathrm{M}} \ln m_{\mathrm{M}} + \nu_{\mathrm{X}} \ln m_{\mathrm{X}} + \nu_{\mathrm{M}} \ln \gamma_{\mathrm{M}} + \nu_{\mathrm{X}} \ln \gamma_{\mathrm{X}} + \nu_0 \ln a_{\mathrm{w}} \tag{16.7}$$

where m_{M}, m_{X}, γ_{M} and γ_{X} represent the molalities and activity coefficients of the cations and anions, respectively, and a_{w} is the water activity defined as

$$\ln a_{\mathrm{w}} = -\phi M_{\mathrm{w}} \sum_i m_i \tag{16.8}$$

where ϕ is the osmotic coefficient and $M_{\mathrm{w}} = 1.801528 \cdot 10^{-2}$ kg mol^{-1} is the molar mass of water. The calculation of solubility equilibria requires the knowledge of activity and osmotic coefficients as a function of composition and temperature. The equations used in the Pitzer model to represent the excess Gibbs energy of electrolyte solutions are reviewed in reference [24]. The equations incorporate an extended Debye-Hückel limiting law and a virial expansion representing short range ionic interactions. Expressions for activity and osmotic coefficients are obtained from appropriate derivatives of the Gibbs free energy equation. The virial coefficients are treated as empirical quantities and are evaluated from experimental data. The adjustable parameters are the binary and ternary interaction parameters. Binary interaction parameters are evaluated from thermodynamic data of binary solutions. Ternary interaction parameters are obtained from experimental data of simple mixtures containing only three different ions, i.e. mixtures of two salts with a common ion. Once all binary and ternary interaction parameters are evaluated, the equations may be used to predict activity and osmotic coefficients in mixed electrolyte solutions of any composition.

To use the model for solubility calculations the equilibrium constants K_{MX} and their temperature dependence are also required. Details of the determination of the model parameters (interaction parameters, solubility products) are provided in references [15, 26, 30]. Figure 16.2 depicts the results of the model parameterization for the solubilities in the ternary system $NaCl$–$NaNO_3$–H_2O system. The symbols represent the literature data that were used for the determination of the ternary model

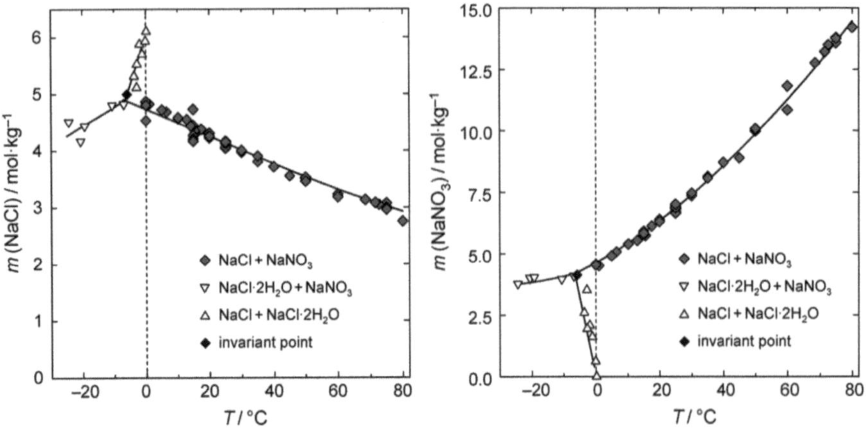

Fig. 16.2 Solubilities of NaCl and $NaNO_3$ at the isothermal invariant points in the ternary NaCl–$NaNO_3$–H_2O system; *symbols* represent literature data [29], *lines* are calculated solubilities

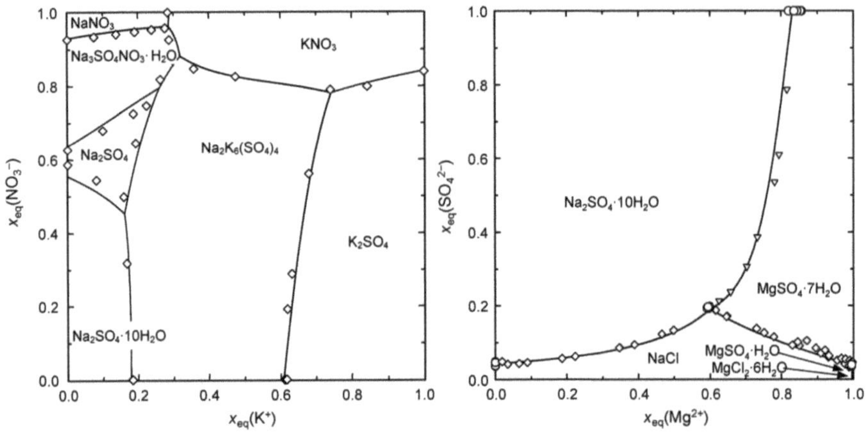

Fig. 16.3 Calculated solubilities (—) compared to experimental data [30] in the reciprocal systems Na^+–K^+–NO_3^-–SO_4^{2-}–H_2O (*left*, 25 °C) and Na^+–Mg^{2+}–Cl^-–SO_4^{2-}–H_2O (*right*, 0 °C); x_{eq} are charge equivalent fractions of cations and anions in saturated solutions (Redrawn from references [26] and [30])

parameters. There is very good agreement between calculated and experimental solubilities confirming that the model equations allow for an excellent reproduction of the activities in mixed solutions even at very high ionic strength.

While binary and common ion ternary experimental data are used to determine all model parameters, data from more complex systems are used to validate the model. Therefore, the model can be tested by comparing predicted solubilities with the available data of mixed systems containing more than three ions. Figure 16.3 shows a comparison of calculated and experimental solubility data in the systems Na^+–K^+–NO_3^-–SO_4^{2-}–H_2O and Na^+–Mg^{2+}–Cl^-–SO_4^{2-}–H_2O. The solubilities are given as Jaenecke projections in terms of the charge equivalent cation and anion mole fractions, i.e. $x_{eq}(K^+)$ and $x_{eq}(NO_3^-)$ in Fig. 16.3a, and, $x_{eq}(Mg^{2+})$ and $x_{eq}(SO_4^{2-})$ in Fig. 16.3b. There is very good agreement between the model calculations and the experimental data. It should be noted that the calculated solubilities are true model predictions since solubility data of reciprocal systems were not used in the parameterization of the model. The lines in the interior of the square diagram are the saturation curves for solutions coexisting with two solid phases. They define the crystallization fields of the stable solids, e.g. $NaNO_3$, KNO_3, $Na_2SO_4 \cdot 10H_2O$, Na_2SO_4, $NaNO_3 \cdot Na_2SO_4 \cdot H_2O$ and $Na_2SO_4 \cdot 3K_2SO_4$ in the Na^+–K^+–NO_3^-–SO_4^{2-}–H_2O system. The intersections of the lines in Fig. 16.3 are the invariant points of the system representing the compositions of solutions in equilibrium with three solid phases, which is the maximum number according to the Gibbs phase rule.

Equilibrium models have been successfully used to predict the crystallization behavior of mixed salt systems and the critical environmental conditions that cause damage in historic monuments [e.g. 31]. A major problem in the modeling of the crystallization behavior of salt mixtures arises from kinetic effects. The formation of several salts is kinetically hindered such that metastable phases form in the crystallization pathway. A prominent example is Na_2SO_4âĹŹ$7H_2O$, which is often formed instead of the stable compound Na_2SO_4âĹŹ$10H_2O$ [32, 33]. In principle, phase equilibria involving metastable phases can be treated using the same thermodynamic model [15, 30]. However, an equilibrium model cannot, of course, predict the conditions under which a system follows a metastable crystallization pathway.

Comprehensive reviews of the Pitzer ion interaction approach and the form of the equations used for the calculation of volumetric quantities can be found elsewhere [27]. The total volume V of a multiple-solute electrolyte solution is given by

$$V = n_1 V_1^0 + \sum_i n_i \overline{V}_i^0 + V^{ex} \tag{16.9}$$

where $V_1{}^{o}$ is the molar volume of water, $\overline{V}_i^0$ are the partial molal volumes of ions i in solution at infinite dilution, n_1 and n_i are number of moles of water and dissolved ions, respectively, and V^{ex} is the excess volume of the mixed solution. Using the molal concentration scale

$$V = 1000 v_w + \sum_i m_i \overline{V}_i^0 + V^{ex} \tag{16.10}$$

where m_i are the molalities of the ions in solution, and v_w is the specific volume of water. The excess volume V^{ex} is the pressure derivative of the excess Gibbs energy. Hence, the ion interaction equations for volume calculations are obtained by taking the appropriate derivation of the interaction equation for the excess Gibbs energy [27]. These equations contain the pressure derivatives of the binary and the ternary interaction parameters. The calculation of total volumes in porous media

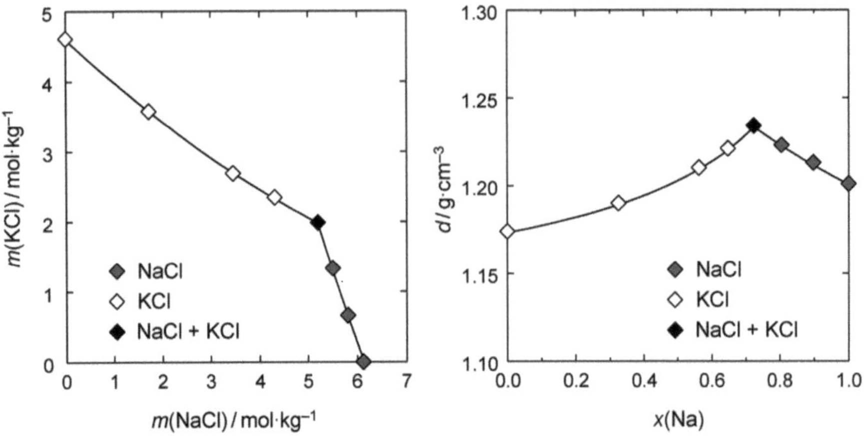

Fig. 16.4 Calculated and experimental solubilities in the mixed system NaCl–KCl–H_2O (*left*) and densities *d* of saturated solutions (*right*) at 20 °C, *x*(Na) is the mole fraction of NaCl in the mixed solutions, *symbols* represent experimental data [30]

also requires the knowledge of mineral densities. The model parameters for the density model were taken from an earlier study [34]. A comparison of calculated and experimental densities in mixed solutions of the NaCl–KCl–H_2O system at 20 °C is depicted in Fig. 16.4. The left diagram shows the solubilities in the ternary system; the right diagram is a plot of the densities of the saturated solutions shown in the left diagram. There is excellent agreement between calculated and experimental data.

The treatment so far considers solubility equilibria in fully saturated porous media. In very small pores and in capillary bridges, however, solubilities are affected by the vapor–solution and crystal–solution interfaces. The additional influences are the result of the low liquid phase pressure [22, 35] and curved crystal–liquid interfaces [18]. Liquid phase pressure and curvature of the crystal–liquid interface affect the thermodynamic solubility products. There is also an influence of pressure on the activity coefficients that needs to be considered [36]. A detailed treatment of these influences is beyond the scope of this short review. Preliminary results of our attempts to extend our model including the calculation of freezing temperatures of pure water and electrolyte solutions in mesoporous materials and the calculation of solubilities in small unsaturated pores are provided in two recent reports [37, 38].

References

1. Goudie A, Viles H (1997) Salt weathering hazards. Wiley, Chichester
2. Doehne E (2002) Salt weathering: a selective review. Geol Soc Lond Spec Publ 205:51–64
3. Steiger M, Charola AE, Sterflinger K (2011) Weathering and deterioration. In: Siegesmund S, Snethlage R (eds) Stone in architecture. Springer, Berlin, pp 227–316

4. Evans IS (1970) Salt crystallization and rock weathering: a review. Rev Geomorph Dyn 19:153–177
5. Scherer GW (1999) Crystallization in pores. Cem Concr Res 29:1347–1358
6. Flatt RJ (2002) Salt damage in porous materials: how high supersaturations are generated. J Cryst Growth 242:435–454
7. Scherer GW (2004) Stress from crystallization of salt. Cem Concr Res 34:1613–1624
8. Steiger M (2005) Crystal growth in porous materials – I: the crystallization pressure of large crystals. J Cryst Growth 282:455–469
9. Becker GF, Day AL (1916) Note on the linear force of growing crystals. J Geol 24:313–333
10. Correns CW, Steinborn W (1939) Experimente zur Messung und Erklärung der sogenannten Kristallisationskraft. Z Krist A 101:117–135
11. Coussy O (2004) Poromechanics. Wiley, Chichester
12. Espinosa-Marzal RM, Scherer GW (2010) Mechanisms of damage by salt. Geol Soc Lond Spec Publ 331:61–77
13. Espinosa-Marzal RM, Hamilton A, McNall M, Whitaker K, Scherer GW (2011) The chemo-mechanics of crystallization during rewetting of limestone impregnated with sodium sulfate. J Mater Res 26:1472–1481
14. Chatterji S, Jensen AD (1989) Efflorescence and breakdown of building materials. Nordic Concr Res 8:56–61
15. Steiger M, Asmussen S (2008) Crystallization of sodium sulfate phases in porous materials: the phase diagram Na_2SO_4–H_2O and the generation of stress. Geochim Cosmochim Acta 72:4291–4306
16. Shahidzadeh-Bonn N, Desarnaud J, Bertrand F, Chateau X, Bonn D (2010) Damage in porous media due to salt crystallization. Phys Rev E 81:066110
17. Espinosa Marzal RM, Scherer GW (2008) Crystallization of sodium sulfate salts in limestone. Environ Geol 56:605–621
18. Steiger M (2005) Crystal growth in porous materials – II: influence of crystal size on the crystallization pressure. J Cryst Growth 282:470–481
19. Tuller M, Or D (2001) Hydraulic conductivity of variably saturated porous media: film and corner flow in angular pore space. Water Resour Res 37:1257–1276
20. Sghaier N, Prat M, Ben NS (2006) On the influence of sodium chloride concentration on equilibrium contact angle. Chem Eng J 122:47–53
21. Shahidzadeh-Bonn N, Rafaï S, Bonn D, Wegdam G (2008) Salt crystallization during evaporation: impact of interfacial properties. Langmuir 24:8599–8605
22. Bouzid M, Mercury L, Lassin A, Matray J-M, Azaroual M (2011) In-pore tensile stress by drying-induced capillary bridges inside porous materials. J Colloid Interface Sci 355:494–502
23. Steiger M, Linnow K, Juling H, Gülker G, El Jarad A, Brüggerhoff S, Kirchner D (2008) Hydration of $MgSO_4 \cdot H_2O$ and generation of stress in porous materials. Cryst Growth Des 8:336–343
24. Pitzer KS (1991) Ion interaction approach: theory and data correlation. In: Pitzer KS (ed) Activity coefficients in electrolyte solutions. CRC Press, Boca Raton, pp 75–153
25. Harvie CE, Møller N, Weare JH (1984) The prediction of mineral solubilities in natural waters: the Na–K–Mg–Ca–H–Cl–SO_4–OH–HCO_3–CO_3–CO_2–H_2O system to high ionic strengths at 25°C. Geochim Cosmochim Acta 48:723–751
26. Steiger M, Kiekbusch J, Nicolai A (2008) An improved model incorporating Pitzer's equations for calculation of thermodynamic properties of pore solutions implemented into an efficient program code. Constr Build Mater 22:1841–1850
27. Monnin C (1989) An ion interaction model for the volumetric properties of natural waters: density of the solution and partial molal volumes of electrolytes to high concentration at 25°C. Geochim Cosmochim Acta 53:1177–1188
28. Criss CM, Millero FJ (1996) Modeling the heat capacities of aqueous 1–1 electrolyte solutions with Pitzer's equations. J Phys Chem 100:1288–1294
29. Silcock HL (1979) Solubilities of inorganic and organic compounds, vol 3. Ternary and multicomponent systems of inorganic substances, vol 3. Pergamon Press, Oxford

30. Steiger M, Linnow K, Ehrhardt D, Rohde M (2011) Decomposition reactions of magnesium sulfate hydrates and phase equilibria in the $MgSO_4$–H_2O and Na^+–Mg^{2+}–Cl^-—$SO_4{}^{2-}$–H_2O systems with implications for mars. Geochim Cosmochim Acta 75:3600–3626
31. Sawdy A, Price C (2005) Salt damage at Cleeve Abbey, England. Part I: a comparison of theoretical predictions and practical observations. J Cult Herit 6:125–135
32. Hamilton A, Hall C, Pel L (2008) Sodium sulfate heptahydrate: direct observation of crystallization in a porous material. J Phys D Appl Phys 41:212002
33. Saidov TA, Espinosa-Marzal RM, Pel L, Scherer GW (2012) Nucleation of sodium sulfate heptahydrate on mineral substrates studied by nuclear magnetic resonance. J Cryst Growth 338:166–169
34. Steiger M (2000) Total volumes of crystalline solids and salt solutions. In: Price CA (ed) An expert chemical model for determining the environmental conditions needed to prevent salt damage in porous materials, European commission, research report 11., pp 53–63
35. Zilberbrand M (1999) On equilibrium constants for aqueous geochemical reactions in water unsaturated soils and sediments. Aquat Geochem 5:195–206
36. Monnin C (1990) The influence of pressure on the activity coefficients of the solutes and on the solubility of minerals in the system Na-Ca-Cl-SO_4-H_2O to 200 °C and 1 kbar, and to high NaCl concentration. Geochim Cosmochim Acta 54:3265–3282
37. Steiger M (2006) Freezing of salt solutions in small pores. In: Konsta-Gdoutos MS (ed) Measuring, monitoring and modelling concrete properties. Springer, Dordrecht, pp 661–668
38. Steiger M, Linnow K (2009) Phase equilibria in mesoporous materials. In: Franke L, Deckelmann G, Espinosa-Marzal R (eds) Simulation of time dependent degradation of porous materials. Cuvillier Verlag, Göttingen, pp 183–195

Chapter 17
Shaping the Interface – Interactions Between Confined Water and the Confining Solid

Dag Kristian Dysthe

Abstract The thermodynamics and kinetics of stressed solids with aqueous solutions in the grain interfaces are summarized. The main focus is on pressure solution and the connection to force of crystallization (crystallization pressure) is shown. The largest driving force of pressure solution, the work term, is linear in stress, all other contributions to the driving force are more than 3 orders of magnitude smaller. The minor driving forces (strain energy, surface tension, surface charges) are important for the transformation and stability of the interface structure and thereby the kinetics of pressure solution. Interface structure and stability may change strain rates up to 7 orders of magnitude.

17.1 Introduction

Fluids confined between solid interfaces less than a micrometer apart are ubiquitous in nature [25]. In grain aggregates like soils, sediments and rocks the existence of thin water films at grain contacts is a source of enhanced transformation of the media. At the grain contact that confines a thin water film the surfaces of the solid grains are in a stress state that differs from the grain surface facing water in the pore space. If the water in the pore space is in equilibrium with the solid grain surfaces the solid will dissolve in the contact and the material will diffuse through the water film to the pore space. Inversely, if the water in the pore space is supersaturated in some species they may diffuse into the grain contact and crystallize in the water film. In the first case the grain aggregate will compact under external stress (or due to it's own weight), in the second case the grain aggregate may break under local stresses generated by the crystallization. The stability of the solid-water-solid

D.K. Dysthe (✉)
Physics of Geological Processes, Department of Physics, University of Oslo, PO Box 1048, Blindern, 0316 Oslo, Norway
e-mail: d.k.dysthe@fys.uio.no

L. Mercury et al. (eds.), *Transport and Reactivity of Solutions in Confined Hydrosystems*, NATO Science for Peace and Security Series C: Environmental Security,
DOI 10.1007/978-94-007-7534-3_17,

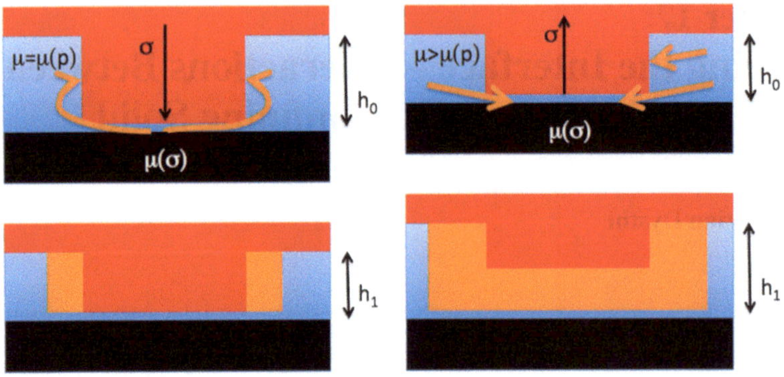

Fig. 17.1 The inverse processes *pressure solution* (*left*) and *force of crystallization* (*right*). **Left**: The water (*bule or light layered grey*) in the pore space is in equilibrium with the solids at the fluid pressure p. The grains are forced together causing a stress σ in the grain contact. Since $\mu(\sigma) > \mu(p)$ the solid grains (*red or dark grey*) will dissolve in the contact, diffuse out through the thin water film and crystallize in the pore space (*orange or intermediate grey*) and $h_1 < h_0$. **Right**: The water (*light layered grey*) in the pore space is supersaturated with respect to some solute. If $\mu > \mu(\sigma)$ there will be diffusion of the solute into the grain contact. Crystallization in the grain contact will induce a stress σ that may push the grains further apart such that $h_1 > h_0$

interface affects both the kinetics of this process and the continued existence of the water film in the grain contact. If the interface is unstable and water is expelled from the contact, the aggregate solidifies and deformation stops. The questions of thermodynamics and kinetics of the solid-water-solid interface at grain contacts is therefore fundamental to understanding for example soil formation by weathering and lithification of soils and sediments (Fig. 17.1).

As direct observations and measurements of natural grain boundaries are difficult, the transport properties of grain boundaries are often characterized by the product between an effective thickness of the interface, Δ_{gb}, and an effective diffusion coefficient D. The fluid film thickness, Δ_{gb}, can vary from about 1 nm (four molecular water layers) to possibly 100 nm depending on the nature of the mineral, fluids and the normal stress [10]. The diffusion coefficient, D, of ions in water at room temperature is of the order 10^{-9} m^2s^{-1} and may be reduced by one to two orders of magnitude [1, 9] by the structure of a confined fluid film, depending on the film thickness. Consequently, the product of fluid film thickness and diffusion coefficient Δ_{gb} D is of the order 10^{-20} to 10^{-16} m^3s^{-1}. These values are consistent with the experimental results.

In the seminal work of Weyl [24] on pressure solution and the force of crystallization, he used two models, one assuming an exponential relation and one assuming independence between disjoining pressure and surface separation. Weyl [24] had some evidence of the existence of a fluid film, but at the time it was still considered hypothetical. He considered the fluid film to be continuous with the bulk fluid, with the same thermodynamic properties like the dependence of solubility on pressure. The fluid pressure inside the fluid film was taken to be the surface normal stress transmitted to the solid surfaces. Furthermore the assumption of continuity, i.e. no jump in pressure and mass conservation resulted in the solution of a continuously varying fluid film thickness in pressure solution (saturated bulk solution) and a growth rim of a certain width during crystallization (supersaturated bulk solution). This continuous stress distribution model of Weyl has been shown to be in contradiction with experimental results [22].

To summarize so far, the fluid-like water confined between the solid surfaces:

- transmits normal stress between the solid surfaces;
- allows molecular diffusion at rates comparable to a normal liquid;
- should only with the greatest care be treated as continuous with the fluid in the pore space and is perhaps better treated as a separate thermodynamic phase.

Actual grain contacts are not necessarily atomically flat and several conceptual models of rough grain contacts have been proposed. These models all consider the deformation of the solid surfaces: the grain contacts are formed by the non-equilibrium processes of pressure solution and force of crystallization. We will in the following focus on pressure solution and the different thermodynamic and kinetic contributions to the development of grain interface structure.

17.2 Thermodynamics of Pressure Solution

In this section we will consider the effect of non-hydrostatic stress and the driving force of pressure solution. The development of this theory, starting with Gibbs [13] has resulted in many different approaches, both equilibrium and non-equilibrium, to deriving the driving forces of pressure solution. Luckily, they all end up with essentially the same driving forces. Except for temperature, T, we will use capital letter symbols for extensive quantities and lower case symbols for intensive quantities. Specific quantities are molar quantities, for example the molar volume is $v = V/N$, where N is the number of moles in volume V. We will use subscript s for solid and subscript l for liquid.

17.2.1 Chemical Potential of Solids in Contact with Their Solution

We choose to start by the Gibbs, G, and Helmholtz, F, free energies of a solid of volume V subject to a hydrostatic pressure p:

$$G = pV + F. \tag{17.1}$$

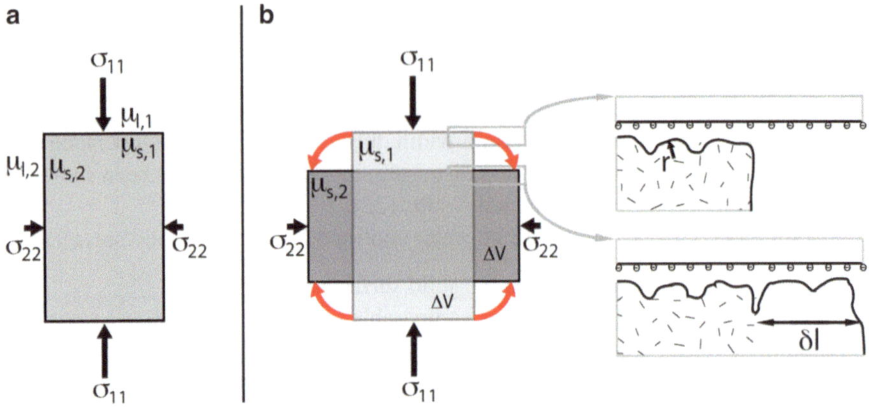

Fig. 17.2 Driving force of pressure solution: (**a**) Surface normal stresses on a grain and the chemical potentials of the solid and liquid at the different interfaces; (**b**) Grain deformation by transfer of a volume $\Delta V = l^2 \delta l$ under the same stresses resulting in the deformation work $W = \Delta V(\sigma_{11} - \sigma_{22})$, enlargement on the right: contributions of surface curvature, r, to the Helmholtz energy, f_γ, the "electrochemical" contribution of the surface charges, f_{sc}, and the contribution of dislocations, $f_{d,}$ in the original crystal. If a steady state deformation is reached, the typical surface curvatures and surface charges must be constant and the only energy change on deformation is proportional to the change in contact area $\Delta A = l\delta l$ (l is the characteristic dimension of the crystal). If the precipitated material is without dislocations and unaffected by the stress component σ_{11} these energy terms, f_d and f_{el} are proportional to the displaced volume ΔV

Dividing by the number of moles, the chemical potential at surface i of a solid (Fig. 17.2a) may be expressed as:

$$\mu_{s,i} = \sigma_{ii} v_s + f_s, \quad f_s = f_{el} + f_d + f_\gamma + f_{sc}, \tag{17.2}$$

where f_s is the specific Helmholtz free energy of the solid and σ_{ii} is the stress normal to surface i, defined as positive when pointing into the solid, thus $p = \frac{1}{3}\sum \delta_{ij}\sigma_{ij}$. The five terms contributing to the chemical potential are summarized in the following:

1. The work related to surface normal stress accounts for the largest contribution to the chemical potential of the solid adjacent to the surface: $\mu_{sp,i} = v_s\sigma_{ii}$, $\mu_{sp} = v_s p$
2. The elastic term of the specific Helmholtz free energy accounts for the non-hydrostatic strain energy:

$$f_{el} = v_s\left(\sigma_{ij} - p\right) d\varepsilon_{ij} = \frac{v_s\left(\sigma_{ij} - p\right) d\sigma_{ij}}{(1-\nu)\,E} \tag{17.3}$$

where the latter equality is valid for isotropic solids with Young's modulus E and Poisson's ratio ν. The ratio between the energy due to hydrostatic pressure and energy due to elastic strain (for the same magnitude of hydrostatic pressure and non-hydrostatic stress) is $\frac{\mu_{sp}}{f_{el}} = (1-\nu)\frac{E}{\sigma_{ii}} \leq 1{,}000$ for quartz because the Young's modulus is 1,000 times larger than the yield stress. Thus for most rock materials and stress conditions, the energy related to surface normal stress is much larger than the non-hydrostatic strain energy term.

3. The plastic strain energy, f_d, of a crystalline solid is stored in defects, such as dislocations. For a large enough density of dislocations, for example greater than 10^{10} dislocations per cm^2 for quartz, the free energy of a crystal can be significantly increased. The energy of a dislocation can be divided in two terms: the energy of the core of the dislocation itself, and the strain energy due to the deformation of the crystalline lattice (Wintsch and Dunning, 1985). This induces an increase of the solubility of quartz. For example, at 400 °C and 300 MPa, a value of 10^{10} dislocations per cm^2 in quartz may induce an increase of solubility close to 1 %. For non-deformed quartz crystals, the density of dislocations is usually smaller than 10^9 cm^{-2}, which is not sufficient to obtain a significant increase of the chemical potential of the solid.
4. The surface energy contribution, $f_\gamma = \frac{2v_s\gamma}{r}$, to the Helmholtz energy may be viewed as an enhanced pressure contribution to the chemical potential. The Laplace pressure difference over a curved interface is $\Delta p = \frac{2\gamma}{r}$, thus for typical interfacial energies of mineral-water interfaces of 0.1–1 J/m^2 [21] a radius of curvature of 10^{-6} m corresponds to pressures in the range 0.2–2 MPa. This contribution to the Helmholtz energy affects a volume of typically the surface area A times the typical radius of curvature, r, of surface roughness: rA, thus the effects can only be found on small length scales, like for the organization of the grain boundary.
5. The last term, f_{sc}, is also a surface contribution to the Helmholtz energy accounting for the effect of surface charges and ionic environment affecting the local solubility and strength of intermolecular bonds of the solid. The term f_{sc} will only contribute to the change of energy of the system if the state of the interface (ion concentrations etc.) changes during the process or if the area of contact changes. The solid volume affected by this surface effect is the area times some small length scale, ξ, typically a few molecular layers thick, i.e. $\sim 10^{-9}$ m. This may be very important on a local grain boundary scale when the interface structure is modified, but during the deformation of whole grains the total change in surface energy is small compared to volume energies like $\mu_{s,p}, f_{el}$ and f_d.

Although the four latter terms are much smaller than the first term, they may be very important in determining the local shape and organization of the grain boundary.

17.2.2 Driving Force of Pressure Solution and Force of Crystallization (Crystallization Pressure)

The classical way of calculating the thermodynamic driving force of a process is to consider changes to a well defined thermodynamic system. Fig. 17.2b portrays the most important contribution in pressure solution, the work W performed by the surroundings on a solid grain when a volume ΔV is moved from face 1 to face 2:

$$W = \Delta V\left(\sigma_{11} - \sigma_{22}\right) \tag{17.4}$$

Identifying this work as the change in free energy of the grain and dividing by the mass of the volume ΔV displaced one obtains the "work term of the driving force" of pressure solution:

$$\Delta\mu_W = \mu_{sp,1} - \mu_{sp,2} = v_s\left(\sigma_{11} - \sigma_{22}\right) \tag{17.5}$$

which we may identify as originating from term 1 of the chemical potential of the solid in the list above. Considering Fig. 17.2b, the total change in free energy of the solid grain is

$$\begin{aligned}\Delta G = {} & \Delta V\left(\mu_{sp,1} + f_{el,1} + f_{d,1} - \mu_{sp,2} - f_{el,2} - f_{d,2}\right) + \Delta A r\left(f_{\gamma,1} - f_{\gamma,2}\right) \\ & + \Delta A \xi\left(f_{sc,1} - f_{sc,2}\right)\end{aligned}$$

$$\begin{aligned}\Delta\mu_{tot} &= \left(\mu_{sp,1} + f_{el,1} + f_{d,1} - \mu_{sp,2} - f_{el,2} - f_{d,2}\right) + \frac{r}{l}x\left(f_{\gamma,1} - f_{\gamma,2}\right) + \frac{\xi}{l}\left(f_{sc,1} - f_{sc,2}\right) \\ &= \mu_{s,1} - \mu_{s,2}\end{aligned} \tag{17.6}$$

This demonstrates that the surface tension and surface charge effects are negligible as driving forces unless the grain size, l, is very small. The smaller terms (2–5 in the list above) of the chemical potential may, however contribute significantly to the organization of the interfaces and thereby to the kinetics of the process.

How does this relate to chemical potential differences in the fluid phase? The fluid chemical potential is

$$\mu\left(\mathrm{p,T,c}\right) = \mu_0\left(\mathrm{p,T}\right) + \mathrm{RTln}\left(\mathrm{a}\right) = \mu_0\left(\mathrm{p,T}\right) + \mathrm{RTln}\left(\chi\mathrm{c}\right) \tag{17.7}$$

where R is the universal gas constant, T the temperature, $a = \chi\mathrm{c}$ is the activity of the solute, where χ is the activity coefficient ($\chi = 1$ for an ideal solution) and c the concentration of the dissolved solid material in the fluid, and subscript 0 signifies the reference state of $c = 0$. The work term of the driving force for pressure solution (Eq. (17.5)) yields:

$$\Delta\mu_W = \left(\sigma_{11} - \sigma_{22}\right)v_s = RTln\frac{a\left(\sigma_{11}\right)}{a\left(\sigma_{22}\right)} \tag{17.8}$$

Here we have expressed the solid chemical potentials in terms of the fluid chemical potential, which is equivalent to using the solid–fluid equilibrium condition $\mu_{l,i}(p,T) = \mu_{s,i}(p,T)$. We have also assumed that the surface normal stress in the grain interface is equivalent to the fluid pressure of a "normal", bulk fluid. Eq. (17.8) is the same as equation (16.1) in the chapter by Steiger in this book where $(\sigma_{11}–\sigma_{22})$

is the effective pressure (relative to the pore pressure $p = \sigma_{22}$) that a supersaturated solution with activity $a = a(\sigma_{22})$ may generate. This demonstrates that the thermodynamics of pressure solution and force of crystallization (crystallization pressure) are the same, only the direction of the process is reversed. The chapter by Espinosa-Marzal of this book describes observations of the state of a nanoconfined aqueous solution. Those observations question whether it is correct to describe the confined fluid as a "normal", bulk fluid, simply substituting the fluid pressure for surface normal stress. Until further work has been done to clarify the thermodynamics of the nanoconfined fluid we find it safer to stick with the formulation based on the chemical potential of the solid.

We have already mentioned that pressure solution is a three steps process. During the process the system is out of equilibrium and one must consider the chemical potential along the entire path of the process. The first step is the subdivision shown in Fig. 17.2b into the three steps and three driving forces.

Dissolution:

$$\Delta\mu_D = \mu_{s,1} - \mu_{l,1} \tag{17.9}$$

Transport:

$$\Delta\mu_T = \mu_{l,1} - \mu_{l,2} \tag{17.10}$$

Precipitation:

$$\Delta\mu_P = \mu_{l,2} - \mu_{s,2} \tag{17.11}$$

For a closed system, all three steps are necessary and one regains $\Delta\mu_{tot}$ of equation (17.6) where the driving force is determined by the solid grain only. For an open system, $\mu_{l,2}$ may be determined by long distance advection or diffusion of material and one may consider only the first two steps of the process. Alternatively one may choose to split the transport step as consisting of two parts, grain contact diffusion (short range) and long range transport.

17.3 Kinetics of Pressure Solution

There is much insight to be gained in considering the limiting cases that one of the three steps of pressure solution is much slower than the other two processes. These limiting cases may for closed systems be translated into conditions on the driving forces:

- Dissolution limited kinetics $\mu_{l,1} = \mu_{l,2} = \mu_{s,2} \Rightarrow \Delta\mu_D = \mu_{s,1} - \mu_{s,2}$
- Transport limited kinetics $\mu_{s,1} = \mu_{l,1},\ \mu_{l,2} = \mu_{s,2} \Rightarrow \Delta\mu_T = \mu_{s,1} - \mu_{s,2}$
- Precipitation limited kinetics $\mu_{s,1} = \mu_{l,1} = \mu_{l,2} \Rightarrow \Delta\mu_P = \mu_{s,1} - \mu_{s,2}$

Using these assumptions, the driving force for pressure solution is identical for the three cases; however the rate at which this occurs will depend on the limiting step. Pressure solution of quartz is known to be dissolution limited at low temperature, below 150 °C [2] but is controlled by diffusion at higher temperature [14], that of soluble salts like NaCl [23] to be transport limited, however, special impurities that attach to the surfaces of calcite [26] and quartz [2, 3] may render pressure solution of these minerals precipitation limited. We will consider the two first steps in the following sections. The kinetics of precipitation in this context does not distinguish itself from precipitation and crystal growth in other geological context and we will not discuss it any further.

The main processes we consider here relate to two grains with a single grain boundary. A sediment or a rock consists of an aggregate of grains and the deformation of the sediment or rock is the sum of discrete dissolution-transport-precipitation processes. Consider an aggregate of spherical grains of average radius l. Let the grain boundary radius be a and let the grain boundary fluid thickness be $\Delta_{gb.}$ Then the average grain boundary region has an area of π a^2 facing the grains and an area $2\pi\ a\Delta_{gb}$ facing the surrounding fluid. The fluxes of dissolution, j_D, and transport, j_T, (moles per unit time per unit area) are related by conservation of mass, $j_D\pi\ a^2 = j_T 2\pi\ a\ \Delta_{gb}$. The strain rate of the grain aggregate is then:

$$\dot{\varepsilon} = \frac{1}{l}\frac{dl}{dt} = \frac{1}{l}v_s j_D = \frac{1}{l}v_s j_T \frac{2\Delta_{gb}}{a} = \frac{2v_s j_T \Delta_{gb}}{l^2\,(a/l)} \tag{17.12}$$

The relationship (17.12) is the basis for rate laws for aggregate deformation, whatever expression of fluxes we choose.

17.3.1 Rate of Dissolution

We have already established that the driving force of dissolution is $\Delta\mu_D = \mu_{s,1} - \mu_{1,1}$. There are several ways of treating the kinetics of dissolution using diffuse or sharp interface conditions in a macroscopic description or by a more explicit description of the discrete steps of the chemical reactions of dissolution.

In non-equilibrium thermodynamics the dissolution flux j_D (moles per unit time per unit area) is considered to be linear in the jump condition $\Delta\mu_D$ at the sharp interface [12, 16]:

$$j_D = -k_\mu \Delta\mu_D = -k_\mu v_s\,(\sigma_{11} - \sigma_{22}) \tag{17.13}$$

where j_D is the dissolution flux and k_μ a dissolution rate constant. This relation is also inherent in diffuse interface (as opposed to sharp interface) models like the Cahn-Hilliard equation [4] where transport in both bulk and through the interface is treated as diffusion.

From an experimental tradition where concentrations and rates are measured it is more common to assume the dissolution rate to be proportional to the concentration jump:

$$j_D = -k_c \Delta c = -k_c c\left(\sigma_{22}\right)\left[\frac{\chi\left(\sigma_{11}\right)}{\chi\left(\sigma_{22}\right)} e^{\frac{\left(\sigma_{11}-\sigma_{22}\right)v_s}{RT}} - 1\right] \tag{17.14}$$

Here we have rearranged Eq. (17.8) to express the difference in fluid concentrations:

$$\Delta c = c\left(\sigma_{11}\right) - c\left(\sigma_{22}\right) = c\left(\sigma_{22}\right)\left[\frac{\chi\left(\sigma_{11}\right)}{\chi\left(\sigma_{22}\right)} e^{\frac{\left(\sigma_{11}-\sigma_{22}\right)v_s}{RT}} - 1\right] \approx \frac{c\left(\sigma_{22}\right)\left(\sigma_{11}-\sigma_{22}\right)v_s}{RT} \tag{17.15}$$

It is worth discussing in what range of stresses $(\sigma_{11} - \sigma_{22})$ the last approximation is valid and Eqs. (17.13) and (17.14) yield approximately the same result. The approximation $e^x - 1 \approx x$ is wrong by a factor 2 for $x \approx 1.25$. Thus for quartz at 300 K we may calculate the reasonable range of stresses for the linear approximation to be

$$\frac{\left(\sigma_{11}-\sigma_{22}\right)v_s}{RT} < 1.25 \;\Rightarrow\; \left(\sigma_{11}-\sigma_{22}\right) < 0.13\,GPa \tag{17.16}$$

This is not very far from the yield stress of natural quartz (0.15–0.2GPa). The difference between Eqs. (17.13) and (17.14) is therefore not very big at stresses where pressure solution can be assumed to dominate over plastic strain as a deformation mechanism. It is still of interest to get a better understanding of the underlying basis for the assumptions. We will therefore discuss dissolution kinetics based on the basic concept of rates of change limited by an activation barrier. This is the basis of many statistical mechanical models and of the Arrhenius equation, Eyring rate theory or transition state theory of chemical reactions. The following example is not rigorous and is included to illustrate a possible rationale of three aspects: non-linear rheology, electrochemical effects and confined fluid films.

Consider a liquid in a state where dissolved solid molecules are characterized by μ_l and a solid at chemical potential $\mu_s > \mu_l$. Assume that there is an interfacial phase or state between the solid and liquid where the solid molecules going from solid to dissolved state (or inverse) have to be "activated" to the molar energy $g_A > \mu_s$. This assumption of an activated transition state differs from the non-equilibrium thermodynamic assumption of a sharp or diffuse interface with a *linear* transport coefficient. The probability $P_{s\text{-}l}$ of some molecule in the solid to cross the activation barrier is

$$P_{s-l} = \frac{N_s}{Z_s} e^{-\left(g_A - \mu_s\right)/RT} \tag{17.17}$$

where Z_s is the partition function of the solid and N_s is the number of molecules in the solid adjacent to the interface. The probability of the opposite jump is

$$P_{l-s} = \frac{N_l}{Z_l} e^{-(g_A-\mu_l)/RT} \tag{17.18}$$

where N_l is the number of molecules in the liquid adjacent to the interface. The dissolution rate is proportional to the difference in probabilities:

$$j_D \propto \frac{N_l}{Z_l} e^{-(g_A-\mu_l)/RT} \left(\frac{N_s Z_l}{N_l Z_s} e^{(\mu_s-\mu_l)/RT} - 1 \right) \propto e^{-g_A/RT} \left(e^{(\sigma_{11}-\sigma_{22})v_s/RT} - 1 \right) \tag{17.19}$$

Here we have assumed that $N_s Z_l / N_l Z_s \sim 1$, that the system is dissolution limited and regained an expression of the dissolution rate of the same form as equation (17.14). Thus in cases where dissolution studies have shown that there is a significant activation barrier, one may have to consider dissolution controlled pressure solution as a non-Newtonian rheology. The apparent contradiction between Eqs. (17.14), (17.19) and (17.13) is equivalent to the result of deGroot and Mazur (p205) that "the linear relations of thermodynamics of irreversible processes hold for chemical reactions when the condition A << RT is satisfied" [5]. More recent work ([17], chapter 7) demonstrates that one may obtain non-linear flux-force relationships in the framework of irreversible thermodynamics if one integrates the entropy production along the reaction pathway (that includes the energy barrier). A proper treatment of transition state theory of dissolution in confined fluids and under stress will be necessary to understand the electrochemical effects of ions in solution and surface charges [19].

To sum up, inserting Eqs. (17.13) and (17.14) into (17.12), the dissolution limited pressure solution strain rate of an aggregate is

$$\begin{aligned} \dot{\varepsilon} &= \frac{v_s j_D}{l} = \frac{k_c v_s c(\sigma_{22})}{l} \left[\frac{\chi(\sigma_{11})}{\chi(\sigma_{22})} e^{-\frac{(\sigma_{11}-\sigma_{22})v_s}{RT}} - 1 \right] \\ &\approx \frac{k_c {v_s}^2 (\sigma_{11}-\sigma_{22})}{l} \frac{c(\sigma_{22})}{RT} = \frac{k_\mu {v_s}^2 (\sigma_{11}-\sigma_{22})}{l} \end{aligned} \tag{17.20}$$

The limit of the linear approximation is commented above.

17.3.2 Rate of Transport

The transport flux, j_T, by diffusion may be expressed as a diffusion coefficient, D_K, times the gradient in chemical potential:

$$
\begin{aligned}
\mathrm{j_T} &= -D_K \nabla \mu_{\mathrm{T}} = -D_K \frac{\nu_s\left(\sigma_{11}-\sigma_{22}\right)}{r} \\
&= -D_F \nabla c, \quad D_F = D_K \frac{RT}{c}\left(1+\frac{d \ln \chi}{d \ln c}\right)
\end{aligned} \tag{17.21}
$$

where r is the transport length. The "kinetic" diffusion coefficient, D_K is related to self diffusion of molecules in the fluid and varies only slowly with concentration, also close to the spinodal and critical points of the solution [8]. The "Fick diffusion coefficient", D_F, however depends on the thermodynamic factor $\left(1+\frac{d \ln \chi}{d \ln c}\right)$ that equals 1 for ideal solutions and is zero at the spinodal of a mixture [8, 20]. This means that diffusion limited pressure solution rate is always a linear function of stress. If we had used the concentration formulation with the Fick diffusion coefficient D_F inserting Eq. (17.15) for Δc we would have been led to believe that $\mathrm{j_T}$ was exponential in stress. The expression (17.21) for the transport flux may be inserted into equation (17.12) to obtain the strain rate of an aggregate, but there are still some parameters to be determined.

When material dissolves through the grain boundary region it has to be transported in this grain boundary region to a less stressed region. The important phenomenological parameters for this transport are the diffusion coefficient of the material in the grain boundary region, D, the effective grain boundary fluid thickness Δ_{gb} and the limiting transport length, r. Figure 17.3 summarizes the four main models of the grain boundary and the "realizations" of the transport parameters associated with each model. The model names have varied somewhat through time and our naming scheme is specific to the parameter realizations tabulated in Fig. 17.3. The oldest model, the thin film (TF) model of [24] assumes that the entire grain boundary is smooth and kept apart by a fluid film as described in the Introduction. The limiting transport length, r, is therefore the grain boundary radius, the thickness of the grain boundary fluid is $\Delta_{gb} = \Delta_{gb,ff} \sim 1$–$10(-100)$nm and the diffusion coefficient $D = D_{ff} = (10^{-1}$–$10^{-2})D_{bulk}$, where D_{bulk} is the bulk fluid diffusion coefficient. The next model we will mention, which is a cartoon model of the formal treatment of dynamic grain boundary structure proposed by [18], assumes a dynamic roughness (DR) of the grain boundary. The real contacts supporting stress are small and change from place to place as old contacts dissolve and new contacts form. These points of contact do not limit the transport, the limiting transport length is the grain boundary radius and the fluid that limits the transport is not under stress, thus $\Delta_{gb} = \Delta_{gb,r} \sim 0.1$–$10\ \mu$m and $D = D_{bulk}$. The third model, the Island-Channel (IC) model, that has been ascribed to Gratz [6, 14] and assumes that transport is limited by diffusion in thin fluid films between contacting islands, thus $r = d_{isl}$, $D = D_{ff}$ and $\Delta_{gb} = \Delta_{gb,ff}$. The island size d_{isl} is some constant, presumably in the range 0.1–10 μm. The fourth model, the Dynamic Island Channel (DIC) model, based on quantitative experimental data for NaCl [11] is much like the IC model except that the island size increases with time t since the last stress change (be it an earthquake or a local grain rearrangement) as $d_{isl}(t) = t^{1/3}$which results in a dramatic strain hardening.

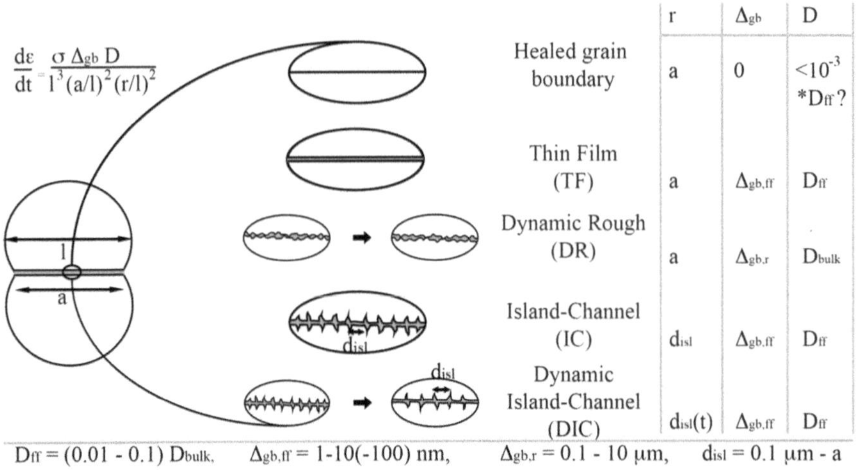

Fig. 17.3 Five cartoon models of the grain boundary during pressure solution and the parameter combinations that distinguish them. The rate limiting transport length, r, is either the entire grain boundary radius, a, or an average island size, $d_{isl} \sim (0.1\text{–}10)\mu\text{m}$, that is either a constant or that increases with time, $d_{isl}(t) \sim t^{1/3}$. The grain boundary width, Δ_{gb}, is either the fluid film thickness $\Delta_{gb,ff} \sim 1\text{–}10$ (possibly 1–100)nm or an effective width of a rough contact, $\Delta_{gb,r} \sim 0.1\text{–}10\ \mu\text{m}$. The diffusion coefficient is either that of the bulk fluid, typically $D_{bulk} \sim 10^{-9}\ \text{m}^2/\text{s}$ @25C, or that of a confined fluid film, $D_{ff} \sim (10^{-1}\text{–}10^{-2})D_{bulk}$. Inserting these values into the strain rate equation one recognizes that the strain rate may vary up to **7 orders of magnitude** depending on the type of interface

We may summarize all four models using the fundamental parameters D, Δ_{gb} and r and rewrite equation (17.12) using the transport distance r for a and using equation (17.21):

$$\dot{\varepsilon} = \frac{2v_s j_T \Delta_{gb}}{l^2\,(r/l)} \approx \frac{2cv_s \Delta_{gb} D \Delta\mu}{RT l^3 \left(\frac{r}{l}\right)^2} \propto \frac{\sigma\, \Delta_{gb} D}{l^3 \left(\frac{r}{l}\right)^2 \left(\frac{a}{l}\right)^2} \tag{17.22}$$

In the last proportionality we have used the work term of the driving force of pressure solution (equation (17.5)) for $\Delta\mu$ with $\sigma_{11} - \sigma_{22} \sim \sigma/(a/l)^2$ where a is the load bearing grain contact radius and σ is the global effective stress. The last 15 years much theoretical work has focused on the stability of the grain boundary in order to give a rational basis for the cartoon models presented here, but there is still much work to be done to reach that goal.

17.4 Conclusions

Pressure solution and force of crystallization are important in Nature. We have shown that the largest driving force is linear in stress, all other contributions to the driving force are more than 3 orders of magnitude smaller. The minor driving forces (strain energy, surface tension, surface charges) are important for the transformation and stability of the interface structure. Interface structure/stability may change strain rates up to 7 orders of magnitude.

Acknowledgements I wish to thank Jean-Pierre Gratier and Francois Renard with whom I have worked for years on these topics and who have pushed me to go critically through this theory.

References

1. Alcantar N, Israelachvili J, Boles J (2003) Forces and ionic transport between mica surfaces: implications for pressure solution. Geochim Cosmochim Acta 67:1289–1304
2. Bjørkum PA, Oelkers EH, Nadeau PH, Walderhaug O, Murphy WM (1998) Porosity prediction in quartzose sandstones as a function of time, temperature, depth, stylolite frequency, and hydrocarbon saturation. AAPG Bull 82:637–648
3. Bjørlykke K, Hoeg K (1997) Effects of burial diagenesis on stresses, compaction and fluid flow in sedimentary basins. Mar Pet Geol 14:267–276
4. Cahn JW, Hilliard JE (1958) Free energy of a nonuniform system (I): interfacial energy. J Chem Phys 28
5. de Groot SR, Mazur P (1984) Non-equilibrium thermodynamics. Dover publication, New York
6. den Brok S (1998) Effect of microcracking on pressure-solution strain rate: the Gratz grain-boundary model. Geology 26:915–918
7. Dewers T, Ortoleva P (1990) A coupled reaction/transport/mechanical model for intergranular pressure solution stylolites, and differential compaction and cementation in clean sandstones. Geochim Cosmochim Acta 54:1609–1625
8. Dysthe DK, Hafskjold B (1995) Inter- and intradiffusion in liquid mixtures of methane and n-decane. Int J Thermophys 16:1213–1224
9. Dysthe DK, Podladchikov Y, Renard F, Feder J, Jamtveit B (2002a) Universal scaling in transient creep. Phys Rev Lett 89 (24): art. no. 246102
10. Dysthe DK, Renard F, Porcheron F, Rousseau B (2002b) Fluid in mineral interfaces – molecular simulations of structure and diffusion. Geophys Res Lett 29: art. no. 1109
11. Dysthe DK, Renard F, Feder J, Jamtveit B, Meakin P, Jøssang T (2003) High resolution measurements of pressure solution creep. Phys Rev E 68(1):011603. doi:10.1103/PhysRevE.68.011603
12. Gamsjäger E (2007) A note on the contact conditions at migrating interfaces. Acta Mater 55:4823–4833
13. Gibbs JW (1877) On the equilibrium of heterogeneous substances. Trans Conn Acad 3:108–248, 343–524
14. Gratz AJ (1991) Solution-transfer compaction of quartzites: progress toward a rate law. Geology 19:901–904
15. Gratier JP, Guiguet R, Renard F, Jenatton L, Bernard D (2009) A pressure solution creep law for quartz from indentation experiments. J Geophys Res Solid Earth 114(3):B03403. doi:10.1029/2008JB005652
16. Kjelstrup S, Bedeaux D (2008) Non-equilibrium thermodynamics of heterogeneous systems. Ser Adv Stat Mech 16
17. Kjelstrup S, Bedeaux D, Johannessen E (2010) Non-equilibrium thermodynamics for engineers. World Scientific, New Jersey
18. Lehner FK, Bataille J (1985) Nonequilibrium thermodynamics of pressure solution. Pure Appl Geophys 122:53–85

19. Kristiansen K, Valtiner M, Greene GW, Boles JR, Israelachvili JN (2011) Pressure solution – the importance of the electrochemical surface potentials. Geochim Cosmochim Acta 75:6882–6892
20. Naumann EB, He DQ (2001) Nonlinear diffusion and phase separation. Chem Eng Sci 56:1999–2018
21. Parks GA (1984) Surface and interfacial free energies of quartz. J Geophys Res 89:3997–4008
22. Røyne A, Dysthe DK (2012) Rim formation on crystal faces growing in confinement. J Cryst Growth 346:89–100
23. Spiers CJ, Brzesowsky RH (1993) Densification behaviour of wet granular salt: theory versus experiment. In: Seven symposium of salt., pp 83–91
24. Weyl PK (1959) Pressure solution and the force of crystallization: a phenomenological theory. J Geophys Res 64:2001–2025
25. Wogelius RA, Dysthe DK (2006) Confined fluids in the Earth's crust – properties and processes. Chem Geol 230:175–181
26. Zhang XM, Spiers CJ, Peach CJ (2010) Compaction creep of wet granular calcite by pressure solution at 28 degrees C to 150 degrees C. J Geophys Res Solid Earth 115

Chapter 18
Geochemistry of Capillary Hydrogeochemical Systems in Arid Environments

Arnault Lassin, Lionel Mercury, and Mohamed Azaroual

Abstract In arid environments, porous media are unsaturated with water which is submitted to capillary constraints. The present chapter focuses on the geochemical impacts of such physical constraints and how theoretical analysis can help interpreting field or laboratory observations. The basic principles of capillary geochemistry suggest that the fate of contaminants, either organic or inorganic, can be significantly impacted in terms of reactive mass transfer in addition to flow and transport processes. All these mechanisms are closely interconnected, what makes the description of the behavior of such systems very complicated. An important work still has to be done in order to achieve such a goal: a number of mechanisms are not taken into account in the current state of development of the capillary geochemistry, namely mechanisms that occur in the thinnest confining geometries, like disjoining pressures, surface forces, etc.

18.1 Introduction

Porous media containing water in dry or arid contexts are only partially saturated, which means that water is strongly bounded to the solid matrix either in a capillary state or as thin hydration films on mineral surfaces. These particular features modify the thermo-physical properties of the confined water compared to the free bulk

A. Lassin (✉) • M. Azaroual
Institut des Sciences de la Terre d'Orléans, UMR 7327 Université d'Orléans - CNRS - BRGM, BRGM, 3 avenue Claude Guillemin, 45060 Orléans, Cedex 2, France
e-mail: a.lassin@brgm.fr

L. Mercury
Institute of Earth Sciences, University of Orléans, Orléans, France
e-mail: lionel.mercury@univ-orleans.fr

L. Mercury et al. (eds.), *Transport and Reactivity of Solutions in Confined Hydrosystems*, NATO Science for Peace and Security Series C: Environmental Security,
DOI 10.1007/978-94-007-7534-3_18,

water (see previous chapters). This, in turn, leads to dramatic modification of the geochemical behavior of water-rock-gas systems compared to what is classically observed in saturated porous media. Then questions arise: (i) how much different and specific are these behaviors? (ii) What are their effective roles on mass and heat transfers from the soil surface to the underlying aquifers and on the water quality of these aquifers? (iii) In particular, what can be their role in the fate of pollutants and pesticides? (iv) Could they also play a role in the response to the global warming and the fate of greenhouse-effect gases?

In order to interpret observations made on the field or in the laboratory (Sect. 18.2), theoretical developments have been performed (Sect. 18.3) for dealing with such systems and for starting bringing some tentative answers to questions above (Sect. 18.4).

18.2 Observations from Field and Laboratory Experiments

The first major field observations leading to the idea that the peculiar state of water in unsaturated porous media could shift chemical equilibria were performed by Yves Tardy and his collaborators [1–3]. These authors pointed out the coexistence of local mineralogical heterogeneities in old lateritic profiles, where thermodynamic equilibrium was expected. More precisely, they observed that hydrated minerals (like gibbsite or goethite) preferentially precipitate in large pores while anhydrous or less hydrated minerals (e.g. kaolinite, hematite) rather formed in smaller pores.

Recently, specific experiments have been performed in order to explore more deeply the underlying mechanisms. Bouzid et al. [4] observed the formation of differently shaped salt crystals in the pores of polycarbonate filters, depending on the ambient relative humidity (RH) and the water repartition/dynamics in the porosity. Using NMR technics, Rijniers et al. [5] measured the solubility of sodium carbonate in a natural porous medium. They observed that this solubility varied as a function of pore size on model porous materials of known pore sizes. The authors recorded that the solubility of sodium carbonate increased with decreasing pore size while sodium sulfate was not sensitive to pore size. Lassin et al. [6] monitored the partial pressure of CO_2 (pCO_2) at equilibrium with samples of an argillaceous rock at various RH. They observed that the equilibrium pCO_2 was decreasing with decreasing RH.

All these observations indicate that capillary and/or confined porous media exhibit specific complex chemical behaviors compared to bulk water-mineral-gas systems. Therefore, in parallel to these experimental works, theoretical developments have been performed, as described in the following for capillary systems.

18.3 Theoretical Background

18.3.1 Thermodynamics of Chemical Reactivity in Capillary Systems

The chemical behavior of capillary systems has been studied for more than 10 years, and the theoretical framework is now well established for calculating the standard thermodynamic properties of chemical compounds submitted to capillary pressure [7–9]. Briefly, the Young-Laplace and Kelvin laws imply that pore water at physical equilibrium with a given relative humidity is subjected to a pressure lower than in the surrounding gas phase. For sufficiently low relative humidity, the internal pressure of water – below the gas-water meniscus – can be negative. The corresponding impact on the thermodynamic properties of water, dissolved species and minerals, and therefore on operant chemical reactions can be calculated using the Thermo-ZNS calculation code [9]. The latter includes the equations of state (EoS) of water after Wagner and Pruss [10] and aqueous species after Helgeson et al. [11] and Tanger and Helgeson [12]. The extended use of these equations to capillary systems and negative pressures can be discussed (see Chap. 13 [13]). Their limit of validity, in terms of pressure, is not known and specific studies would be necessary to check or improve existing equations.

However, using the state-of-the-art EoS, one can show that capillary constraints can significantly impact water thermodynamic and electrostatic properties like the dielectric constant, ε, and its derivatives with respect to pressure and temperature. As an illustration, the Born function $Z\ (= -1/\varepsilon)$ is plotted on Fig. 18.1 as a function of temperature, at different pressures. This affects the thermodynamic properties of aqueous species too, which depend on temperature and pressure, but also on the dielectric properties of the solvent, namely water [11]. The apparent standard Gibbs free energy of Na^+, $SO_4^=$ and Cl^- are plotted on Fig. 18.1, respectively, as a function of pressure, at different temperatures. They show very different behaviors from one aqueous species to another, and from positive to negative pressures. Since, for every chemical reaction, the mass action law involves activity of aqueous species rather than concentration, another parameter related to aqueous species may be affected by capillary pressure. This is the activity coefficient by the means of the Debye-Hückel coefficients A_γ and B_γ [9] which are functions of water density and dielectric constant, and of temperature.

Consequently, the thermodynamic impact of capillary pressure on homogeneous chemical equilibria (*i.e.* in water) cannot be foreseen easily. Obviously, this also concerns heterogeneous equilibria (*i.e.* mineral-water and gas-water systems). Furthermore, in addition to the single pressure effect, the chemistry of capillary systems offers specific features called the anisobar contexts (i.e., where phases can undergo different pressures). They apply particularly to gas-aqueous solution

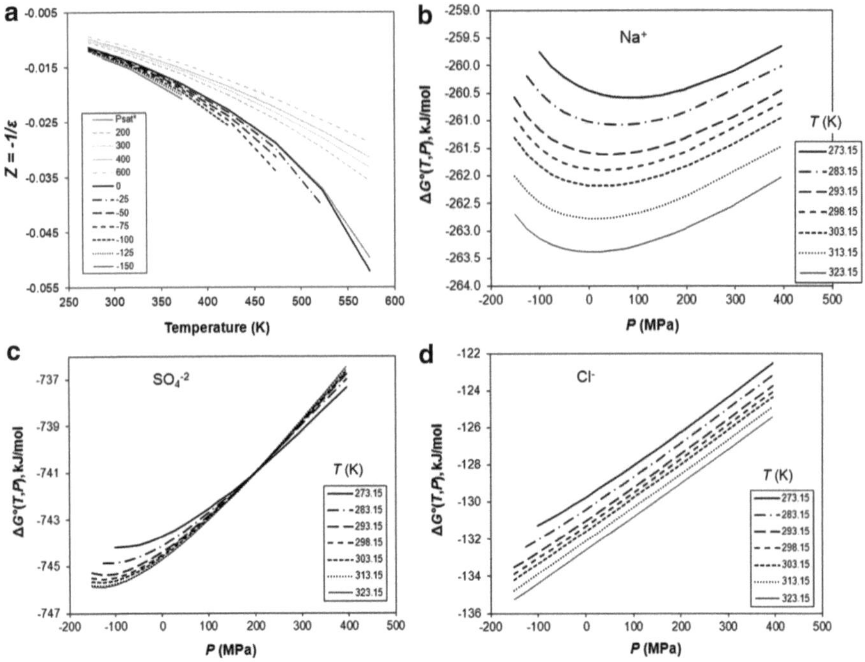

Fig. 18.1 Temperature and pressure dependence of various thermodynamic and electrostatic properties of water and aqueous compounds: (**a**) Born function Z of water, and apparent standard Gibbs free energy (**b**) of Na^+, (**c**) of $SO_4^=$, and (**d**) of Cl^-

equilibria since a gaseous compound cannot undergo negative pressures whereas its aqueous equivalent undergoes the internal capillary water pressure. A similar situation can be envisioned for minerals initially present in the rock matrix. They are submitted to the local pressure constraint when the porosity is saturated with water and the same pressure applies to these minerals during desaturation regardless of water content. On the other hand, if the capillary aqueous solution is supersaturated with respect to a mineral, the latter should precipitate at the capillary water pressure. Some evidence for this isobar/anisobar situation was obtained experimentally [4] as explained in Sect. 18.4. According to the theoretical developments described above, thermodynamic properties and thus solubility of a mineral differ according to the isobar or anisobar context. As a general trend, anisobar conditions are less favorable to mineral stability than isobaric ones, which is physically consistent. The difference between these two contexts depends on the volumetric properties of the mineral. In particular, hydrated mineral phases are generally more affected by pressure than anhydrous ones. On the other hand, gases are expected to be more soluble in capillary conditions.

18.3.2 Thermodynamic Database

In order to simulate the behavior of hydro-geochemical systems, we use the well-known PhreeqC-2 geochemical calculation code [14]. It allows calculating the aqueous speciation of solutions, saturation indices of minerals, partial pressures of gases, amounts of precipitated or dissolved minerals, etc. The use of this code at temperatures and pressures far from the liquid–vapor saturation curve is possible if appropriately adapted thermodynamic databases are available. Such databases including both isobar and anisobar mineral dissolution/precipitation reactions were generated using the Thermo-ZNS code indicated above. The reference thermodynamic database implemented in Thermo-ZNS is the Thermoddem database [15] which provides information on critically selected standard thermodynamic properties of chemical compounds and reactions necessary for their calculation at various temperatures and pressures.

18.4 Interpretations

Once the theoretical framework is defined, experimental observations evoked previously can be interpreted quite simply. In this section we focus on experiments from Bouzid et al. [4] and from Lassin et al. [6]. In the former work, the pores of a polycarbonate filter were initially filled with a NaCl aqueous solution slightly under-saturated with respect to halite. During the progressive drying of the filter, different stages were observed and, applying the principles of capillary geochemistry, Bouzid et al. [4] could provide the interpretation summarized hereafter (Fig. 18.2).

(a) Precipitation of (cubic) halite while porosity is still saturated with water. Part of the volume of water which has evaporated is occupied by the salt. In some cases the corners of the precipitated salt can touch the pore wall, thus modifying the pore geometry.
(b) As evaporation proceeds, de-saturation occurs and capillary water goes in the angular spaces created between the pore wall and the salt. A typical anisobar context thus takes place between water and solids. The less soluble and most supple solid (the polycarbonate pore wall) deforms while the most soluble and less supple solid (halite) dissolves, its destabilization being predicted by thermodynamics. The corners of the crystal are eroded.
(c) On the contrary, the isobar polymorph of halite, probably a nano-crystalline or an amorphous halite, is stabilized and thus massively precipitates under the capillary meniscus before water has totally evaporated.

All of these successive stages are fully consistent with the theoretical interpretation of the geochemical mechanisms in capillary context. They cannot be explained by the classical geochemistry applied to water-saturated porous systems.

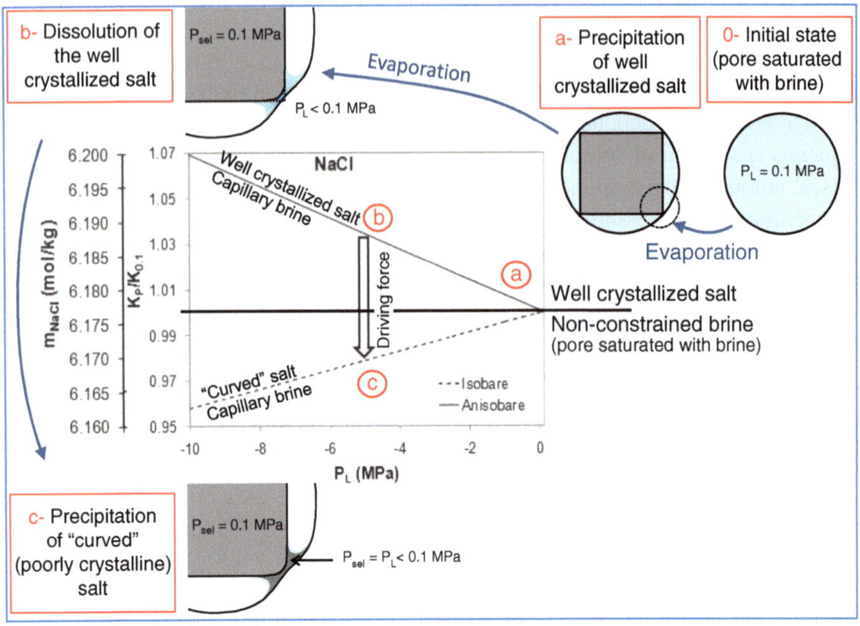

Fig. 18.2 Successive reactive steps during the evaporation of a NaCl brine from the porosity of a polycarbonate filter and calculated solubility product of halite as a function of capillary pressure, both in the isobar and the *anisobar* contexts. (**a**) Precipitation of well crystallized cubic halite; (**b**) establishment of capillary conditions and dissolution of the cubic (*anisobar*) halite angles; (**c**) precipitation of isobar poorly crystallized halite under the capillary meniscus (Modified after [4])

The second series of experiments [6] recorded the CO_2 partial pressure at equilibrium with a clay rock in a closed system at various relative humidity values. The rock (the Callovo-Oxfordian argillite) and its pore water are well known. A geochemical model has been developed, involving mineral dissolution/precipitation and cation exchange reactions at thermodynamic equilibrium [16], and is used as a reference for describing this water-rock system. In particular, this model calculates the correct in situ pCO_2. By simply applying the principles of capillary geochemistry and without any parameter adjustment, Lassin et al. [6] could reproduce with a good accuracy the observed decrease of the equilibrium pCO_2 with decreasing relative humidity, from 8 down to less than 1 mbar and from 100 down to 50 %, respectively. The authors took into account the evaporation of water from porosity and the consequent increase in salinity. They considered that anisobar alumino-silicates could only dissolve according to the time scale of the experiments (from 6 to 18 months). On the other hand, more reactive minerals, like carbonates, were initially present as anisobar phases and could evolve to their isobar polymorphs. Here again, the observed equilibrium pCO_2 could be reproduced only if principles of capillary geochemistry were applied.

18.5 Concluding Remarks

These results were obtained in very different porous hydro-geochemical systems thus illustrating the wide applicability of the capillary geochemistry approach. Consequently, a new interpretation of the observations of Rijniers et al. [5], amongst others, could be done in the light of the principles of capillary geochemistry. Applications to other domains can therefore be envisioned quite straightly. They are related to industrial activities like CO_2 (or more generally: non-wetting phases) geological storage as well as nuclear waste repository in argillaceous rocks. They are also related to natural environments which have been impacted by human activity, in particular for studying the fate of inorganic and organic pollutants in arid environments.

Results above also illustrate the significant impact that capillary pressure can have on geochemical equilibria in arid environments. However, the obvious question to ask concerns the volume of water involved and the consequences on the mass/heat transfers through the porous medium. According to Pettenati et al. [17] significant amounts of water can remain held in the porosity for low relative humidity, depending on the pore network structure and petrophysical properties of the soil.

On the other hand, the different chemical compounds present in the pore solution can play a significant role on the air-solution interfacial tension and on the wettability of the pore walls depending on their affinity to interfaces and on their concentration in water. This may in turn affect the water content versus relative humidity relationship, but also the relative permeabilities in multiphase systems as well as the hysteresis and memory effects, etc., and thus the mass/heat transfers.

The principles on which capillary geochemistry is based are fairly consistent with approaches focused on other processes like crystallization pressure (see Chap. 16 [18]) and solution pressure phenomena (see Chap. 17 [19]). In such processes, anisobar conditions are also encountered in confined porous media, with strong relationships with pore water chemistry and shapes of interfaces. These similarities suggest that a single and general formalism able to describe all the above systems can be developed. An important work remains to achieve such a goal since a number of mechanisms are not taken into account in the current state of development of the capillary geochemistry. In particular, this approach only deals with bulk properties of capillary water. It does not take yet into account mechanisms that occur in the thinnest confining geometries, like disjoining pressures or surface forces (see Chap. 1 [21]). Works are under progress in this direction (see chapters of [22–24]).

Acknowledgments This work has received support from the French Agency for Research (Agence Nationale de la Recherche, ANR) through the grants CONGE BLAN-61001 and Labex Voltaire ANR-10-LABX-100-01.

References

1. Tardy Y, Nahon D (1985) Geochemistry of laterites. Stability of Al-goethite, Al-hematite and Fe^{3+}-kaolinite in bauxites and ferricretes. An approach to the mechanism of concretion formation. Am J Sci 285:865–903
2. Trolard F, Tardy Y (1987) The stabilities of gibbsite, boehmite, aluminous goethites and aluminous hematites in bauxites, ferricretes and laterites as a function of water activity, temperature and particle size. Geochim Cosmochim Acta 51:945–957
3. Tardy Y, Novikoff A (1988) Activité de l'eau et déplacement des équilibres gibbsite-kaolinite dans les profils latéritiques. CR Acad Sci Paris 306:39–44
4. Bouzid M, Mercury L, Lassin A, Matray J-M, Azaroual M (2011) In-pore tensile stress by drying-induced capillary bridges inside porous material. J Colloid Interface Sci 355:494–502
5. Rijniers L, Huinink H, Pel L, Kopinga K (2005) Salt crystallization as damage mechanism in porous building materials – a nuclear magnetic resonance study. Magn Reson Imaging 23(2):273–276
6. Lassin A, Marty N, Henry B, Trémosa J, Gailhanou H, Madé B, Altmann S, Gaucher EC (2013) Equilibrium partial pressure of CO_2 in the Callovo-Oxfordian argillite as a function of relative humidity. Procedia Earth Planet Sci 7:459–462
7. Mercury L, Tardy Y (2001) Negative pressure of stretched liquid water. Geochemistry of soil capillaries. Geochim Cosmochim Acta 65(20):3391–3408
8. Mercury L, Azaroual M, Zeyen H, Tardy Y (2003) Thermodynamic properties of solutions in metastable systems under negative or positive pressures. Geochim Cosmochim Acta 67:1769–1785
9. Lassin A, Azaroual M, Mercury L (2005) Geochemistry of unsaturated soil systems: aqueous speciation and solubility of minerals and gases in capillary solutions. Geochim Cosmochim Acta 69:5187–5201
10. Wagner W, Pruss A (2002) The IAPWS formulation 1995 for the thermodynamic properties of ordinary substance for general and scientific use. J Phys Chem Ref Data 31:387–535
11. Helgeson HC, Kirkham DH, Flowers GC (1981) Theoretical prediction of the thermodynamic behavior of aqueous electrolytes at high pressures and temperatures: IV. Calculation of activity coefficients, osmotic coefficients, and apparent molal and standard and relative partial molal properties to 600° C and 5 kb. Am J Sci 281:1249–1516
12. Tanger JC IV, Helgeson HC (1988) Calculation of the thermodynamic and transport properties of aqueous species at high pressures and temperatures: revised equations of state for the standard partial molal properties of ions and electrolytes. Am J Sci 288:19–98
13. Imre AR (2013) Stability and negative pressure in bulk and confined liquids. In: Mercury L, Tas N, Zilberbrand M (eds) Transport and reactivity of solutions in confined hydrosystems. Springer, Dordrecht, pp 147–153
14. Parkhurst DL, Appelo CAJ (1999) User's guide to PHREEQC (version 2). USGS water resources investigations report 99–4259, 312 p
15. Blanc P, Lassin A, Piantone P, Azaroual M, Jacquemet N, Fabbri A, Gaucher EC (2012) Thermoddem: a geochemical database focused on low temperature water/rock interactions and waste materials. Appl Geochem 27:2107–2116. Available at http://thermoddem.brgm.fr/
16. Gaucher EC, Tournassat C, Pearson FJ, Blanc P, Crouzet C, Lerouge C, Altmann S (2009) Geochim Cosmochim Acta 73:6470–6487
17. Pettenati M, Mercury L, Azaroual M (2008) Capillary geochemistry in non-saturated zone of soils. Water content and geochemical signatures. Appl Geochem 23:3799–3818
18. Steiger M (2013) Crystal growth and phase equilibria in porous materials. In: Mercury L, Tas N, Zilberbrand M (eds) Transport and reactivity of solutions in confined hydrosystems. Springer, Dordrecht, pp 185–194

19. Dysthe D (2013) Shaping the interface – interactions between confined water and the confining solid. In: Mercury L, Tas N, Zilberbrand M (eds) Transport and reactivity of solutions in confined hydrosystems. Springer, Dordrecht, pp 195–208
20. Or D, Tuller M (1999) Liquid retention and interfacial area in variably saturated porous media: upscaling from single-pore to sample-scale model. Wat Resour Res 35:3591–3605
21. Espinosa-Marzal RM (2013) Interactions in water across interfaces: from nano to macro-scale perspective. In: Mercury L, Tas N, Zilberbrand M (eds) Transport and reactivity of solutions in confined hydrosystems. Springer, Dordrecht, pp 1–14
22. Bergonzi I, Mercury L (2013) Infrared-Thermodynamics conversion as a function of temperature: towards confined water. In: Mercury L, Tas N, Zilberbrand M (eds) Transport and reactivity of solutions in confined hydrosystems. Springer, Dordrecht, pp 43–53
23. Mercury L, Shmulovich KI (2013) Experimental superheating and cavitation of water and solutions at spinodal-like negative pressures. In: Mercury L, Tas N, Zilberbrand M (eds) Transport and reactivity of solutions in confined hydrosystems. Springer, Dordrecht, pp 155–167
24. Tas N, Brunets N, van Honschoten J, Haneveld J, Jansen H (2013) Static and dynamic capillarity in silicon based nanochannels. In: Mercury L, Tas N, Zilberbrand M (eds) Transport and reactivity of solutions in confined hydrosystems. Springer, Dordrecht, pp 29–41

Chapter 19
Evaporation from a Porous Medium in the Presence of Salt Crystallization

Marc Prat

Abstract The interplay between salt transport, crystallization and evaporation from a porous medium is a topic rich in interesting open problems. This is illustrated through the consideration of a few of them from experiments with model porous media. We notably discuss the factors controlling the localization of crystallization spots at the evaporative surface of a porous medium and the impact of surface crystallization on evaporation kinetics.

19.1 Introduction

Evaporation in a presence of dissolved salt is a key process in several applications such as the injection of CO_2 in saline aquifers, the soil salinization problem and the preservation of our cultural heritage (frescos, statues, monuments, etc). The latter application is directly related to the damages that can be induced in a porous material or at its surface by the salt crystallization, i.e. [1]. The damage problem is generally studied in relation with the so-called crystallization pressure, e.g. [2, 3], which is the key concept in order to evaluate the stresses generated within the pores by the crystallization. According to the current theory, the crystallization pressure notably depends on the local supersaturation, that is the amount of dissolved salt in excess compared to the amount corresponding to the thermodynamic equilibrium between the crystal and the solution. It is therefore crucial to predict the salt concentration transient distribution within the system during the evaporation process in order to be able to address the associated poremechanical problem or more modestly to predict where the damages are likely to be generated in the materials.

M. Prat (✉)
Institut National Polytechnique de Toulouse (INPT), Institut de Mécanique des Fluides de Toulouse (IMFT), UMR 5502 CNRS-UPS, Allée Camille Soula, F-31400 Toulouse, France
e-mail: mprat@imft.fr

L. Mercury et al. (eds.), *Transport and Reactivity of Solutions in Confined Hydrosystems*, NATO Science for Peace and Security Series C: Environmental Security,
DOI 10.1007/978-94-007-7534-3_19,

Even in the absence of coupling with poremechanical effects, the study is made complicated by the various couplings existing between evaporation, salt transport and crystallization. The main objective of the present article is to illustrate the interplay between different phenomena (in the absence of structural damages). We indeed believe that it is crucial to develop a much better understanding of the trio: evaporation/transport/crystallization. A few obvious basic questions are: Is evaporation in the presence of salt faster, identical or slower than for pure water? How does the effect of salt on evaporation depend on the initial salt concentration? Does the salt crystallize at the surface of the porous medium or within the porous medium? These are the main points that will be addressed in what follows. The idea is to give an overview of some key results. Details can be found in the references. Also, we do not discuss in details evaporation with pure water, see [4] in the present book for an introduction to evaporation in porous media. The results we discuss were obtained with sodium chloride, which can be considered as a simple salt, notably because it forms an anhydrous crystal.

19.2 Efflorescence/Subflorescence

Salt crystals appear in a solution when the dissolved salt concentration C reaches the crystallization concentration C_{sat}. The latter is generally greater than the saturation concentration corresponding to the thermodynamic equilibrium between the crystal and the solution. Depending on various factors, the crystallisation concentration can be reached within the porous medium, leading to formation of subflorescence or at the surface of the porous medium, which leads to formation of efflorescence (Fig. 19.1). In classical drying experiments, the advection effect induced in the porous medium by the evaporation process is sufficient to lead to a concentration build-up at the surface of the porous medium, e.g. [6, 7], and thus to efflorescence.

Fig. 19.1 Efflorescence and subflorescence in a model porous medium [5]. Subflorescence refers to crystallized salt structures inside the porous medium whereas efflorescence refers to crystallized salt structures at the surface of the porous medium

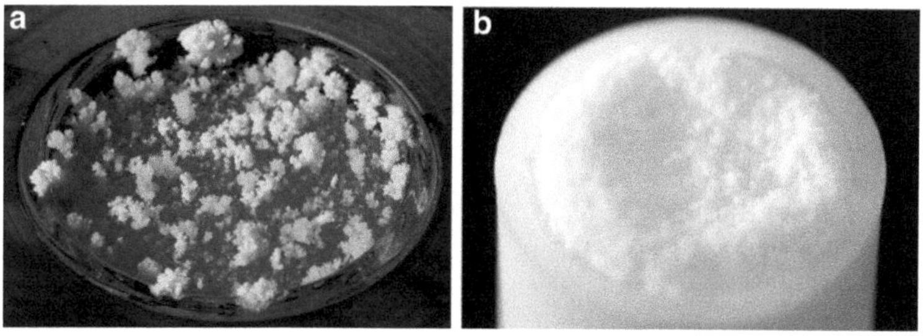

Fig. 19.2 Example of cauliflower (or *patchy*) efflorescence (**a**) and crusty efflorescence (**b**)

Subflorescence can form at later stages when the liquid is not connected anymore to the outer open surface of the porous medium or more generally when evaporation takes place not at the surface but inside the porous medium.

19.3 Cauliflower Efflorescence and Crusty Efflorescence

As reported in [8] and illustrated in Fig. 19.2, two main types of efflorescence are distinguished: cauliflower, also referred to as patchy, and crusty. The crusty efflorescence forms a crust covering entirely the surface (the surface of the underlying porous surface is not visible where the crust is present) whereas the cauliflower efflorescence is discrete, characterized by well individualized salt structures. As can be seen from Fig. 19.2a a significant fraction of the surface of the porous medium remains free of efflorescence when the efflorescence is patchy. The factors controlling which type of efflorescence is going to form are a subject of investigation, [8, 9]. Patchy efflorescence will be favoured when the pores are relatively big and the evaporation rate is small whereas a crust is expected when the pores are smaller and/or the evaporation rate is high. Also, it should be mentioned that distinguishing only two types of efflorescence is a simplification. It is likely that the typology of efflorescence is richer, e.g. [10]. In our experiments, these two main types can be clearly distinguished, notably through their impact on evaporation (see Sect. 19.6 below).

19.4 Efflorescence Is Porous. Efflorescence Growth Mechanism

Efflorescence is porous. This has been clearly established for the patchy efflorescence using simple capillary invasion experiments (see Fig. 19.3) and X-ray microtomography [11]. The dissolved salt is transported through the pore network of the efflorescence structures up to their upper regions where the evaporation fluxes

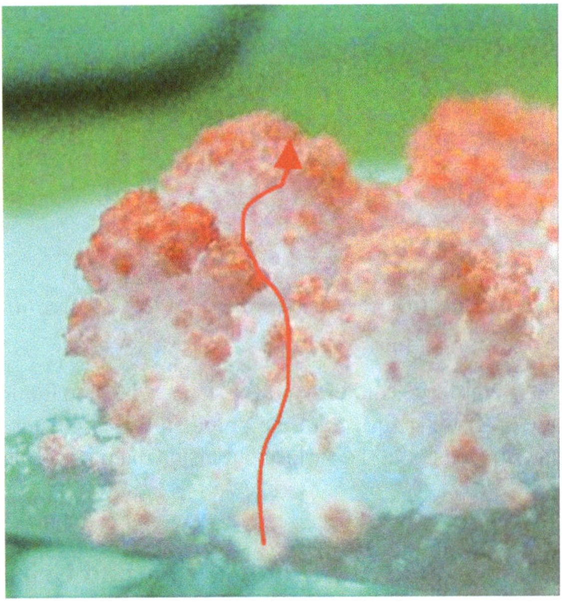

Fig. 19.3 Patchy efflorescence is porous. A *red* dyed aqueous solution put in contact with a dry efflorescence rises by capillary forces up to the *top* of efflorescence where higher evaporation fluxes induce a preferential deposition of the dye

inducing the transport are higher. This leads to the preferential crystallization in the upper regions of the salt structures. As a result the growth of efflorescence occurs from above by successive deposition of salt crystals from above. This is therefore an evaporation flux controlled growth, which has some obvious but not yet fully studied connections with classical Laplacian growth processes (the vapour partial pressure is governed by a Laplacian operator in the diffusive limit), see for example [12] for an introduction to Laplacian growth.

The case of the crusty efflorescence is much less clear. The model proposed in [9] for explaining qualitatively the transition from patchy to crusty as the mean pore size of the underlying porous medium is decreased is based on the assumption that the crusty efflorescence is porous. It has been also seen that the vapour can be transported through a salt crust [13]. However, the details of the crust formation and crust growth are a widely open subject. Further experimental investigations are needed.

19.5 Factors Affecting the Localisation of Cauliflower Efflorescence

One striking feature of the patchy efflorescence is the formation of discrete crystallization spots at the porous surface (see Fig. 19.2a). This is interpreted as the signature of the existence of local maxima in the salt concentration field at the porous surface prior and up to the formation of first crystals. The existence of spatial fluctuations in the salt concentration field is explained by the spatial fluctuations in the velocity field existing in the liquid phase. Because the advection effect is significant, the spatial fluctuations in the velocity field induce fluctuations in the concentration field. The factors affecting the salt concentration distribution and thus the localization of crystallization spots at the surface are therefore intimately related to the factors affecting the velocity distribution within the pores. As discussed in [9, 14] and [15], those factors are the disordered nature of a porous medium (= random distribution of pore sizes), the evaporation flux distribution at the surface, e.g. [14], and the possible Darcy' scale heterogeneities of the medium, e.g. [15]. The more detailed analysis reported in [9] leads to distinguish the impact of surface disorder and internal disorder. The surface disorder refers to the spatial fluctuations in the pore opening at the surface, which induce fluctuation in the evaporation rate at the surface of each meniscus, and thus velocity fluctuations in the adjacent pores. The internal disorder refers to the fluctuations in the pore size within the porous medium, which induce fluctuations in the velocity field, and thus in the salt concentration field.

As discussed in [9] and [15], salt concentration gradients are important only in a region of size $\xi(t)$ adjacent to the evaporative surface. This is a consequence of the advection effect on the salt transport, see also [6]. As a result only the internal disorder of this region of size $\xi(t)$ is important. The disorder located further away from the evaporative surface can generally be ignored.

The next question is why the patchy efflorescence continues to grow under the form of well individualized salt structures. As discussed in [9] and [14], this is explained by the screening of the porous surface free of efflorescence located between the already growing efflorescence structures and the redirection of dissolved salt present in the porous medium toward the growing salt structures. The screening means that the evaporation flux at the porous medium surface between the growing efflorescence structures tends to zero, which "kills" the advection effect and thus the salt concentration build-up at the surface.

19.6 Impact of Efflorescence Type on Evaporation

As reported in [8], the type of efflorescence, i.e. crusty or patchy, has a great impact on evaporation. The patchy efflorescence does not affect significantly the evaporation process and can even enhance the evaporation rate compared to pure water. By contrast the crusty efflorescence can greatly affect the evaporation process and even blocks the evaporation. It has also been observed [8] that the interplay between drying and the efflorescence formation leads to a non-monotonous variation of the drying rate with the initial salt concentration when the efflorescence is patchy but not when the efflorescence is crusty. This has to do with a porous medium "coffee ring" effect due to evaporation fluxes higher at the periphery of the sample, see [8] for more details. The ring effect is illustrated in Fig. 19.4.

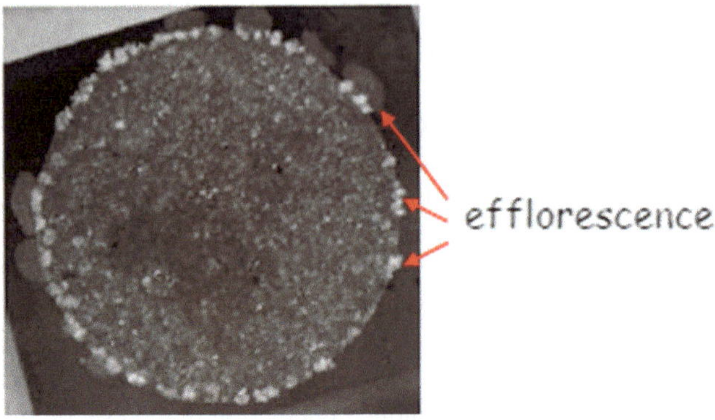

Fig. 19.4 Top view of the evaporative surface of a porous sample and illustration of the efflorescence ring effect. Efflorescence first forms at the periphery of the porous sample owing to evaporation fluxes greater at the periphery

19.7 Conclusion

Evaporation from a porous medium in the presence of dissolved salt is a particularly interesting problem because of the complex interplay between evaporation, salt transport and crystallisation. This field is widely open and many questions are still to be answered. The detailed understanding of the crusty efflorescence growth and the patchy/crusty transition are just two examples of interesting questions. A more advanced understanding of the transport/crystallization problem is needed not only because the crystallization can greatly affect the evaporation but also because the salt concentrations that can be reached in the solution in contact with growing crystals has a direct impact on the crystallisation pressure, and therefore the possible associated poromechanical effects.

As for evaporation with pure water, e.g. [4], studies in recent years have mainly focussed on microporous materials (pore size equal or greater than 1 μm). Very little is in fact known for the systems involving nanopores (~ pore sizes lower than 100 nm).

Acknowledgement I am very thankful to the students and colleagues I have worked with over the years on the "salt problem": S. Ben Nasrallah, P. Duru, H. Eloukabi, F. Hidri, M. Marcoux, N. Sghaier, S. Veran-Tissoires. Their contribution is greatly appreciated.

References

1. Goudie A, Viles H (1997) Salt weathering hazards. Wiley, Chichester
2. Scherer GW (2004) Stress from crystallization of salt. Cem Concr Res 34:1613–1624
3. Coussy O (2006) Deformation and stress from in-pore drying-induced crystallization of salt. J Mech Phys Solid 54:1517–1547

4. Assouline S, Narkis K, Veran-Tissoires S, Marcoux M, Prat M (2013) Impact of heterogeneity on evaporation from bare soils. In: Mercury L, Tas N, Zilberbrand M (eds) Transport and reactivity of solutions in confined hydrosystems. Springer, Dordrecht, pp 65–74
5. Eloukabi H, Sghaier N, Prat M, Ben Nassrallah S (2011) Drying experiments in a hydrophobic model porous medium in the presence of a dissolved salt. Chem Eng Technol 34(7):1085–1094
6. Huinink HP, Pel L, Michels MAJ (2002) How ions distribute in a drying porous medium: a simple model. Phys Fluid 14(4):1389–1395
7. Sghaier N, Prat M, Ben Nasrallah S (2007) On ions transport during drying in a porous medium. Transp Porous Media 62:243–274
8. Eloukabi H, Sghaier N, Ben Nasrallah S, Prat M (2013) Experimental study of the effect of sodium chloride on drying of porous media: the crusty-patchy efflorescence transition. Int J Heat Mass Transf 56:80–93
9. Veran-Tissoires S, Prat M (Submitted) On the formation of patchy salt efflorescence at the evaporative surface of a porous medium
10. Rodriguez-Navarro C, Doehne E (1999) Salt weathering: influence of evaporation rate, crystallisation rate and crystallisation pattern. Earth Surf Process Landf 24:191–209
11. Sghaier N, Prat M (2009) Effect of efflorescence formation on drying kinetics of porous media. Transp Porous Media 80:441–454
12. Meakin P (1998) Fractals, scaling and growth far from equilibrium. Cambridge University Press, Cambridge
13. Nachshon U, Shahraeeni E, Or D, Dragila M, Weisbrod N (2011) Infrared thermography of evaporative fluxes and dynamics of salt deposition on heterogeneous porous surfaces. Water Resour Res 47:W12519
14. Veran-Tissoires S, Marcoux M, Prat M (2012) Discrete salt crystallization at the surface of a porous medium. Phys Rev Lett 108:054502
15. Veran-Tissoires S, Marcoux M, Prat M (2012) Salt crystallization at the surface of a heterogeneous porous medium. Europhys Lett 98:34005

Chapter 20
Micro-CT Analysis to Explore Salt Precipitation Impact on Porous Media Permeability

Noam Weisbrod, Uri Nachshon, Maria Dragila, and Avrami Grader

Abstract The concern for water scarcity as a future reality for a growing proportion of the planet is driving the need to understand in greater detail the role of water in ecosystem sustainability and resiliency. Pertinent to this conversation is the role that salt precipitation and salinization has on soil resources. This paper addresses this task by presenting results of a detailed study on the reduction in gas permeability of a porous media by subflorescent salt precipitation. Using high-resolution CT scanning and detailed Lattice Boltzman Modeling, the three dimensional distribution of salt was mapped and its effect on permeability was calculated. It was found that salt precipitate mass increases towards the soil surface and that the effect of subsurface precipitation on gas permeability was significant, potentially impacting gas exchange processes between atmosphere and vadose zone. In addition, results from the LBM were compared to estimates using the Koseny-Carman equation. A good correlation between these two methods indicates that the Koseny-Carman equation may be sufficient to calculate changes in permeability, providing a more accessible tool relative to the more complicated LBM approach. This study has made it clear that knowledge of the spatial distribution of salt precipitation is essential for any estimation of gas permeability changes due to subflorescence.

N. Weisbrod (✉) • U. Nachshon
Department of Environmental Hydrology & Microbiology, Zuckerberg Institute for Water Research, Jacob Blaustein Institutes for Desert Research, Ben-Gurion University of the Negev, Sede Boker Campus, Midreshet Ben-Gurion, Beersheba 84990, Israel
e-mail: weisbrod@bgu.ac.il

M. Dragila
Department of Crop and Soil Science, Oregon State University, 3017 Agriculture and Life Sciences Building, Corvallis, OR 97331, USA

A. Grader
INGRAIN, Houston, TX, USA

L. Mercury et al. (eds.), *Transport and Reactivity of Solutions in Confined Hydrosystems*, NATO Science for Peace and Security Series C: Environmental Security,
DOI 10.1007/978-94-007-7534-3_20,

20.1 Introduction

Sodium Chloride (NaCl) is a very common salt in nature, and while it has many benefits it is also noted for its detrimental effects in agricultural and urban environments. Of increasing concern is the salinization of soil and water supplies that impact the future availability of our natural resources. For these reasons it is important to improve our understanding of its precipitation dynamics, its effect on matrix properties, and its consequent impact on evaporation from porous media. NaCl is known for its tendency to precipitate as an efflorescent salt crust above matrix surfaces during evaporation [6–8], however, in many cases it may also accumulate within matrix pores as a subflorescent formation. Such precipitation may occur under conditions of a receding evaporation front. During the drying process of porous media the evaporation front recedes into the matrix [4], resulting in downward movement of salt crystallization sites.

Of specific interest here is the complex relationship between subsurface precipitation and its subsequent impact on evaporation. Reduction of evaporation rates from porous media due to salt precipitation within (subflorescence) and above (efflorescence) the pores is a known phenomenon, well discussed in Nachshon et al. [6, 7]. The reduction in evaporation rates is explained as the results of an increase in matrix resistivity to vapor flow. Nachshon et al. [6, 7] examined and quantified the impact of NaCl efflorescent salt crust formation on matrix resistivity for vapor flow. It was found that the precipitated surface crust reduces evaporation rates by tens of percentages, since vapor has to diffuse through the salt crust layer that has a vapor diffusion coefficient that is approximately one order of magnitude lower than the porous media. While Nachshon et al. [6] explored the effect of efflorescence on vapor transport; the effect of subflorescence on vapor transport remained an open question. This paper reports on a first attempt to investigate the link between NaCl subflorescent salt precipitation during evaporation and matrix resistivity to vapor flow. To quantify this relationship at the pore scale, matrix porosity and air permeability are examined during NaCl subflorescent precipitation.

20.2 Theory

20.2.1 Permeability

Fluid flow through porous media is affected by the media and fluid properties, as expressed by Fick's law [5],

$$Q = -KA\nabla P \tag{20.1}$$

where, Q (m^3/s) is liquid discharge through the cross sectional area of flow A (m^2) in the porous medium, K is hydraulic conductivity (m/s), and ∇P is the pressure gradient that generates the fluid flow. The hydraulic conductivity is function of liquid and porous media properties, defining as [5],

$$K = \frac{k\,(\rho g)}{\mu} \tag{20.2}$$

where ρ is liquid density (kg/m^3), g is acceleration due to gravity (m/s^2), μ is liquid dynamic viscosity (kg/ms) and k is the intrinsic permeability of the medium (m^2 or Darcy, where 1D ≈ 10^{-12} m^2) [5]. k is the only parameter affected by the medium properties, which includes porosity (ϕ) and the mean grain size of the medium = (d (m)). Kozeny, then Carman, followed a bit later by Fair and Hatch formulated several empirical equations to correlate porous media properties to permeability. However, the most accepted correlation is the Kozeny-Carman equation that correlates ϕ, d, and k [1],

$$k = \frac{d^2}{180}\frac{\phi^3}{(1-\phi)^2} \tag{20.3}$$

Quantification of k for porous media in the field is practically impossible; it is traditionally measured in the laboratory on regularly shaped rock samples by forcing a fluid through the rock and recording the resulting fluid flux and pressure drops (e.g., [3]). However, new, state of the art technologies enable measurement of porous media permeability by numerically simulating fluid flow through a direct digital representation of a real pore space obtained by high-resolution 3D imaging. Such imaging and simulations can be rapidly and massively conducted on physical samples of irregular shapes and sizes that are impossible to handle in the physical laboratory. Moreover, in the case of salt precipitation in the media, using liquids for characterization is impossible as it may dissolve part of the precipitated salts. In the last two decades, the lattice Boltzmann method (LBM) has matured to become an efficient numerical scheme for simulations of fluid flow [10]. The slow viscous flow needed for permeability estimation is simulated by the LBM, which mathematically mimics the Navier–Stokes equations of viscous flow by treating the fluid as a set of particles with specific interaction rules. LBM provides great advantages over directly solving the equations of flow in that it directly handles the boundary conditions on a complex realistic pore surface. Full description of the LBM, for permeability estimation can be found at Tolke et al. [9]. As will be described later in greater detail, the LBM was used to calculate permeability by using porosity values obtained from high resolution CT images of porous media samples.

20.2.2 Salt Precipitation and Matrix Permeability

Evaporation of saline solution from porous media results in salt precipitation at the evaporation front, above the capillary fringe. Salt species and media moisture conditions both control location of precipitation. NaCl exhibits a preference to precipitate at the liquid–vapor interface. Therefore, the location of NaCl precipitation will depend on the location of the evaporation front. Efflorescent salt crust effect on matrix permeability is relatively easy to model as it creates an additional layer above the porous media, with different permeability. To compute the effective permeability (k_{eff}) of the combined matrix (k_1) and salt crust (k_2) system, the individual layers are summed using an analog to the resistivity of electrical circuits in series [2],

$$k_{eff} = \frac{L_1 + L_2}{\frac{L_1}{k_1} + \frac{L_2}{k_2}} \tag{20.4}$$

where, L_1 and L_2 are the thickness (m) of the matrix and salt crust, respectively. This method was used successfully to quantify the permeability of salt crusts in Nachshon et al. [6].

However, model and estimation of k_{eff} for subflorescent salt precipitation is much more complex, and practically impossible, due to the disperse nature of salt distribution within the pores. The Klinkenberg method is capable of measuring permeability of bulk samples, but is not able to provide information on the permeability profile of the sample which is necessary to understand the relationship between precipitation morphology and its effect on permeability. Nachshon et al. [6] successfully used the Klinkeberg method to measure permeabilities of a Berea sand stone giving values for permeability prior to and following salt precipitation. In this work we present a different approach to better understand salt precipitation effects on permeability while the sample as it being affected by the precipitated salt. We have combined the LBM with high resolution CT scans of the sample to measure salt precipitation and it effect on matrix porosity and permeability in a nondestructive way.

20.3 Materials and Methods

The CT scans to produce the 3D imagining of the porous media, and the LBM method to estimate the matrix permeability and porosity estimation from the CT scans were performed at INGRAIN, Huston (http://www.ingrainrocks.com/). CT scans were done using micro-CT (XRADIA, California, USA), at energy level of 80 kV/8 W, with voxel sizes of 3.83 μm. The sample consisted of dune sand with average grain diameter of 225 μm. Grains size distribution of the sand is presented in Fig. 20.1. The sample was created by placing the sand grains in a plastic cylinder with an inner diameter of 5 and 36 mm height. A consolidated sand core was placed at the bottom to prevent the sand from migrating downward and was sealed at the bottom so air cannot enter from the bottom. Hence, air only enters the sample from the top during the evaporation stage. Water with 5 % NaCl by weight was added to the sample from the top and allowed to distribute itself throughout the core. Then the sample was allowed to dry, while salt moved by capillarity to the top of the

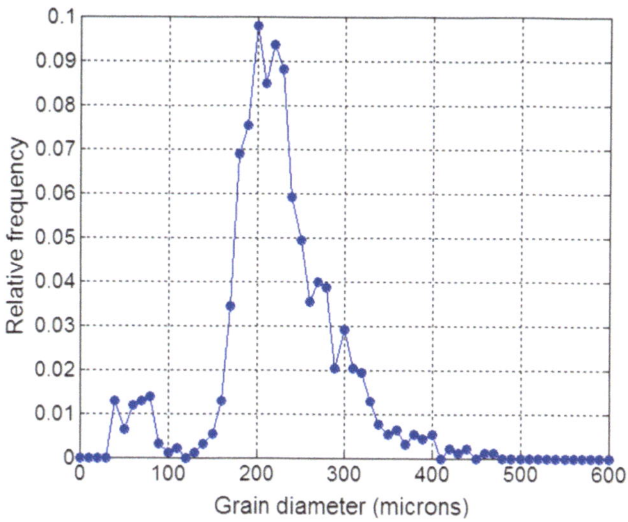

Fig. 20.1 Grain size distribution of the sample

core and crystallized. Prior to sample saturation (dry sample), after saturation, and after free evaporation for a couple of days, the sample was scanned. A subsample of the scanned image comprising a 1.53 mm cube was used to compute the absolute permeability of the dry sample in the vertical direction. The initial permeability was 17.6 D and the porosity was 31.6 %. Figure 20.2 shows 3D imagining of the scanned sample, where the grains are shown as solids and the pores are shown as empty. Because the sample was initially saturated, salt precipitated above as well as below the sample. However, the analysis presented here focuses only on the effect of subsurface precipitation.

Calculation of porosity and permeability by the LBM method were performed on the constructed micro-CT 3D imagining of the sample (Fig. 20.3a) below the matrix surface, ignoring any efflorescent salt precipitation above matrix. In order to compute permeability of discrete layers within the scanned sample, the LBM was applied to growing scanned sections of the sample (Fig. 20.3b). The LBM was applied to the top 1.72 mm of the sample, in increments that increased in thickness, starting at a depth of 1.72 mm and growing upwards towards the surface, with the upper boundary of the subscan at depths of 0.27, 0.33, 0.38, 0.57, 0.76, 0.96, 1.15, 1.34, 1.53 and 1.72 mm. The LBM produce a value for the k_{eff} of the entire examined section (Fig. 20.3c); by knowing the k_{eff} of a smaller section (the previous scanned section), the permeability of the added layer can be calculated using Eq. 20.4 (Fig. 20.3c, d). Repeating of this process over additional sections enables calculation of the discrete permeability of several layers within the sample.

Fig. 20.2 3D imagining of the 1.53 mm cube subsample used to compute the dry sample permeability and porosity

20.4 Results and Discussion

The dry (original state, no salt precipitation) sample porosity was found to be 31.6 %, with permeability of 17.6 D. The ensuing permeability measurements indicated significant reduction in permeability caused by the precipitated salt. The effective permeability of the entire sample was reduced by ~80 %, from 17.6 to 3.7 D. Figure 20.4 presents calculated permeabilities and porosities, using the LBM on ever increasingly larger sections of the scanned sample from bottom to top. It can be seen that the permeability gradually decreases with increasing length of the scanned sections, as it approaches the matrix surface. The most significant reduction in permeability occurred within the top 0.27 mm, where permeability is decreased by 57 %, from 8.6 to 3.7 D.

As mentioned earlier, one of the main advantages of the LBM calculations is the ability to compute the discrete permeability of a narrow layer within the sample. Figure 20.5 displays the calculated permeabilities of the discrete layers. Permeability of each layer was calculated as presented in Fig. 20.3.

It can be seen that in the lower part of the sample (below ~1 mm), permeabilities and porosities were not affected by the salt, maintaining their initial high values

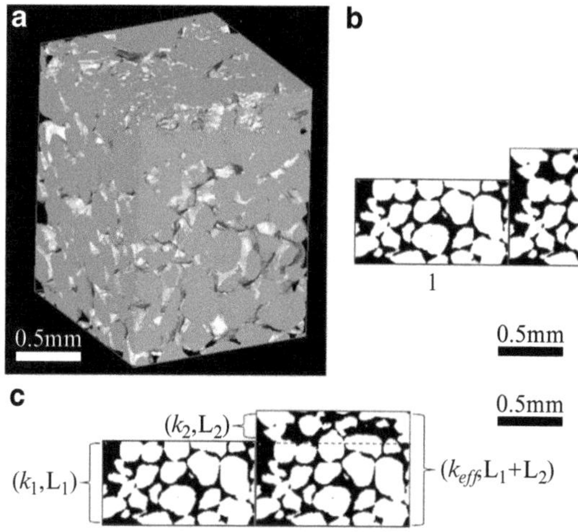

$$k_2 = \frac{L_2}{\frac{L_1 + L_2}{k_{eff}} - \frac{L_1}{k_1}}$$

Fig. 20.3 (**a**) Sample of a 3D image produced from the CT scan. (**b**) 2D samples of the growing sections from the entire sample. The LBM method was applied on every section (e.g., 1 through 4) separately. (**c**) Demonstration of the calculation process to estimate permeability for a discrete thin layer within the sample. First the effective permeability of the small section was found by the LBM. Then the effective permeability of the large section is found by the LBM. The permeability of the large section is defined as k_{eff}, while the permeability of the small section is defined as k_1. k_1 is also the permeability below the dashed line in the large section. The permeability of the thin layer, above the dashed line in the large section is defined as k_2 and it is the only unknown parameter. (**d**) Displays the modification of Eq. 20.4 to compute k_2 when all the other parameters are known

(Figs. 20.4 and 20.5). In the upper parts of the sample, above 1 mm, permeability and porosity are reduced increasingly as the surface is approached, with the most significant reduction in permeability of more than 65 % within the top 0.27 mm. This spatial distribution in permeability reduction is explained by the massive amounts of salt accumulation in the upper parts of the sample, which concurs with the observations presented in [6] and b. Figure 20.6 presents CT images showing the significant salt accumulation near the top of the sample.

The LBM calculated permeabilities and porosities can be compared to the Kozeny-Carman equation (Eq. 20.3). Figure 20.7 displays permeabilities of the discrete layers (Fig. 20.5), calculated by the LBM and Eq. 20.4, and theoretical permeabilities calculated by the Kozeny-Carman equation (Eq. 20.3), for d = 225 μm, which was found to be the average grain diameter.

A good agreement was observed between the LBM calculated permeabilities and the Kozeny-Carman equation (Eq. 20.3), with Pearson's correlation coefficient equal to 0.97, indicating a statistically good correlation. The good match between the LBM and the Kozeny-Carman equation is important as it validates the value of a simple tool, based on the Kozeny-Carman equation, to estimate changes in porous media permeability due to efflorescence salt precipitation. The approach based on the Kozeny-Carman equation is a much simpler calculation, compared to the LBM,

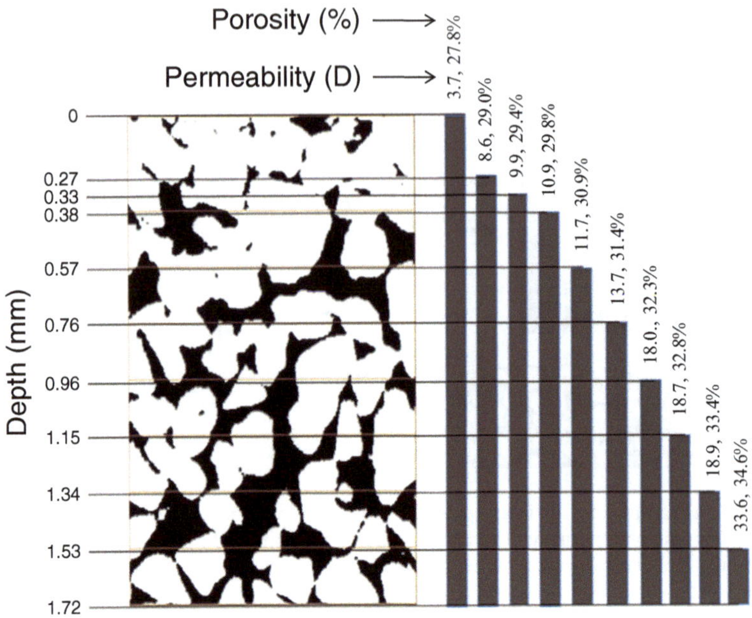

Fig. 20.4 Calculated permeabilities and porosities by the LBM method, for subsamples of the scanned image of increasing thickness, following subflorescent salt precipitation

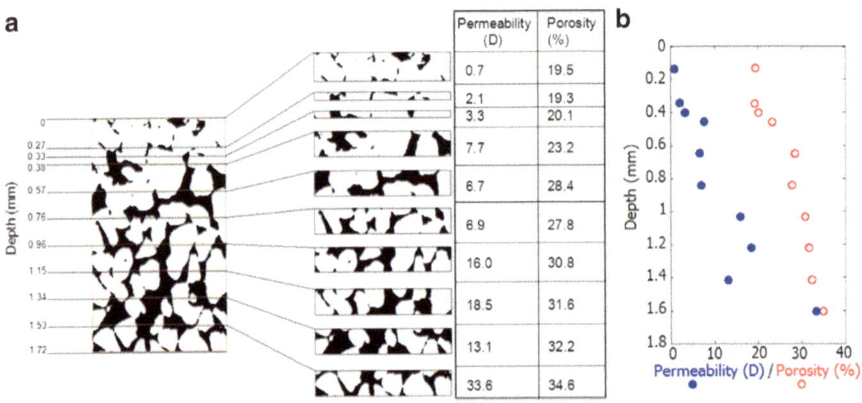

Permeability (D)	Porosity (%)
0.7	19.5
2.1	19.3
3.3	20.1
7.7	23.2
6.7	28.4
6.9	27.8
16.0	30.8
18.5	31.6
13.1	32.2
33.6	34.6

Fig. 20.5 Calculated permeabilities and porosities for discrete layers of the sand sample, following salt precipitation. (a) Permeabilities and porosities per individual layer; and (b) permeability and porosity profile

and this is the major advantage of this method. In order to compute permeability based upon the Kozeny-Carman equation, it is necessary but sufficient to know the media's effective grain diameter and matrix porosity, two parameters that can be attained easily by CT scans. Once these parameters are known, calculation of the

Fig. 20.6 Digital imagining of the CT scanned sample. (**a**) Vertical cross section of the middle of a 3D image. (**b**) Enlargement of the vertical cross section; crumble material is the precipitated salt. (**c**) Three-dimensional renditions of a volume, cropped out of the full sample, showing the *top* and *bottom* of the same volume. *Top* is being clogged by the salt, while the *bottom* is free of salt

effective permeability is simple, based on Eq. 20.3. With regards to extending this equation to quantify the effect of salt precipitation on permeability, it is important to keep in mind that salt is not homogeneously distributed in the matrix during the evaporation process, with most of it accumulated within the top few mm [6, 7]. Therefore, it is necessary to measure the porosity profile of the sample, with as high resolution as possible near the top of the sample. The K-C equation would be used on individual layers, with Eq. 20.4 facilitating calculation of the effective permeability of the entire sample.

A very relevant property of natural soils is media heterogeneity because it will significantly impact the spatial distribution of salt precipitation, and that spatial

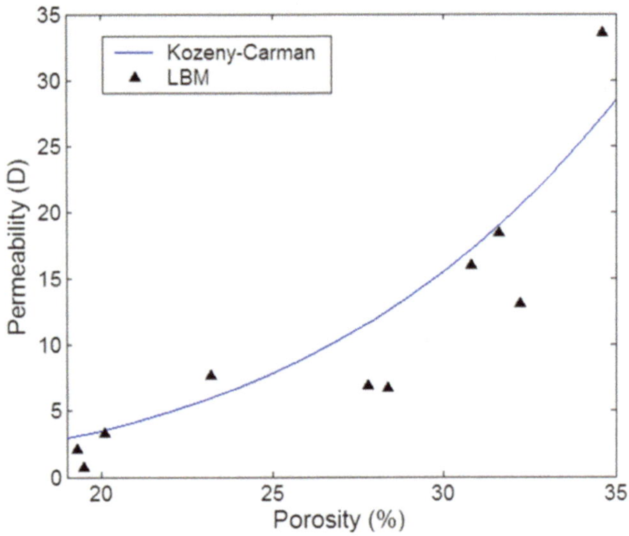

Fig. 20.7 Calculated permeabilities, following evaporation and salt precipitation, by the LBM and by the Kozeny-Carman equation

distribution can have a large effect on gas permeability. It was reported by Nachshon et al. [6] that in highly heterogeneous media subflorescent precipitation preferably deposits in the smaller pores, leaving the larger pores empty of obstruction. They found that during evaporation, while the sample experienced a reduction in permeability of 85 % and an expected permeability reduction (by Eqs. 20.3 and 20.4) of 99 %, the real reduction in permeability using the Klinkenberg test was only 7 %. What this result highlights is that to understand the effect of subflorescent precipitation on system gas exchange it is necessary not only to know the total amount of salt precipitated, but more importantly to understand the internal spatial distribution of the precipitate. Well connected, larger pores, free of salt obstruction can provide a very efficient path for gas exchange, while the remainder of the matrix is isolated from this process.

20.5 Conclusions

The work presented here sheds new light on the effect of NaCl subflorescent salt precipitation, associated with evaporation, on matrix permeability. Changes in vadose zone permeability due to salt precipitation can impact the physical mechanisms controlling exchange of vadose zone and atmosphere gases. These processes include vapor transport, greenhouse gases emissions and soil aeration which are very important for agriculture and many other environmental processes.

In this work, high resolution CT scans and Lattice-Botlzman modeling were used in conjunction to correlate subflorescent salt precipitation associated with evaporation to changes in porosity and permeability of the porous media. Significant changes in permeability, on the order of 60 % were observed due to the reduction in media porosity by NaCl subflorescent precipitation. Most of the salt accumulation occurred near the top of the sample, reducing porosity and permeability mainly in these areas. The permeabilities calculations were conducted by a complex numerical simulation of fluid flow in porous media using the Lattice-Boltzmann method and using for the media structure the 3D images from the high-resolution CT scans. Notably, the relationship salt-affected porosity and permeability calculated by the LBM was also well modeled by the Kozeny-Carman equation. Therefore, due to the good agreement between the much more sophisticated Lattice-Boltzmann method and the Kozeny-Carman equation, it is suggested that the Kozeny-Carman equation could facilitate computation of salt-affected permeability of homogeneous porous media once the porosity is well characterized. The validity of the Kozeny-Carman method for heterogeneous methods should be further explored in the future.

Acknowledgements This work was funded by the Binational Science Foundation (BSF), contract # 2006018 and the U.S. National Science Foundation (NSF) grant #0510825.

References

1. Bear J (1988) Dynamics of fluids in porous media. Courier Dover Publications, New York, p 764
2. Chin CW (2002) Quantitative methods in reservoir engineering. Gulf Professional Publishing, Amsterdam/Boston, p 480
3. Cosenza P, Ghoreychi M, Bazargan-Sabet B, de Marsily G (1999) In situ rock salt permeability measurement for long term safety assessment of storage. Int J Rock Mech Min Sci 36(4):509–526, ISSN 1365–1609, 10.1016/S0148-9062(99)00017-0
4. Fisher RA (1923) Some factors affecting the evaporation of water from soil. J Agric Sci 13(2):121–143
5. Kasenow M (2002) Determination of hydraulic conductivity from grain size analysis. Water Resources Publication, Highlands Ranch, p 97
6. Nachshon U, Weisbrod N, Dragila MI, Grader A (2011) Combined evaporation and salt precipitation in homogeneous and heterogeneous porous media. Water Resour Res 47:W03513. doi:10.1029/2010WR009677
7. Nachshon U, Shahraeeni E, Or D, Dragila MI, Weisbrod N (2011) Infrared thermography of evaporative fluxes and dynamics of salt deposition on heterogeneous porous surfaces. Water Resour Res 47:12. doi:10.1029/2011WR010776
8. Rodriguez-Navarro C, Doehne E (1999) Salt weathering: influence of evaporation rate, supersaturation and crystallization pattern. Earth Surf Process Landf 24(3):191–209
9. Tolke J, Baldwin C, Mu Y, Derzhi N, Fang Q, Grader A, Dvorkin J (2010) Computer simulations of fluid flow in sediment: from images to permeability. Lead Edge 29(1):68–74
10. Walls JD, Sinclair WS (2011) Eagle Ford shale reservoir properties from digital rock physics. First Break 29(6):97–101

Chapter 21
Reactive Transport in Heterogeneous Media

Harvey Scher and Brian Berkowitz

Abstract We analyze the dynamics of reactive transport in heterogeneous media, emphasizing the nature of the chemical reactions and the role of small-scale fluctuations induced by the structure of the porous medium, which is the main component of geological formations. Our goal is the interpretation of the results of laboratory-scale experiments, for which detailed characterization of the system is possible. Modelling approaches have been based on continuum and particle tracking (PT) schemes, which differ in how the fluctuations are incorporated. We choose PT methods wherein space-time transitions are drawn from appropriate probability distributions that have been tested to account for anomalous (non-Fickian) transport. While PT methods have been employed for many years to describe conservative transport, their application to laboratory-scale reactive transport problems in the context of both Fickian and non-Fickian regimes is relatively recent. We concentrate on experimental observations of different types of reactions in heterogeneous media: (1) the dynamics of a bimolecular reactive transport ($A + B \rightarrow C$) in passive (non-reactive) media, and (2) a multi-step chemical reaction, as exemplified in the process of dedolomitization involving both dissolution and precipitation. The fluctuations in a number of the key variables controlling the processes prove to have a dominant role; elucidation of this role forms the basis of the present study. An implication of these findings is that subtle changes in patterns of water flow, as a result of climate change or changes in land use, may have significant effects on water quality.

H. Scher • B. Berkowitz (✉)
Department of Environmental Sciences and Energy Research, Weizmann Institute of Science, Rehovot 76100, Israel
e-mail: Brian.Berkowitz@weizmann.ac.il

L. Mercury et al. (eds.), *Transport and Reactivity of Solutions in Confined Hydrosystems*, NATO Science for Peace and Security Series C: Environmental Security,
DOI 10.1007/978-94-007-7534-3_21,

21.1 Introduction

Fluctuations in key state variables (e.g., concentrations, velocities) caused by multi-scale heterogeneities contribute significantly to the space-time distribution of solutes in natural and reconstructed porous media. In reactive transport the fluctuations extend to the nature of chemical reactions between these solute (or reactant) distributions. Differences in theoretical approaches to modelling reactive transport in heterogeneous media lie largely in the treatment of the quantitative aspects of these fluctuations. We will concentrate on an approach that deals with modelling laboratory-scale experimental observations; it is particle tracking (PT), or more specifically, a PT implementation of the continuous time random walk (CTRW) framework (CTRW-PT). The CTRW-PT is a relatively recent treatment of reactive transport and is the basis of the successful modelling of the experimental observations of bimolecular ($A + B \rightarrow C$) and multi-step chemical reactions. Treatments of these smaller scales, where the chemical reactions occur, play an important role in understanding upscaled processes such as geochemical effects on aquifer water quality induced by changes in precipitation and patterns of water flow, and the feasibility of CO_2 sequestration in geological formations.

Particle tracking (PT) methodologies, first described by Smoluchowski [1] about a century ago, have been implemented widely in various fields of study [e.g., 2–8]. These studies suggest different ways to follow a particle while specifying, probabilistically, reaction rules among particles. Particle tracking techniques are considered to be effective in providing appropriate depictions of some of the salient features of reactive transport scenarios. This is due chiefly to the fact that they are prone to include simple formulations to account for pore-scale fluctuations of relevant quantities [2, 3, 5, 6, 9–12].

A detailed review of particle tracking methods, in the context of random walks, is given in [13]. In [14], a PT modelling approach is applied, within the continuous time random walk (CTRW) framework, to demonstrate the occurrence of non-Fickian transport in a random fracture network. Subsequently, Dentz et al. [15] showed that analytical and numerical solutions of the CTRW equations can be matched well using CTRW-PT simulations. An advantage of the CTRW-PT approach is that it allows considerable flexibility in the choice of probability density functions (pdf's) that govern particle motion in space and time [16]. The solutions of the advection-dispersion equation (ADE) are reproduced using an exponential (in time) pdf together with a spatial pdf having finite moments (e.g., an exponential pdf). Employing instead a truncated power law (TPL) for the temporal pdf results in a full range of non-Fickian behaviors [16, 17].

The PT approach can be shown to converge to a continuum description, so that PT represents in some sense both a "discrete random walk" and a means to solve a continuum equation, without the need to define a mesh in a numerical solution. Indeed, it has been noted [18] that, at least in principle, Eulerian and Lagrangian descriptions of (reactive) transport should yield identical results. However, to capture (at least partially) fluctuations in the local-scale reactant and product concentration distributions, descriptions based on deterministic models of reactive transport, such as the advective-dispersion reaction equation (ADRE), may require high spatial resolution. A critique of this approach is given by [2], who demonstrate agreement between a high resolution numerical solution of the continuum-scale equation and simulations using the stochastic model of [10, 19, 20].

In Sect. 21.2 we briefly review the salient points of CTRW and its major application to non-Fickian transport, which is ubiquitous in heterogeneous media. We introduce PT techniques and show the connection to CTRW. In Sect. 21.3 we focus on the two different problems, which have been the subject of several theoretical and experimental works: (21.1) a bimolecular reactive transport ($A + B \rightarrow C$) scenario in a passive (non-reactive) porous medium, and (21.2) a multi-species reactive system, as exemplified in the context of dedolomitization, where both precipitation and dissolution occur. Section 21.4 contains a discussion of our interpretive results.

21.2 CTRW and Particle Tracking

Random walk models are among the most significant in science and yet are based on a simple structure: a recursion relation wherein the space-time increment of each step is drawn from a common distribution $\psi(\mathbf{s}, t)$ and follows from the previous step,

$$R(s,t) = \sum\nolimits_{s'} \int_0^t \psi(s - s', t - t')\, R(s', t')\, dt' + \delta_{\mathbf{s},0}\delta\left(t - 0^+\right) \tag{21.1}$$

where $R(\mathbf{s}, t)$ is the probability per time for a walker to just arrive at site $\mathbf{s}$ at time t. The CTRW includes a random time for each step, and the main character of the pathways depends on the functional form of $\psi(\mathbf{s}, t)$. $P(\mathbf{s}, t)$ is the probability to be found at $\mathbf{s}$ at time t and is determined by $R(\mathbf{s}, t)$ [17]; it is the normalized concentration and is developed into a nonlocal-in-time pde. This equation reduces to an ADE for an exponential $\Sigma_{\mathbf{s}}\psi(\mathbf{s}, t) \equiv \psi(t)$. The exponential behavior describes a single rate (or transition time) between sites and is appropriate for a homogeneous material. Heterogeneity induces a spectrum of rates, which can be captured by a power-law time dependence, $\psi(t) \sim t^{-1-\beta}$, $\beta < 2$, cf. discussion below. A compact illustration of the capacity of CTRW with a truncated power law $\psi(t)$ to account for highly dispersive breakthrough curves (BTC), with essentially one parameter β, is shown in Fig. 21.1; see [17] for full details.

As outlined in Sect. 21.1, simulation via PT is highly flexible. The character of the transport and reaction can be changed significantly, with the choice of the pdf's affecting the space-time transitions. We use PT to implement the CTRW, i.e., equivalent to solving (21.1), with power law pdf's. This is a relatively new application; it has been demonstrated that the CTRW-PT yields the same results as the solution of the non-local (in time) pde formulation of CTRW [15, 16]. However, in the simulation of reactive transport (see below), the CTRW-PT cannot

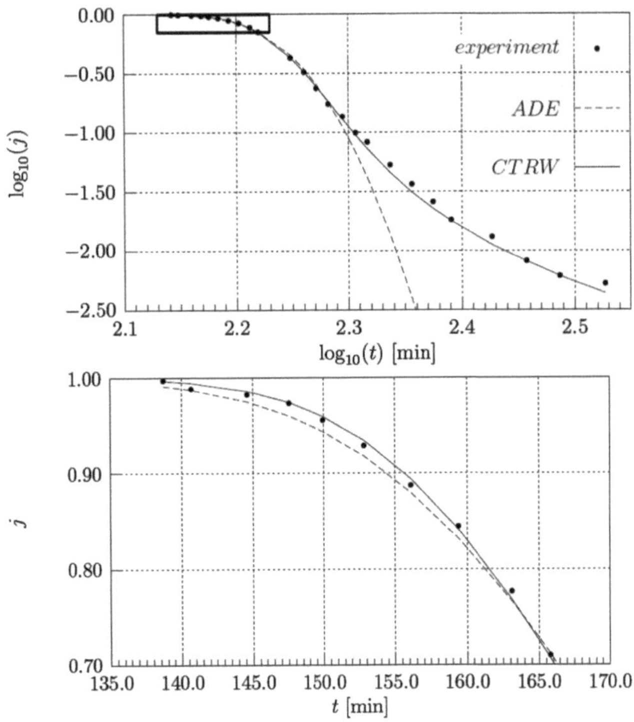

Fig. 21.1 Measured BTC with CTRW/ADE fits for a Berea sandstone core, using a truncated power law function with $\beta = 1.59$. Here the quantity j represents the normalized, flux-averaged concentration. (*top*) Complete BTC. (*bottom*) Region identified by the bold-framed rectangle in top plot. Note the difference in scale units between the plots. Column length equals 0.762 m. Porosity $n = 0.204$. Flux $q = 1.73$ cm^3/min. *Dashed line* is the best ADE model fit. *Solid line* is the best CTRW fit (Data from Scheidegger and Berkowitz [21])

be compared to a corresponding pde because a reaction term for this non-local pde has not yet been constructed. Hence, the PT implementation of non-Fickian transport in combination with multistep and multi-species reactions is unique.

In this CTRW-PT [15, 16], the movement of each particle is governed by the equation of motion:

$$\mathbf{s}^{(N+1)} = \mathbf{s}^{(N)} + \varsigma^{(N)}; t^{(N+1)} = t^{(N)} + \tau^{(N)} \tag{21.2}$$

where a random spatial increment $\varsigma^{(N)}$ and a random temporal increment $\tau^{(N)}$ are assigned to each particle transition from $\psi(\mathbf{s}, t)$ or the decoupled form $p(\mathbf{s})\psi(t)$. For each N step, a velocity v can be derived by $\varsigma^{(N)}/\tau^{(N)}$. During a simulation, particles are followed over a set of sample times. At each one, the locations of all particles are recorded (even in "mid-flight"). As described below, this information is of particular value for proper quantification of reactive transport.

The $p(\mathbf{s})$ is chosen to be an exponential $\lambda_s{}^2\exp(-\lambda_s s)$ throughout (for the angular part see [16]). For the ADE (Fickian) the $\psi(t) = \lambda_t\exp(-\lambda_t t)$ and for the heterogeneous cases $\psi(t) = \Omega\exp(-t/t_2)/(1+t/t_1)^{1+\beta}$ a truncated power law (TPL), where Ω is a normalization coefficient. The TPL is governed by three parameters: a power law exponent β, a characteristic transition time t_1, and a cut-off time to Fickian transport t_2 (see, e.g., [17] for details).

We now include chemical reactions among particles into the CTRW-PT to simulate reactive transport. In one example, for bimolecular reactions discussed in Sect. 21.3, the CTRW-PT methodology is straightforward. Two types of particles are defined, A ("inflowing" particles) and B ("resident" particles), and their movement is governed by the equations of motion (21.2). At each time interval, Δt, all particles in the system are frozen in mid-flight. The minimum distance between A and B particles is determined iteratively until all possible pairs have been checked; the particular details of the A and B particles are updated and recorded. A reaction $A+B \rightarrow C$ (where C replaces the A and B particles, which are removed from the simulation) takes place if this minimum distance is smaller than a prescribed radius of interaction, R (see below). Each C particle is located at the average position between the A and B particles it replaces; the motion of the C particles follows the same transport rules as those specified for A and B. In each time interval, the entire process is repeated until all possible reactions between A and B particles, lying within the interaction radius R at a given time interval, are allowed to occur. The reaction behavior is controlled largely by this radius, which is chosen to be at the pore scale. The time interval, Δt, affects the estimated amount of mixing – and thus the degree of reaction – in the sense that increasing the value of Δt allows more time, before carrying out the calculation step for the reactions, for the plumes of A and B particles to spread and overlap. As a consequence, the choice of Δt affects the number of C particles produced at any given time; obtaining a more continuous temporal assessment of the reactions requires that Δt be sufficiently small.

In another example, one can define different ways to account for reactions in a concentration-based algorithm. Each simulated particle is a representation of a fractional mole amount for one reactant in an experiment. The algorithm will divide the simulated domain into a grid, where for each Δt the concentration of reactants is calculated for each cell. From the concentrations, one can establish the local reaction for each cell without using R as the reaction radius. Using such a scheme, we have incorporated, e.g., precipitation and dissolution with multispecies reactive transport; a detailed account can be found in [22] (discussed in detail below, in Sect. 21.3.2).

21.3 Experimental and PT Modelling of Reactive Transport

21.3.1 Bimolecular Reactive Transport

We consider the experiment on bimolecular reactive transport presented in detail in [23]. Colorimetric reactions between $CuSO_4$ and $EDTA^{4-}$ were measured in a one-dimensional (on average) flow field; the flow cell has dimensions 36 cm (length) × 5.5 cm (width) × 1.8 cm (depth), and was packed with cryolite beads. The flow cell was saturated with $EDTA^{4-}$ (denoted as B particles) and $CuSO_4$ (denoted as A particles) was subsequently injected as a step input. The mixing and reaction zone of the product $CuEDTA^{2-}$ were measured via fluorescence. These experiments are characterized by large values of Pe ($\approx O(10^3)$) and Da ($\approx O(10^{12})$), indicating that advective and reactive processes are significantly faster than diffusion.

Pore-scale fluctuations in concentrations are visible in the experiments in [23]. Clearly, the dispersion is not uniform and the reaction fronts (forward and backward tailing regions) are not uniform or sharp. Rather, as emphasized in [24], colorized "islands" of pixels are distributed broadly and heterogeneously in these regions, demonstrating "isolated" reactions occurring far from the $CuEDTA^{2-}$ plume center of mass; also sharp localized concentration gradients can be observed. This behavior suggests significant non-uniformity in reactant mixing and product spreading. Close inspection of Fig. 21.4 in [23] reveals pronounced tailing effects in the cross-sectional averaged concentration profile which are not captured by an ADE.

There have been several efforts to model these data [9, 23–26]. Gramling et al. [23] use an analytical solution of the ADRE which was derived on the assumption that the concentration of the limiting reactant is instantaneously and completely consumed in the reaction. They observe that this analytical solution significantly over-estimates the space-time evolution and the total production of C in the system. Indeed, the ADRE solution overestimates the peak concentration value by as much as 40 % as compared to the measured one; similarly, the estimated cumulative mass production is up to 20 % larger than the measured production.

In [9, 24] CTRW-PT simulations consider particle transport in a rectangular, two-dimensional domain, with impermeable horizontal boundaries and fixed flux vertical boundaries. The domain is filled initially with a low concentration (c_B) of uniformly distributed B particles. At $t = 0$, A particles are introduced at the inlet boundary with the same concentration c_A as c_B. Technical details regarding insertion of A particles into the domain, treatment of boundary conditions and sensitivity to the selected time interval Δt are reported in [9, 24].

Figure 21.2 contains the experimental data from Figure 5b of [23], showing the temporal evolution of the spatial concentration profile of C particles at three different observation times. The figure reports the results obtained by [24] using a TPL-PT, which is marginally Fickian with the power law exponent $\beta = 1.96$ ($\beta \geq 2$ is Fickian), together with the Fickian ADE-PT. Note that in the CTRW-PT approach, for $\beta > 1$, the mean location of the tracer plume scales as t, and the standard deviation $\sigma \sim t^{(3-\beta)/2}$, whereas for the ADE-PT, $\sigma \sim t^{1/2}$. For discussion of the red line we refer the reader to [26]. The dotted line is the analytical solution adopted in [23] which assumes instantaneous reaction with complete pore-scale mixing. It is significant that [24] calibrated their model parameters against a subset of the measurements, and provided predictions that match the remaining measurements. In these plots, the parameters are tuned manually until a satisfactory visual fit is obtained in [24]. Note that the ADE-PT model is unable to capture the forward and backward tails of the C profiles.

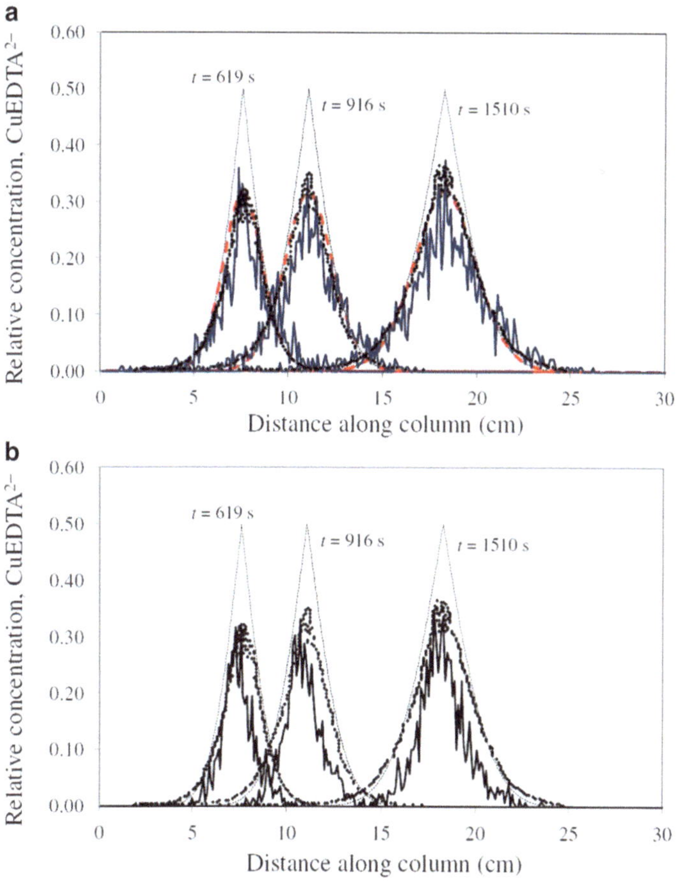

Fig. 21.2 (**a**) Particle tracking simulations with the non-Fickian CTRW [24] (*jagged blue line*), and ADRE solution of [26] (*smooth red line*), compared to the experimental measurements of [23] (*dots*) and the ADRE solution of [23] (*dotted line*), showing the relative concentration of C particles at different times. (**b**) Particle tracking simulations with a Fickian ADE [24] (*jagged line*), compared to the experimental measurements of [23] (dots) and the ADRE solution of [23] (*dotted line*), showing the relative concentration of C particles at different times. In both (**a**) and (**b**), the PT simulations of [24] and ADRE solution of [26] were fit to the measured profile at $t = 619$ s, and the same parameters were used as predictions for the two later times $t = 916$ s and $t = 1{,}510$ s

The time evolution of the measured cumulative mass of $CuEDTA^{2-}$ produced is presented in Fig. 21.3 for the same experiment shown in Fig. 21.2. Data are contrasted against the modelling results of [23, 24, 26]. The agreement between experiments and the models is remarkable, considering the limitations associated with the concept of effective dispersion and reaction rate. One can note that at early times the reaction is relatively fast. The mass production rate, which is the temporal derivative of the curve in Fig. 21.3, then displays a power law decay (see Figure 3a in [24]) with time, being controlled mostly by the slow diffusion of the reacting

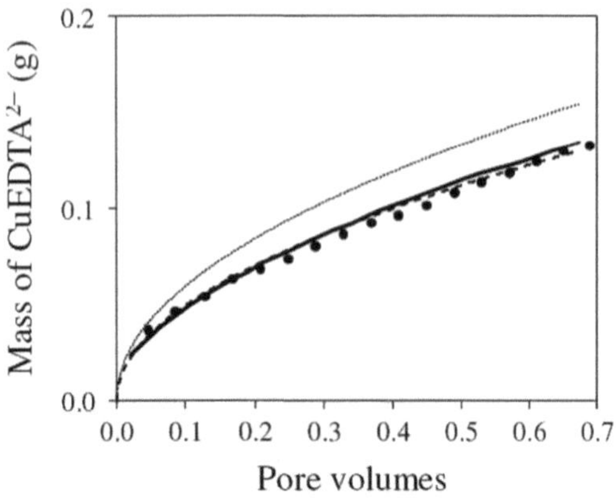

Fig. 21.3 Total mass production given from the particle tracking simulations of [24] (*smooth line*) and ADRE simulations of [26] (*dashed line*), compared to the experimental measurements (*large dots*) and the ADRE standard pore-scale mixed model of [23] (*dotted line*)

species into and out of the low velocity pores, causing incomplete mixing. The PT simulations in Figure 3a of Edery et al. [24] indicate ongoing reactions, even after long time periods, which suggest the occurrence of rare fluctuation-dominated dynamics. The fluctuation-dominated dynamics in the velocity are apparent in a phase diagram that shows, for each reactant, the median velocity of the reacted particles vs. median of the total particle velocity (see Figure 7 of [24]). This diagram shows different distribution functions (ADE-PT, TPL-PT), and illustrates that the (inflowing) *A* reacting particles have an overall higher velocity relative to the mean, or median, velocities of the total set of particles, while the (resident) *B* reacting particles have an overall lower velocity. These systematic deviations from the average velocity are caused by fluctuation-dominated mechanisms. It is also demonstrated that the TPL-PT, which has a heavier weighting on the tails, affects the magnitude of the fluctuations more strongly than the ADE-PT. The mixing dynamics produce a void area between the *A* and *B* profiles. As the void develops only a portion of the *A*, *B* transitions can cross the void. An analytic treatment and further theoretical developments are presented in [24].

21.3.2 Multispecies Heterogeneous Reaction

In this section, we model a different reactive transport scenario – an experiment of dedolomitization [27]. The major change is the necessity of accounting for multi-step, multi-species reactions, which are both reversible and irreversible in a reactive medium. The framework of the reactions is:

$$CaMg(CO_3)_2 + 4H^+ \rightarrow Ca^{2+} + Mg^{2+} + 2H_2CO_3 \tag{21.3}$$

$$H_2CO_3 \leftrightarrow H^+ + HCO_3^- \tag{21.4}$$

$$HCO^{-3} \leftrightarrow H^{+} + {CO_3}^{2-} \tag{21.5}$$

$$Ca^{2+} + {CO_3}^{2-} \leftrightarrow CaCO_3 \tag{21.6}$$

In principle one could set up nonlinear coupled pde's for each reactant and solve the equations with incorporation of the effect of medium heterogeneities or use a PT, Sect. 21.2, for each reactant and specify for each one a local reaction scheme. Both of these alternatives for (21.3), (21.4), (21.5), and (21.6) would be rather daunting. Fortunately, a few simplifications of (21.3), (21.4), (21.5), and (21.6) make the problem tractable.

As detailed in [27], we consider a column packed with crushed sucrosic dolomite, $CaMg(CO_3)_2$ (containing 1.5 % calcium carbonate $CaCO_3$), in the form of particles having a diameter range of 350–500 μm; hence it is one of the abundant species. A mixture of $CaCl_2$/HCl in water is injected across the entire inlet boundary of the column, which results in the dissolution of dolomite and precipitation of $CaCO_3$; Ca^{2+} is also one of the abundant species.

The reaction cascade (21.3), (21.4), (21.5), and (21.6) is irreversibly initiated by the local pH. The two middle reactions, (21.4) and (21.5), are reversible and sensitive to the local pH. We assume these reactions constitute a local carbonate equilibrium and an input of H_2CO_3 produces an output of ${CO_3}^{2-}$, which immediately reacts with the abundant Ca^{2+} to produce $CaCO_3$. The framework (21.3), (21.4), (21.5), and (21.6) simplifies to

$$D + 2h \rightarrow 2A \tag{21.7}$$

$$A \leftrightarrow h + C \tag{21.8}$$

where $D =$ dolomite, $A = H_2CO_3$, $C = CaCO_3$ and $h = 2H^{+}$. In a low pH region, there is dissolution of D, (21.7), and C (in the sucrosic dolomite) $h + C \rightarrow A$, (21.8). In a high pH region there is precipitation of C, $A \rightarrow h + C$, (21.8).To subordinate (21.4), (21.5), (21.6), and (21.8), we require a figure of merit of the carbonate equilibria to produce A as a function of the local pH. We use as the cumulant of this probability pdf that A is produced as the fraction (as a function of pH) of A in the carbonate equilibria

$$a_1\,(\mathrm{pH}) = \frac{10^{-2\mathrm{pH}}}{10^{-2\mathrm{pH}} + 10^{-2\mathrm{pH}-\mathrm{pKa1}} + 10^{-\mathrm{pKa1}-\mathrm{pKa2}}} \tag{21.9}$$

with the standard values $pK_{a1} = 6.35$, $pK_{a2} = 10.33$ [e.g., 28] and a range $0 \le \alpha_1 \le 1$. The reactive transport with (21.3), (21.4), (21.5), and (21.6) is represented by the PT of the species h and A with the initiating dissolution reaction governed by the local pH, with accounting for either (21.7) or (21.8) (arrow to the left). The fluctuations of the transport and the chemical reactions are modeled with the TPL-PT resulting in a profile of h, which produces A with the probabilistic weighting of $\alpha_1(\text{pH})$. In essence, the system of reactions (21.7) and (21.8) is an interchange of A and h with interaction with the abundant D, C and Ca^{2+} as sources and sinks.

There is an additional level of fluctuations due to the modification of the medium by the reaction, i.e., changes in local porosity by precipitation of C and dissolution of D and C. The feedback from the porosity change is the effect on the local flow, which can subsequently change the precipitation pattern. This pattern, which we refer to as "banding", is affected by a number of parameters and is the key to our comparison to the experimental observations. To develop this picture fully, we describe the PT simulation of the h and A dynamics.

The details of the simulation algorithm are contained in [22]; basically we consider a two-dimensional rectangular column medium, with the C and D particles being distributed either uniformly or heterogeneously, divided into cells 1 mm $\times$ 1 mm. The numbers of A and h particles, which are advanced by TPL-PT, in each cell, are updated iteratively during each time step. With a uniform injection of Ca^{2+} and H^+ across the entire inlet boundary, we track the local pH fluctuations along the advancing front – the pH in each cell. The updated pH in each cell, denoted pH_{loc}, determines the value $\alpha_1(\text{pH}_{\text{loc}})$. If a chosen random number U $(0 \leq U \leq 1)$ is less than this value, dissolution occurs; otherwise, precipitation occurs. To facilitate the computation, the number of injected H^+ particles is restricted to 10^4, so that each particle carries a parcel of moles of H^+ to account for the total injected pH, e.g., pH $= 3.5$. The time interval Δt sets (using the average velocity) the average distance each particle can traverse before being allowed to react.

At each Δt, all particles are "frozen" in space and the possible chemical reactions are determined. All cells are examined according to the dissolution/precipitation process described above, with reaction (21.8) proceeding in one of two directions. After the dissolution of C do we allow D particles (dolomite) to take part in the dissolution process. In this case, dissolution of D, following (21.7) requires two h particles and produces two A particles in each reaction. The A particles are subsequently transported similar to the h particles. Similarly do we determine calcium carbonate precipitation.

In a porous medium, a particle will in general move more slowly in a denser cell (densities of dolomite or calcium carbonate particles) or, often bypass it, preferring to advance more easily in a less dense cell. These spatial variations will in turn alter the precipitation/dissolution patterns, causing a feedback between the flow pattern and the density in the cells. To mimic these feedback mechanism effects in our PT simulations, we modify the velocity distribution as a function of the local (cell) density of C and D particles.

The key comparisons to the experimental data of [29] (especially Figure 6 within that paper), to our simulation results are given in the patterns of the spatial distributions of $CaCO_3$ as a function of the pH of the injected fluid and the flow rate Q as well as other variables within this broad range we have produced patterns similar to the observations of [29] with the same sensitivity of these patterns to the various parameters.

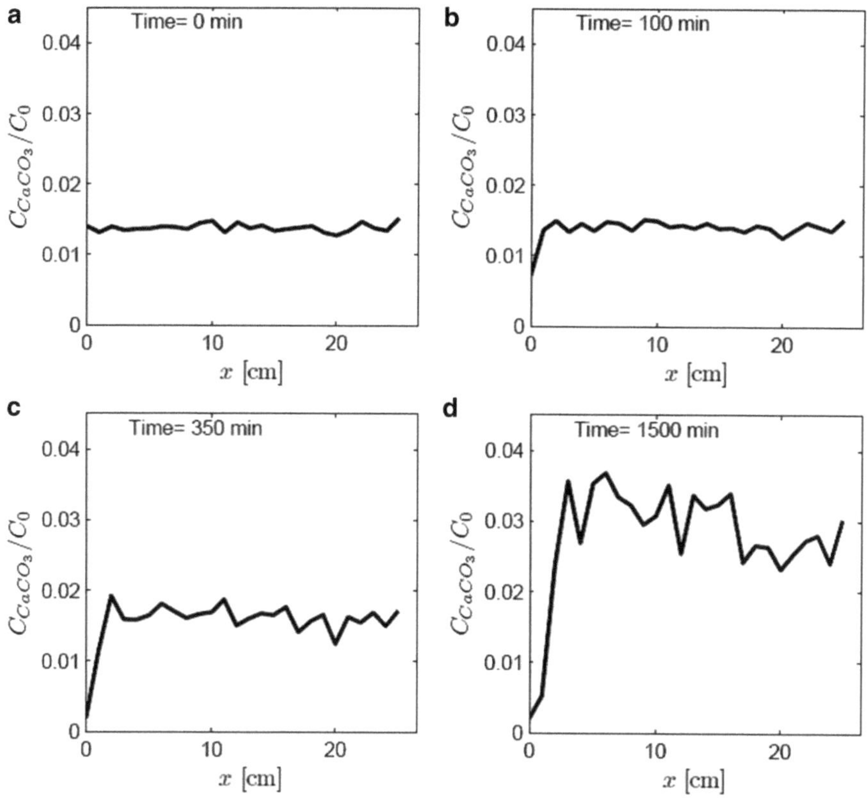

Fig. 21.4 The time evolution of the amount of calcium carbonate relative to the total amount of initial dolomite and calcium carbonate (C_0), for an initially heterogeneous distribution of dolomite and calcium carbonate, with pH = 4.5, $Q = 1.5$ cm^3/min (After [22])

We restrict our results here to a limited set of parameter values. We fix pH = 4.5, 3.5, $Q = 1.5$, 5 cm^3/min, $\Delta t = 6$ s, $\beta = 1.8$ and $R = 0.03$. In Fig. 21.4, observe the strong decrease in the presence of D and C particles near the inlet, at all times. This is due to the rapid drop in pH (from the initial fluid value of 8) with the injection of h particles dissolving C and D particles and producing A particles. Subsequently, downstream each A can precipitate as a C. The overall growth in C is highly irregular in space and time, as in Fig. 21.4 at $t = 1{,}500$ min; three maximum concentration values appear at $x = 7$, 15, and (on average) 22 cm. This corresponds to a banding effect, with a relatively clear temporal evolution. The main difference between this set of bands and the one given in Figure 6 in [29] is the level of "noise". The statistics of C precipitation are less averaged due to the computational demands involved with tracking each H^+ particle discussed above.

We define a precipitation band as the increase in cross-sectional concentration of C particles, along the x-axis, followed by a subsequent decrease farther along the x-axis. The phenomenon of banding, its detailed shape definition, and its position are due to a number of factors. These factors affect the dynamics of the h particles, which play the main role in the production of A, via (21.7). In a low pH environment, C and D are dissolved where the production of A replaces the h, creating a higher pH environment. The A particles that escape from the low pH cells, due to faster-than-average local velocity, cause precipitation and the peaks of C. When precipitation of C occurs, h particles are released and advance downstream, subsequently dissolving regions beyond the peaks. This process, which leads to distinct regions of precipitation and dissolution, occurs because the advancing h and A particles do not move uniformly in space.

We have demonstrated how simplified chemical rules can be combined with detailed simulation of transport to mimic the dynamic patterning of precipitation and dissolution, and evolution of calcium carbonate banding, which result from dedolomitization. The core of the simulations is the novel treatment of the injected H^+ particles as the key reactant. The H^+ particles initiate the dissolution reaction and determine the chemical environment for precipitation of $CaCO_3$. In both reactions, there is also modification of the medium leading to local changes in the permeability field and the flow patterns in the cell. The flow patterns effectively change the local pH levels, and the entire cycle is subsequently updated.

21.4 Discussion and Perspectives

There is a large array of configurations to consider for reactive transport in heterogeneous media, each with distinct subtleties. We specialize to the one common to the laboratory experiments discussed in this paper, namely, the step injection of chemicals into a prepared reservoir of reactants. Distinct features associated with such experiments include the transport or spreading of the reactants, the nature of the chemical reactions, the evolution of the mixing zone, and the feedback of a medium affected by the reactions. We assess the role of fluctuations of pore-scale state variables, induced by the structure of the porous medium, on these features. The accurate characterization of fluctuations necessitates departure from an average quantity. The nature of the average in turn becomes an important issue, e.g., is the particular procedure adopted for averaging concentrations and reactions suppressing the effect of fluctuations? We discussed above these aspects for two prototype scenarios, for which detailed measurements at the laboratory scale are available. We focused on modelling techniques based on a PT approach that involves an implementation of CTRW that is not a particle tracking method per se, in the sense that particles are not advanced by advection along streamlines and by diffusion/dispersion by a random component. Rather, the CTRW-PT involves the use of appropriate pdfs describing the space-time transitions, which have been shown in extensive testing to account for anomalous (non-Fickian) transport in a variety of laboratory and field observations. The method provides a good model to account for the effect of small-scale fluctuations on the advancing profile(s). Hence, the intimate connection between fluctuations in the transport and fluctuations in the reactions is clear.

References

1. Smoluchowski M (1906) Zur kinetischen theorie der brownschen molekularbewegung und der suspensionen. Ann Phys 21:756–780
2. Palanichamy J, Becker T, Spiller M, Köngeter J, Mohan S (2007) Multicomponent reaction modelling using a stochastic algorithm. Comput Vis Sci 12(2):51–61. doi:10.1007/s00791-007-0080-y
3. Fabriol R, Sauty JP, Ouzounian G (1993) Coupling geochemistry with a particle tracking transport model. J Contam Hydrol 13:117–129
4. Sun NZ (1999) A finite cell method for simulating the mass transport process in porous media. Water Resour Res 35:3649–3662
5. Cao Y, Gillespie D, Petzold L (2005) Multiscale stochastic simulation algorithm with stochastic partial equilibrium assumption for chemically reacting systems. J Comput Phys 206:395–411
6. Lindenberg K, Romero AH (2007) Numerical study of $A + A \rightarrow 0$ and $A + B \rightarrow 0$ reactions with inertia. J Chem Phys 127(17):174506. doi:10.1063/1.2779327
7. Srinivasan G, Tartakovsky DM, Robinson BA, Aceves AB (2007) Quantification of uncertainty in geochemical reactions. Water Resour Res 43:W12415. doi:10.1029/2007WR006003
8. Yuste SB, Klafter J, Lindenberg K (2008) Number of distinct sites visited by a subdiffusive random walker. Phys Rev E 77(3):032101
9. Edery Y, Scher H, Berkowitz B (2009) Modeling bimolecular reactions and transport in porous media. Geophys Res Lett 36:L02407. doi:10.1029/2008GL036381
10. Gillespie D (1977) Exact stochastic simulation of coupled chemical reactions. J Phys Chem 25(81):2340–2361
11. Sahimi M, Gavals GR, Tsotsis TT (1990) Statistical and continuum models of fluid–solid reactions in porous media. Chem Eng Sci 45(6):1443–1502
12. Sahimi M (1993) Flow phenomena in rocks: from continuum models to fractals, percolation, cellular automata, and simulated annealing. Rev Mod Phys 65(4):1393–1534
13. Delay F, Ackerer P, Danquigny C (2005) Simulating solute transport in porous or fractured formations using random walk particle tracking: a review. Vadose Zone J 4:360–379. doi:10.2136/vzj2004.0125
14. Berkowitz B, Scher H (1998) Theory of anomalous chemical transport in fracture networks. Phys Rev E 57(5):5858–5869
15. Dentz M, Cortis A, Scher H, Berkowitz B (2004) Time behavior of solute transport in heterogeneous media: transition from anomalous to normal transport. Adv Water Resour 27(2):155–173. doi:10.1016/j.advwatres.2003.11.002
16. Dentz M, Scher H, Holder D, Berkowitz B (2008) Transport behavior of coupled continuous-time random walks. Phys Rev E 78:041110. doi:10.1103/PhysRevE.78.041110
17. Berkowitz B, Cortis A, Dentz M, Scher H (2006) Modeling non-Fickian transport in geological formations as a continuous time random walk. Rev Geophys 44:RG2003. doi:10.1029/2005RG000178
18. Gardiner CW (2004) Handbook of stochastic methods for physics, chemistry and the natural sciences, vol 13, 3rd edn, Springer series in synergetics. Springer, New York
19. Gillespie D (1976) A general method for numerically simulating the stochastic time evolution of coupled chemical reactions. J Comput Phys 22:403–434
20. Gillespie D (1977) Concerning the validity of the stochastic approach to chemical kinetics. J Stat Phys 16(3):311–318
21. Scheidegger AE (1959) An evaluation of the accuracy of the diffusivity equation for describing miscible displacement in porous media. In: Proceedings of the theory of fluid flow in porous media conference. University of Oklahoma, Norman, pp 101–116
22. Edery Y, Scher H, Berkowitz B (2011) Dissolution and precipitation dynamics during dedolomitization. Water Resour Res 47:W08535. doi:10.1029/2011WR010551
23. Gramling CM, Harvey CF, Meigs LC (2002) Reactive transport in porous media: a comparison of model prediction with laboratory visualization. Environ Sci Technol 36(11):2508–2514

24. Edery Y, Scher H, Berkowitz B (2010) Particle tracking model of bimolecular reactive transport in porous media. Water Resour Res 46:W07524. doi:10.1029/2009WR009017
25. Rubio AD, Zalts A, El Hasi CD (2008) Numerical solution of the advection reaction–diffusion equation at different scales. Environ Model Software 1(23):90–95
26. Sànchez-Vila X, Fernàndez-Garcia D, Guadagnini A (2010) Interpretation of column experiments of transport of solutes undergoing an irreversible bimolecular reaction using a continuum approximation. Water Resour Res 46:W12510. doi:10.1029/2010WR009539
27. Singurindy O, Berkowitz B (2003) Evolution of hydraulic conductivity by precipitation and dissolution in carbonate rock. Water Resour Res 39:1016. doi:10.1029/2001WR001055
28. Manahan SE (2000) Environmental chemistry, 9th edn. CRC Press, Boca Raton
29. Singurindy O, Berkowitz B (2003) Flow, dissolution, and precipitation in dolomite. Water Resour Res 39:1143. doi:10.1029/2002WR001624

Chapter 22
Extraction of Water from the Atmosphere in Arid Areas by Employing Composites "A Salt Inside a Porous Matrix"

Larisa Gordeeva and Yuri Aristov

Abstract This communication is addressed to sorptive extraction of water from the atmosphere in arid areas. The method includes (a) sorption of water vapour in an adsorber in the night-time when the air relative humidity is comparatively high, and (b) desorption of the stored water and its subsequent collection in a condenser in the day-time. New materials adapted to this process are highly welcome. Composites "a salt inside a porous matrix" (CSPMs) have enhanced water sorption capacity and their properties may be intently varied in a wide range. In this communication, we make a preliminary analysis of CSPMs application for extraction of water from the atmosphere. Firstly, a general scheme of the water extraction is described. Then, we form a mental representation of an ideal solid sorbent that is optimal for the extraction of water from the atmosphere. Finally, we discuss how to design a real CSPM with properties meeting the formulated requirements, what are roles of the salt and the matrix, etc.

22.1 Introduction

A sustainable water supply is of vital importance for the humanity development. If all the freshwater on the planet were divided equally among the global population, there would be 5,000–6,000 m^3 of water available for everyone, every year. This global calculation gives an impression of abundance. However, the world's freshwater resources are distributed very unevenly, as is the world's population [1]. The water-stressed basins are situated mainly in northern Africa, the Mediterranean region, the Middle East, the Near East, southern Asia, northern China, Australia, the USA, Mexico, north eastern Brazil and the west coast of South America [2, 3].

L. Gordeeva • Y. Aristov (✉)
Boreskov Institute of Catalysis, Lavrentiev av. 5, 630090 Novosibirsk, Russia
e-mail: aristov@catalysis.ru

L. Mercury et al. (eds.), *Transport and Reactivity of Solutions in Confined Hydrosystems*, NATO Science for Peace and Security Series C: Environmental Security,
DOI 10.1007/978-94-007-7534-3_22,

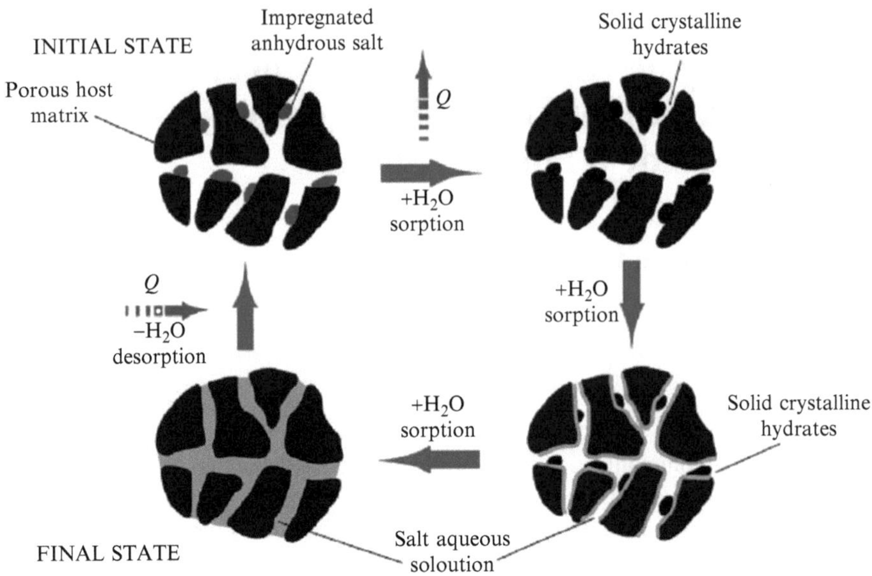

Fig. 22.1 Schematics of water sorption on CSPMs [6]

At the same time, the earth's atmosphere is a huge reservoir of moisture. The total amount of water in the atmosphere of $13 \cdot 10^3$ km^3 greatly exceeds the decade's global demand. In the extremely arid region of Africa (Sahara desert) the water content amounts of 4–8 g per m^3 of air that corresponds to the vapour partial pressure of 5–10 mbar. In several arid regions, the humidity can reach 15 g per 1 m^3 of air. Even in the Sahara and Namib deserts the absolute humidity is on a par with that for Oregon and Siberia. In contrast to rainfall, which is characterised by the seasonal irregularity, the atmospheric humidity is the only continuously available water resource for the most desert regions. Sorptive extraction of the water from the atmosphere could significantly alleviate the problem of water shortage in arid regions.

Although this approach is known from ancient times, nowadays it has attracted much attention again. New opportunities might be opened by means of developing new materials with advanced sorption properties adapted to the application involved. Composites "a salt inside a porous matrix" (CSPMs) have enhanced water sorption capacity and their sorption properties may be intently varied in a wide range to fit requirements of a particular application [4, 5]. Typical CSPM is a two-component material that consists of a host matrix and an inorganic salt placed inside the matrix pores (Fig. 22.1). A salt S is the main sorbing component. It reacts with vapour forming a hydrate $S \cdot nH_2O$ according to the reaction

$$S + n\,H_2O = S \cdot n\,H_2O. \tag{22.1}$$

Further water sorption leads to the hydrate dissolution/liquefying and its transformation to an aqueous salt solution inside the pores [4–7]. As a result of the volume filling of pores (Fig. 22.1), a large water sorption capacity can be obtained (up to 0.6–1.0 g/g) which is superior to that of common adsorbents, like silica gels and zeolites. The equilibrium pressure of the hydrate formation is a function of solely temperature

$$\ln P^*(T) = -\Delta G^\circ/\mathrm{R}T = -\Delta H^\circ/\mathrm{R}T + \Delta S^\circ/\mathrm{R}, \tag{22.2}$$

where ΔG° is the variation of the standard Gibbs energy of reaction (22.1). According to the Gibbs phase rule, the system "salt-H_2O" is mono-variant, therefore at a fixed temperature a salt completely transforms to appropriate hydrate at a certain pressure $P^*(T)$ given by Eq. (22.2). Hence, in the Clausius-Clapeyron diagram, the equilibrium of the hydrate formation/decomposition is depicted by a single isoster.

In this communication, we make a preliminary feasibility analysis of CSPMs application for extraction of water from the atmosphere. Firstly, a general scheme of this process is described. Then, we form a mental representation of an ideal solid sorbent that is optimal for the sorptive extraction of water from the atmosphere. Finally, we discuss how to design a real CSPM with properties meeting the formulated requirements, what are roles of the salt and the matrix, etc.

22.2 General Scheme of the Sorptive Extraction of Water from the Atmosphere

For the sorptive water extraction from the atmosphere, it is important that the absolute air humidity in arid areas undergoes seasonal changes but is relatively stable during a day. On the contrary, the relative humidity varies greatly through a day: it is minimal in the day-time and rises sharply at night when the ambient temperature falls down. The general process scheme consists of two stages (Fig. 22.2). The first stage is water sorption in the night-time at high relative humidity. Ambient air passes throughout a bed of solid sorbent and gets dry. The latter stage – water desorption and subsequent water condensation – is carried out in the day-time. Heat needed for the water desorption must be supplied from an external heat source: in hot and sunny climates a solar energy can be used. Finally, the water desorbed is collected in a condenser. A reservoir buried underground at the depth exceeding 1 m can serve as the condenser. This condenser supplied with a heat exchanger at the bottom can maintain inside the temperature of 10–20 °C even in the day-time, when the desorption and condensation stages proceed.

The efficiency of the water extraction depends on the climatic conditions and is expected to increase at higher day-time absolute humidity and larger difference between day and night temperatures and, consequently, between the day and night relative humidity. Indeed, the mean diurnal temperature variation may reach 15–20 °C, therefore the relative pressure of water vapour may change from 0.2–0.4 to 0.5–0.8 through a day [8].

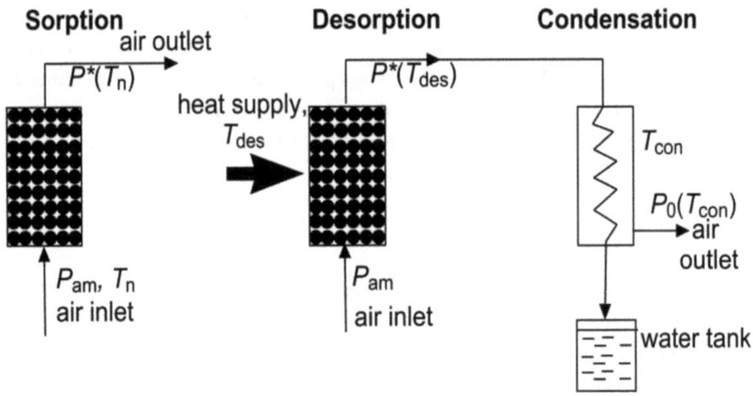

Fig. 22.2 General scheme of the water extraction from air (Adapted from Aristov et al. [4]. Copyright (2013), with permission from Elsevier)

22.3 Requirements for the Optimal Adsorbent/Salt

22.3.1 General Considerations

For analyzing the process of sorptive water extraction we use the Clausius-Clapeyron diagram (Fig. 22.3). This P-T chart presents liquid-vapour equilibrium of pure H_2O and solid-vapour equilibrium of the hydrate formation/decomposition (reaction 22.1). As the salt is the main sorbing component, the latter P-T line represents, in a first approximation, the sorption equilibrium between the CSPM, in whole, and water vapour. Below, we make qualitative points about desirable properties of an optimal solid sorbent (hydration reaction) and then present several case studies.

The water sorption occurs at the ambient vapour pressure P_{am} and the average night temperature T_n and desorption proceeds at the same pressure P_{am} and desorption temperature T_{des}. Consequently, reaction (22.1) has to occur at the equilibrium temperature $T^*(P_{am})$ obeying the relation $T_n < T^*(P_{am}) < T_{des}$. In other words, the hydration isoster of the optimal sorbent (or the equilibrium line of the optimal hydration reaction) has to lie between the points (P_{am}, T_n) and (P_{am}, T_{des}) (Fig. 22.3). During sorption, the water pressure in the outlet air is approaching the equilibrium pressure $P^*(T_n)$ over the salt-hydrate system (Fig. 22.3) and the degree of water extraction from the inlet air flux is $\delta_{sorb} = \Delta P_{ads}/P_{am} = [P_{am} - P^*(T_n)]/P_{am}$. Evidently, this degree increases at lower pressure $P^*(T_n)$ in the outlet air flux. Thus, the affinity of optimal sorbent to water should be high enough to get low $P^*(T_n)$.

At the second stage, desorption proceeds by heating the sorbent up to T_{des} that is the temperature level of heat used for desorption (regeneration). At this temperature, the equilibrium water pressure in the outlet air equals $P^*(T_{des})$ (Fig. 22.3). The water desorbed has to be collected in the condenser at temperature T_{con}. The degree of water collection is $\delta_{col} = \Delta P_{con}/P^*(T_{des}) = [P^*(T_{des}) - P_{con}]/P^*(T_{des})$. Evidently,

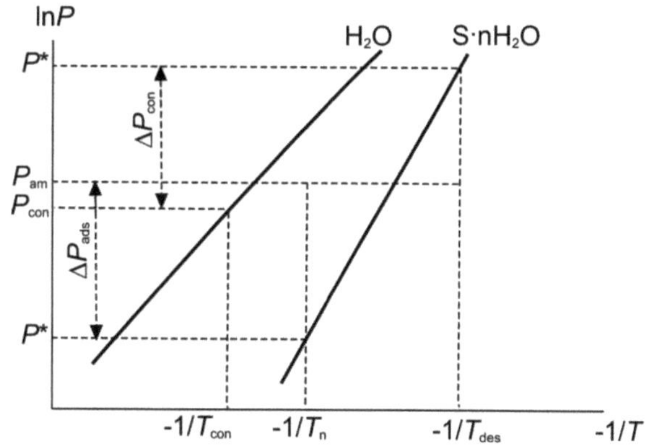

Fig. 22.3 Requirements for the optimal adsorbent/salt: equilibrium vapour pressure over H_2O and a salt hydrate $S \cdot nH_2O$

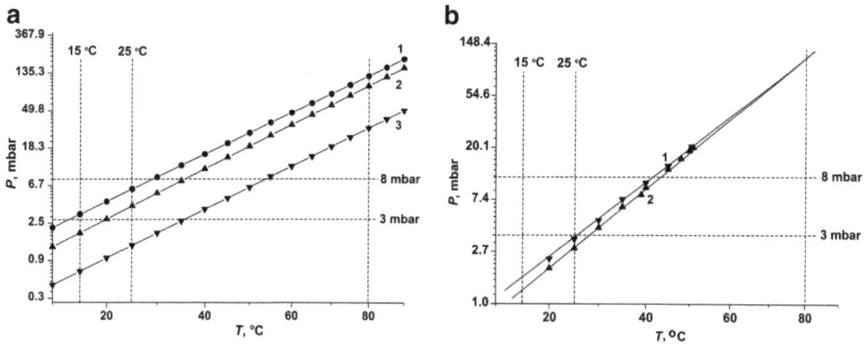

Fig. 22.4 Equilibrium water pressure of transitions between hydrates $CaCl_2 \cdot nH_2O$ at $n = 4–6$ (line 1), 2–4 (line 2), 1–2 (line 3) (**a**) and $Ca(NO_3)_2 \cdot nH_2O$ $n = 2–3$ (line 1) and 0–2 (line 2) (**b**)

this degree increases at higher pressure $P^*(T_{des})$ in the outlet air and lower condensation pressure P_{con}. Hence, at the desorption stage, the affinity of the optimal solid sorbent to water should not be too high that would otherwise reduce the pressure $P^*(T_{des})$. Searching for appropriate compromise may be done by considering the P-T diagram of a particular solid sorbent (see the case studies below). For these studies, the conditions at the sorption and desorption stages are fixed as follows (the so-called standard conditions): $P_{am} = 8$ mbar, $T_n = T_{con} = 15$ °C ($P_{con} = 17$ mbar) and $T_{des} = 80$ °C which are typical of many arid areas [8].

22.3.2 *Case Study 1: The System "$CaCl_2 – H_2O$"*

Four stable crystalline hydrates of $CaCl_2$ are known: $CaCl_2 \cdot H_2O$, $CaCl_2 \cdot 2H_2O$, $CaCl_2 \cdot 4H_2O$, and $CaCl_2 \cdot 6H_2O$ [9]. Appropriate equilibrium lines are presented on Fig. 22.4a. During the sorption stage, at $T = 15$ °C and $P_{am} = 8$ mbar, the salt reacts with water forming a hydrate $CaCl_2 \cdot 6H_2O$, first, at the adsorber inlet.

Then, the sorption front moves towards the adsorber exit. The equilibrium vapour pressure over this hydrate at $T = 15$ °C is 3.2 mbar that results in the degree of water extraction $\delta_{sorb} = 0.59$. As a matter of fact, it is larger because at the adsorber outlet the degree of the salt hydration n is lower than 6. For instance, at $n = 4$, the equilibrium pressure over $CaCl_2 \cdot 4H_2O$ $P^*(15\ °C) = 1.9$ mbar and $\delta_{sorb} = 0.76$; at $n = 2$, $P^*(15\ °C) = 0.69$ mbar and $\delta_{sorb} = 0.91$. The water capacity of $CaCl_2 \cdot 6H_2O$ is 0.97 g H_2O/g $CaCl_2$.

During the desorption stage, the hexa-hydrate decomposes at 29 °C giving $CaCl_2 \cdot 4H_2O$, which, in its turn, decomposes at 35 °C to form $CaCl_2 \cdot 2H_2O$, and the latter – at 53 °C yielding $CaCl_2 \cdot H_2O$ (Fig. 22.4a). To remove the last H_2O molecule, temperature over 80 °C is required which is poorly available by solar heat from a simple flat receiver. The equilibrium vapour pressure over $CaCl_2 \cdot 6H_2O$ at $T = 80$ °C is 126 mbar that results in the degree of water collection $\delta_{col} = 0.87$ at $T_{con} = 15$ °C.

What happens if the ambient air contains less moisture, e.g. $P_{am} = 3$ mbar. In this case, the affinity of this salt to water is not sufficient to bind $6H_2O$ molecules, and the hydrate $CaCl_2 \cdot 4H_2O$ forms (Fig. 22.4a). Its water capacity is 0.65 g/g. Less efficient is also the desorption/condensation stage, because for $CaCl_2 \cdot 4H_2O$ $P^*(80\ °C)$ is 84 mbar and $\delta_{col} = 0.80$. According to the diagram of Fig. 22.4a, the extraction efficiency dramatically falls at $P_{am} < 1.9$ mbar when only di-hydrate $CaCl_2 \cdot 2H_2O$ forms that has both low water capacity and $P^*(80\ °C)$.

Another important climatic parameter is the average night temperature T_n that dictates conditions of the sorption stage. It appears that $P_{am} = 8$ mbar and $T_n = 25$ °C are still profitable for the formation of $CaCl_2 \cdot 6H_2O$. Moreover, $CaCl_2$ allows the efficient water extraction from the inlet air and ensures the maximal sorption capacity (0.97 g/g) even at higher sorption temperature T_n up to c.a. 30 °C (Fig. 22.4a).

Increase in the condensation temperature from 15 to 25 °C reduces the degree of water condensation down to $\delta_{col} = 0.75$ and 0.63 in case of the $CaCl_2 \cdot 6H_2O$ and $CaCl_2 \cdot 4H_2O$ formation, respectively. If only $CaCl_2 \cdot 2H_2O$ forms at sorption stage, the condensation completely stops because for this hydrate $P^*(80\ °C) = 31.3$ mbar $< P_{con}(25\ °C) = 31.7$ mbar.

In sum, at the standard climatic conditions, $CaCl_2$ allows efficient extraction of water at the sorption stage and its condensation at the desorption stage. The salt performance drops down dramatically if the vapour pressure in the inlet air reduces below 1.9 mbar, means, in extremely arid zones. The collection efficiency is quite sensitive to the condensation temperature. Hence, the affinity of $CaCl_2$ to water vapour is quite balanced that allows efficient realization of both process stages. Interestingly, this salt was empirically selected to be used in the first lab-scale prototypes of water extraction units employing $CaCl_2$/(carbon Sibunit), $CaCl_2$/(silica gel KSK) [6] and $CaCl_2$/MCM-41 [10]. Testing of these units confirmed the feasibility of the water extraction from the atmosphere with the output of 3–5 tones of fresh water per 10 tones of the dry composite per day [6]. Analysis of the water produced with $CaCl_2$/(carbon Sibunit) composite demonstrated that its organoleptic quality is good and chemical composition is close to that of distilled water. The quality of the water passes the Russian portable water standard.

Table 22.1 Equilibrium temperature $T^*(P_{am})$, pressures $P^*(T_n)$ and $P^*(T_{des})$, mass of water Δw exchanged between various salt crystalline hydrates $S \cdot nH_2O$ and $S \cdot mH_2O$, and degrees of the water extraction δ_{sorb} and collection δ_{col}

Salt	n–m	$T^*(P_{am})$, °C	$P^*(T_n)$, mbar	$P^*(T_{des})$, mbar	δ_{sorb}	δ_{col}	Δw, g/g
$MnCl_2$	1–2	54	0.4	80	0.95	–	0.14
	2–4	26	3.3	242	–	0.92	0.29
$Mg(NO_3)_2$	2–6	54	0.4	80	0.95	0.78	0.49
$CaCl_2$	2–1	54	0.6	35	0.93	–	0.16
	4–2	36	1.9	98	–	–	0.32
	6–4	29	3.2	154	–	0.89	0.32
$CuSO_4$	1–3	58	0.6	32	0.93	–	0.23
	3–5	28	2.7	247	–	0.93	0.23
$CaSO_4$	0–0.5	27	0.3	80	0.96	–	0.07
	0.5–2	56	3.4	233	–	0.93	0.20
Na_2SO_3	0–7	37	1.6	100	0.80	0.83	1.0
Na_2HPO_4	7–12	37	1.6	101	0.80	0.83	0.63
$Ca(NO_3)_2$	0–2	36	1.2	–	0.85	–	0.22
	2–3	39	1.5	80	–	0.78	0.11

22.3.3 Case Study 2: The System "$Ca(NO_3)_2 - H_2O$"

Two stable crystalline hydrates of $Ca(NO_3)_2$ are known ($Ca(NO_3)_2 \cdot 2H_2O$ and $Ca(NO_3)_2 \cdot 3H_2O$) that at $P_{am} = 8$ mbar are formed at very close temperatures $T^*(P_{am}) = 39$ and 36 °C, respectively (Fig. 22.4b). During the sorption stage at $T_n = 15$ °C and $P_{am} = 8$ mbar, the salt reacts with water forming $Ca(NO_3)_2 \cdot 3H_2O$. The equilibrium vapour pressure over $Ca(NO_3)_2 \cdot 2H_2O$ at $T_n = 15$ °C is 1.3 mbar that would result in the degree of water extraction $\delta_{sorb} = 0.84$. The water capacity of $Ca(NO_3)_2 \cdot 3H_2O$ is 0.5 g/g. It is worth-noting that the water extraction remains constant even at $P_n = 8$ mbar and $T_n = 35$ °C. Thus, $Ca(NO_3)_2$ might be suitable for hotter climate than $CaCl_2$. At $P_{am} = 3$ mbar and $T_n = 15$ °C, the water extraction falls down to $\delta_{sorb} = 0.46$. Thus, at the standard climatic conditions, $Ca(NO_3)_2$ is somewhat less efficient than $CaCl_2$.

During the desorption stage, the equilibrium vapour pressure over $Ca(NO_3)_2 \cdot 3H_2O$ at $T_{des} = 80$ °C is 110 mbar that results in the degree of water collection at $T_{con} = 15$ °C $\delta_{col} = 0.84$. At $T_{des} \leq 48$ °C the condensation stops completely because $P^*(48\ °C) = 17$ mbar $\approx P_{con}(15\ °C) = 17$ mbar. The increase in the condensation temperature from 15 to 25 °C reduces the degree of water condensation down to 0.71.

Similar analysis has been performed for a variety of salts which thermodynamic data on the hydrates formation are available in literature. A number of salts and their hydrates appear to obey the formulated requirements (Table 22.1).

Thus, several salts are found to be promising for extraction of water from the atmosphere as they allow effective realisation of both sorption and desorption stages. In practice, the use of a bulk salt is limited by the following factors:

1. The formation of salt hydrate is accompanied by the intense reorganization of the crystalline structure. Therefore, the reaction may be inhibited and a certain extra-cooling below the equilibrium temperature is necessary to initiate the hydration process in bulk [11]. This inhibition may lead to a pronounced synthesis/decomposition hysteresis as well.
2. Reaction between the salt crystallites and vapour results in the formation of a crystalline hydrate on the surface of the salt. Further reaction requires the diffusion of vapour through the layer of the hydrate that could be a very slow process.
3. Swelling of the salt during the hydrate formation leads to its mechanical destruction and dust formation.
4. At large n, the hydrated salt may liquefy.

Just to overcome these shortages, it was suggested embedding the salt inside the pores of a host matrix which is the main CSPM concept [4, 6]. Below we discuss how the salt properties may change inside the matrix pores.

22.4 Change in the Salt Properties Inside the Matrix Pores

The confined salt may change the sorption properties due to its dividing into tiny pieces (the size effect) either/and its interaction with the matrix (the guest-host effect). The deep understanding of both effects may assist in intent variation of CSPMs properties to adapt them to particular conditions of water extraction or collection, etc.

22.4.1 Effect of the Pore Size

The main role of the host matrix is to disperse the salt over a large surface and prevent its aggregation. It is found that the salt dispersing may change its equilibrium with water vapour: e.g. the temperature of $Ca(NO_3)_2$ hydration increases [12] and the equilibrium vapour pressure over $CaCl_2$ decreases [13] in smaller pores (Fig. 22.5a, b). These effects are due to the fact that smaller salt particles have higher affinity to water vapour. Hence, CSPMs with smaller pores may be useful for extracting water from more dry air. Simultaneously, larger temperature is necessary for removal of the sorbed water.

In whole, making use of matrices with various pore size assists in fine tuning of the salt sorption equilibrium and managing the outlet pressures $P^*(T_n)$ and $P^*(T_{des})$. Important that a scale of the pore size effect is 5–15 °C that is in agreement with typical variation of the average night temperature in a variety of arid climates.

To analyze the increase in the hydration temperature in smaller pores (means, for smaller salt particle) we consider the Gibbs free energy of the reagents $\Delta G_r^\circ(T)$

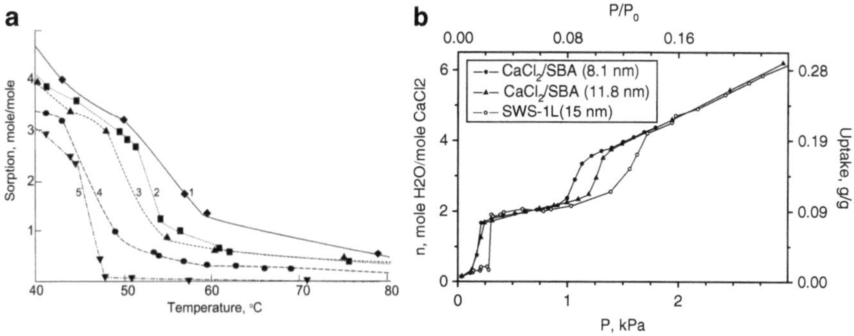

Fig. 22.5 (**a**) Isobars of vapour sorption by the $Ca(NO_3)_2$/silica with the silica pore size 6 (1), 9 (2), 12 (3), 15 (4) nm, and by a bulk $Ca(NO_3)_2$ (5), $P(H_2O) = 17$ mbar [12]; (**b**) – isotherms of vapour sorption by the $CaCl_2$/SBA-15 with the pore size of 8.1 and 11.8 nm, and SWS-1L = $CaCl_2/SiO_2$(KSK) $T = 50$ °C (Reprinted from Glaznev et al. [13]. Copyright (2013), with permission from Elsevier)

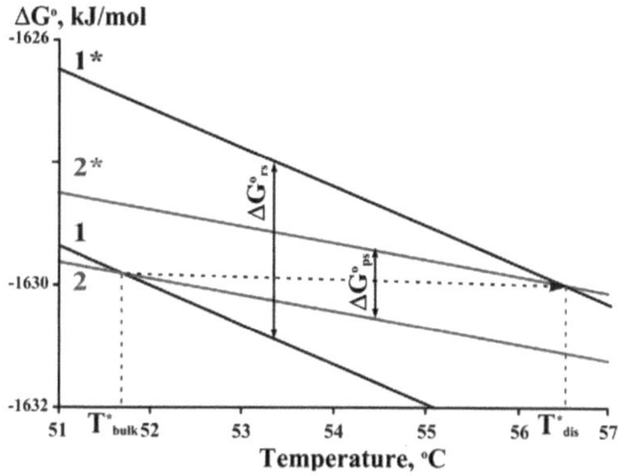

Fig. 22.6 The Gibbs free energy for $Ca(NO_3)_2$ hydration in bulk (1,2) and dispersed states (1*,2*): 1,1* – reagents; 2, 2* – products [14]. $\Delta G_{rs}{}^\circ - \Delta G_{ps}{}^\circ = 1.7$ kJ/mol

and products $\Delta G_p{}^\circ(T)$ of reaction (22.1) assuming $S = Ca(NO_3)_2$ (Fig. 22.6) [14]. The intersection of the curves $\Delta G_r{}^\circ(T)$ and $\Delta G_p{}^\circ(T)$ corresponds to the temperature T^* at which $\Delta G^\circ(T^*) = \Delta G_p{}^\circ(T^*) - \Delta G_r{}^\circ(T^*) = 0$, hence, the system is at equilibrium. E.g., in bulk, the theoretical hydration temperature $T^*(P = 17\ \text{mbar}) = 51.7$ °C (Fig. 22.6). It is higher than the experimental value $T^*(P = 17\ \text{mbar}) \approx 47$ °C (Fig. 22.5a) that is probably due to the effect of hydration inhibition. The difference of c.a. 5 °C presents the aforementioned under-cooling below the equilibrium temperature. In the dispersed state, both $\Delta G_r{}^\circ$ and $\Delta G_p{}^\circ$ are larger than in bulk due to the appropriate contributions of the surface energy of the reagents $\Delta G_{rs}{}^\circ$ and products $\Delta G_{ps}{}^\circ$, and these increments may change the equilibrium temperature of the hydration process (Fig. 22.6). For instance, the hydration temperature should increase if the surface energy of the salt is larger than that of the hydrate ($\Delta G_{rs}{}^\circ > \Delta G_{ps}{}^\circ$). Our graphical analysis shows that the theoretical augmentation of the $Ca(NO_3)_2$ hydration temperature is a linear function

$\Delta T = A\ (\Delta G_{rs}{}^{\circ} - \Delta G_{ps}{}^{\circ}) = 2.88$ K/(kJ/mol) $(\Delta G_{rs}{}^{\circ} - \Delta G_{ps}{}^{\circ})$. At typical values of $(\Delta G_{rs}{}^{\circ} - \Delta G_{ps}{}^{\circ}) = 1\text{–}5$ kJ/mol, the latter relation results in $\Delta T = 5\text{–}15$ °C that is in agreement the experimental findings [14].

22.4.2 *Effect of the Subsidiary Salt*

One more way to adjust the CSPM hydration temperature to particular climatic conditions may be realized by adding a subsidiary salt S2 that modifies properties of the basic salt S1 [15, 16]. It was demonstrated that the addition of metal bromides $MeBr_n$ to chlorides $MeCl_n$ results in the formation of a solid solution with the equilibrium hydration pressure $P^*(T)$ lower than for $MeCl_n$ (not presented). Consequently, the composite (S1 + S2)/(silica gel) may have better performance at drier atmospheric air. By varying the relative bromine content, solid sorbents with a required hydration temperature can be prepared.

22.4.3 *Effect of the Salt Content*

The salt content in the composite is another important parameter, it mainly affecting the CSPM sorption capacity: the water sorption gradually rises at larger salt content [13, 17, 18]. It is expected if keeping in mind that the salt is the main sorbing element of the CSPM. Although a large salt content is profitable, it should be in any way restricted to ensure that at the maximal daily relative humidity, the salt solution, that may form, remains inside the pores (see [5] for more details). Water sorption properties of CSPMs are also sensitive the concentration and pH of the impregnating solution [19], the calcination temperature [20], etc.

22.5 Conclusions

A sustainable water supply is of vital importance for the humanity development. The atmospheric humidity is the only continuously available water resource for the most desert regions. Sorptive extraction of water from the atmosphere is very attractive and could significantly alleviate the problem of water shortage in the most of the arid regions. New solid sorbents with advanced properties adapted to this application are welcome. In this communication, we make an analysis of applying the new composites “a salt inside a porous matrix” (CSPMs) for extracting water from the atmosphere. The general scheme of the water extraction consists of the two stages: the sorption phase and desorption/condensation phase, each imposing specific requirements for an optimal solid sorbent. The requirements are considered by using the Clausius-Clapeyron diagrams of “water-vapour” and “salt hydrate-vapour" equilibriums. Several salts have been selected as promising for extraction of water from the atmosphere as they allow effective realisation of both sorption and desorption stages. Finally, we discuss how the salt sorption properties may change due to its confinement inside the matrix pores. The conclusion is drawn that it is possible to intently vary the hydration temperature and the water sorption capacity in wide range by intelligent selection of the matrix pore size and the salt content as well as by employing a subsidiary salt that modifies properties of the basic one. All these tools may assist in adapting CSPMs to particular conditions of water extraction or collection that may dramatically differ in various climatic zones all over the world.

References

1. Food and Agricultural Organization of the United Nations (2007) World water day 22nd March 2007. Coping with water scarcity – challenge of the twenty-first century. p. 9. http://www.fao.org/nr/water/docs/escarcity.pdf. Accessed 30 Jan 2013
2. Arnell NW (2004) Climate change and global water resources: SRES emissions and socio-economic scenarios. Glob Environ Chang 14:31–52
3. Vörösmarty CJ, Green P, Salisbury J et al (2000) Global water resources: vulnerability from climate change and population growth. Science 289:284–288
4. Aristov YI, Tokarev MM, Gordeeva LG et al (1999) New composite sorbents for solar-driven technology of fresh water production from the atmosphere. Sol Energy 66:165–168
5. Gordeeva LG, Aristov YI (2012) Composites "salt inside porous matrix" for adsorption heat transformation: a current state of the art and new trends. Int J Low Carbon Technol 7(4):288–302
6. Aristov YI (2007) New family of solid sorbents for adsorptive cooling: material scientist approach. J Eng Thermophys 16:63–72
7. Gordeeva LG, Restuccia G, Freni A et al (2002) Water sorption on "LiBr in porous carbons". Fuel Process Technol 79:225–231
8. Arakawa H (ed) (1969) World survey of climatology, vol 8, Climates of Northern and Eastern Asia. Elsevier Scientific Publishing Company, Amsterdam; Takahashi K, Arakawa H (eds) (1981) World survey of climatology, vol 9, Climates of Southern and Western Asia. Elsevier Scientific Publishing Company, Amsterdam; Griffiths JF (ed) (1972) World survey of climatology, vol 10, Climates of Africa. Elsevier Scientific Publishing Company, Amsterdam
9. Watcher (ed) (1992) Kirk-Othmer encyclopaedia of chemical engineering, vol 4, 4th edn. Wiley, New York
10. Ji JG, Wang RZ, Li LX (2007) New composite adsorbent for solar-driven fresh water production from the atmosphere. Desalination 212:176–182
11. Andersson JY (1986) Kinetic and mechanistic studies of reactions between water vapour and some solid sorbents. Department of Physics, Chemistry. The Royal Institute of Technology, Stockholm
12. Simonova IA, Aristov YI (2005) Sorption properties of calcium nitrate dispersed in silica gel: the effect of pore size. Rus J Phys Chem 79:1307–1311
13. Glaznev I, Ponomarenko I, Kirik I et al (2011) Composites $CaCl_2$/SBA-15 for adsorptive transformation of low temperature heat: pore size effect. Int J Refrig 34:1244–1250
14. Simonova IA (2009) Composite water sorbents $Ca(NO_3)_2$/silica and $LiNO_3$/silica. Ph.D. thesis, Boreskov Institute of Catalysis, Novosibirsk
15. Gordeeva LG, Grekova AD, Krieger TA et al (2009) Adsorption properties of composite materials (LiCl + LiBr)/silica. Microporous Mesoporous Mater 126:262–267

16. Gordeeva L, Grekova A, Krieger T et al (2013) Composites “binary salts in porous matrix” for adsorption heat transformation. Appl Therm Eng 50:1633–1638
17. Daou K, Wang RZ, Xia ZZ (2006) Development of a new synthesized adsorbent for refrigeration and air conditioning applications. Appl Therm Eng 26:56–65
18. Okada K, Nakanome M, Kameshima Y et al (2010) Water vapour adsorption of $CaCl_2$-impregnated activated carbon. Mater Res Bull 45:1549–1553
19. Gordeeva LG, Glaznev IS, Savchenko EV et al (2006) Impact of phase composition on water adsorption on inorganic hybrids “salt/silica”. J Colloid Interface Sci 301:685–691
20. Gordeeva LG, Gubar AV, Plyasova LM et al (2005) Composite water sorbents of the salt in silica gel pores type: the effect of the interaction between the salt and the silica gel surface on the chemical and phase compositions and sorption properties. Rus Kinet Catal 46:736–742

MIX
Papier aus verantwortungsvollen Quellen
Paper from responsible sources
FSC® C105338

If you have any concerns about our products,
you can contact us on
ProductSafety@springernature.com

In case Publisher is established outside the EU,
the EU authorized representative is:
Springer Nature Customer Service Center GmbH
Europaplatz 3, 69115 Heidelberg, Germany

Printed by Libri Plureos GmbH
in Hamburg, Germany